SOFTWARE
ENGINEERING

SUBHAJIT DATTA

OXFORD
UNIVERSITY PRESS

OXFORD
UNIVERSITY PRESS

Oxford University Press is a department of the University of Oxford.
It furthers the University's objective of excellence in research, scholarship,
and education by publishing worldwide. Oxford is a registered trademark of
Oxford University Press in the UK and in certain other countries

Published in India by
Oxford University Press
22 Workspace, 2nd Floor, 1/22 Asaf Ali Road, New Delhi 110002, India

ISBN-13: 978-0-19-569656-1
ISBN-10: 0-19-569656-5

Typeset in Times Roman
by Recto Graphics, Delhi 110 096
Printed in India by Manipal Technologies Limited, Manipal

To
my first teacher of programming
I still dream I would be as good at it as you

Preface

To most of the world, software is entirely invisible; ... To many ... software is just a jumbled mass of incomprehensible letters. To a few, quality software-intensive systems are a thing of beauty, full of drama and patterns ... changing and being changed by interaction with the real world.

Grady Booch

When I was studying electrical engineering as an undergraduate student, it was very easy to check if one knew the basics. Ohm's law and Kirchhoff's laws are the bedrock upon which electrical engineering stands and they are the first principles a student has to learn. So if one knew these laws (at least by heart), it was reasonable to assume he or she had an inkling of the basics of electrical engineering. But what does it mean to *know the basics* of software engineering—to know how to write software programs, to know how object-oriented programming differs from procedural programming, or to know which technology will best serve a particular user's needs?

These are important questions for successful software development. But they do not represent the kernel of software engineering. We are still searching for that kernel—the core set of credos representing the foundation of our discipline, upon which all other principles stand.

But the world will not wait while we search. Stroustrup has summarized—what is now widely recognized —that civilization, as we know it today, runs on software. On one hand, software engineers need to build systems that influence our experience of living in deep and diverse ways. On the other hand, the quest continues for the fundamental principles of software engineering.

For software engineers, this dichotomy represents a great challenge, and a greater opportunity. The aim of this book is to help software engineers confront the challenge and embrace the opportunity.

About the Book

When delving into a subject for the first time, a serious difficulty comes from being swamped with too many details. Such details can only be put in their place when one is able to discern the broad contours of the subject. On introduction to an area as large and evolving as software engineering, students need to focus on gleaning the message, rather than gathering minutiae. Accordingly, *Software Engineering: Concepts and Applications* intends to look at the proverbial forest and not the trees. The main thread of discussion highlights the so-called 'big picture', with pointers to important details when necessary.

A textbook on software engineering can hardly hope to be complete. By the time this book reaches you, new advances in theory and practice may have altered the outlook on some of its discussions. To keep up with this change, a useful textbook must keep on refining itself. The Web is an excellent medium for such refinement. I intend to interact with readers through blogs and other online channels; stay tuned to my website at www.dattas.net and the publisher's website at www.oup.com.

In many textbooks on the subject, software engineering in the title is qualified by 'object-oriented' or 'classical' (or 'object-oriented and classical', by particularly versatile authors). It is somewhat restrictive to view software engineering through the lens of a particular development paradigm. Instead, when we have recognized software engineering as an 'engineering', the emphasis should be more around software development within the typically engineering contexts of planning, management, construction, maintenance, etc. This thinking has been reflected in the organization of *Software Engineering: Concepts and Applications*.

Each chapter begins with a discussion of learning objectives and motivation, and ends with a summary, a section pointing to additional material (often Web resources), exercises, and references. Exercises include review questions to test the understanding of key concepts and reflective questions that call for independent thinking. Reflective questions are usually interpretive in nature and may not have unique right or wrong answers. The chapters with mathematical equations or formulae also have numerical problems; these are expected to be solved by quick, back-of-the-envelope kind of calculations. In the chapters that closely relate to programming, programming exercises require students to analyse and/or write programs or program segments. The example programs given in this book are in Java, but they highlight concepts not necessarily specific to a programming language.

Many chapters also contain exhibits, which serve a variety of purposes. Some of the exhibits present interesting facts—often historical—which provide a larger context to the discussion. Some indicate a point of view not universally accepted or outright controversial, which nonetheless imparts a valuable insight. Some exhibits briefly introduce an established method or technique not within the scope of this book but whose familiarity will help advanced students while some present new research results not yet included in standard textbooks.

Content and Coverage

This book is divided into the following logical sections:
- Part I   Understanding the Realm of Software Engineering: Chapters 1 to 3
- Part II   Planning and Managing Software Development: Chapters 4 to 10
- Part III   Making Software: Chapters 11 to 16

- Part IV Testing, Maintaining, and Modifying Software Systems: Chapters 17 and 18
- Part V   Latest Trends of Software Development: Chapters 19 to 23

A brief outline of each chapter is given below.

Chapter 1, *What is Software Engineering?*, introduces various definitions of software engineering, the problems software engineering confronts, software engineering's response to those problems, the challenges with the response, and what it is like being a professional software engineer.

Chapter 2, *Evolution of Software Engineering*, discusses important trends and milestones in the evolution of software engineering.

Chapter 3, *Basic Ideas and First Principles*, addresses the question: Are there laws of software engineering? Software engineering is then compared and contrasted with other engineering disciplines and characterized by its unique features.

Chapter 4, *Software Development Methodologies*, introduces the software development life cycle, discusses philosophies of sequential and iterative software development, and gives an overview of different development methodologies.

Chapter 5, *Place of Process in Software Development*, highlights the importance of processes and discusses some established processes of software engineering at the individual, team, and project levels.

Chapter 6, *Software Estimation*, underscores the significance of estimation in software development. It discusses standard techniques of estimating by judgement, comparison, and correlation, and illustrates how they can be used in estimating size, time, and effort.

Chapter 7, *Role of Metrics in Software Development*, gives an overview of the range of software engineering metrics, their utility, and how they have evolved over time.

Chapter 8, *Software Project Management*, presents the life cycle of a typical software project, followed by discussions of the principles of software project management, process groups, knowledge areas, and the software project management plan. Additionally, team dynamics and the differences between managing and leading are highlighted.

Chapter 9, *Human Aspects of Software Development*, brings into focus the need to help users know their needs, the co-evolution of the problem and solution domains, the importance of language and communication in software development, human-computer interaction, and the significance of usability for software systems.

Chapter 10, *Role of Automation in Software Development*, introduces the 'why', 'how', and 'what' of automating software development, followed by discussions on the range of automation and specific examples of Test and Design automation.

Chapter 11, *Understanding Software Architecture*, presents architectural views of software, the ways in which software architecture is perceived and defined, the importance of architecture for large-scale software systems, how architecture differs from Design, and a brief overview of architectural patterns.

Chapter 12, *Paradigms of Software Development*, starts by making a case for software's complexity and the strategies for addressing such complexity in software development, and then presents three major development paradigms—algorithmic, object oriented, and aspect oriented.

Chapter 13, *Languages of Software Development*, discusses the use of languages at various levels of software development. Programming languages are classified and the factors guiding the choice of a programming language for a particular case are discussed. Then the need for modelling languages is underscored and the Unified Modelling Language (UML) is introduced. Finally, specification languages are presented in the context of formal methods and illustrated through a simple example using Z.

Chapter 14, *Software Development across Workflows and Phases*, delves into the dimensionality of software development. It describes the software development life cycle in terms of the *workflows* (Requirements, Analysis, Design, Implementation, and Test) and *phases* (Inception, Elaboration, Construction, and Transition). This chapter highlights the central activities in building a software system and explains how these are interrelated.

Chapter 15, *Building a Software System: An Extended Case Study*, builds a software system from scratch, demonstrating how the key concepts of software development are applied in practice.

Chapter 16, *Tricks of the Trade*, presents a set of skills that can be used across software development scenarios to build and maintain flexible, resilient, and useful software.

Chapter 17, *Software Testing, Reliability, and Quality*, presents different Testing philosophies, types of Testing, and debugging techniques. It also explores basic ideas of software reliability and the place of the ISO 9000 series of standards, the Capability Maturity Model (CMM), and Six Sigma in the quest for software quality.

Chapter 18, *Towards Software Evolution*, examines continual modification and maintenance of a software system in the light of software entropy.

Chapter 19, *Software Engineering and the World Wide Web*, discusses the salient features of Web-based software systems and highlights the growing importance of the Web as a software development medium.

Chapter 20, *Towards Enterprise Software Development*, presents key characteristics of enterprise software systems and the challenges unique to their development, maintenance, and enhancement.

Chapter 21, *Global Software Development*, discusses the dynamics of distributed teams and remote customers, and the traits of a global software engineer.

Chapter 22, *Open Source Software Development*, traces the evolution of open source software development, followed by a discussion of the range and limitations of open source systems and the implications of open source for the professional software engineer.

Chapter 23, *Future of Software Development*, extrapolates the trends of the present into the future. In conclusion, a survival toolkit for the professional software engineer is suggested, which may help him/her stay current in the face of relentless change.

Acknowledgements

In addition to continuing the supply of love, understanding, and sumptuous food, my wife Reshmi has drawn the diagrams for this book. And all the while, she has continued her own research. My mother keeps on believing I can do things that I cannot. But her faith helps. My brother, a mathematician, has an interest in reading and writing that has greatly shaped my love for books.

I wish to specially thank the editorial staff of Oxford University Press for guiding me through the project. Without their patience and professionalism, this book would have never been published.

My passion for software engineering has come out of the time I spent as a software engineer: in the trenches of development, in research, and in teaching. There are individuals—too numerous to name—whose originality has made me to look anew at things that I had cast aside as old. They are my teachers.

It is pleasure and a privilege to be a software engineer in the world of today and tomorrow; I welcome you to the quest for discovering why.

Subhajit Datta

Brief Contents

Detailed Contents

Part II Planning and Managing Software Development

PART I

UNDERSTANDING THE REALM OF SOFTWARE ENGINEERING

What is Software Engineering?

Learning Objectives

In this chapter, we begin by exploring some of the foundations of software engineering. Specifically, we consider:

- Various definitions of software engineering
- Characteristics of software
- Problems confronting software engineering
- Its response to the problems
- Challenges with the response and the grand challenge
- What it is like to be a software engineer

1.1 MOTIVATION

Most textbooks on software engineering start with a picture of gloom. Copious references are made to the 'software crisis' (see Chapter 2), with indications that the crisis has not ended yet, and insinuations that it may never end. The monumental cost of software failure is highlighted with facts and figures. Perhaps all of this is meant to emphasize the difficulties of software engineering and the onus on an aspiring software engineer. When I read such books as a student (or at least *started* reading), the first few pages of the first chapter had a rather depressing effect. Given the gory details of the crisis software engineering seemed to be perpetually in, I was not sure I wanted to risk my happiness getting sucked into that vortex of missed deadlines, unhappy customers, and other vicissitudes. But in spite of those ominous openings, fortunately, I ended up being a software engineer. After more than a decade of studying, researching, and practicing software engineering, when I come across similar books now, I find their overtures both odious and misplaced.

Odious, since it is both in bad taste and pedagogically sterile to introduce a student to a discipline by reciting all the privations of the past. It is very important to challenge the student, but it does no good to present the discipline in a foreboding light. While recognizing that many of the facts reflecting on the difficulties of software engineering are true, their introduction in the first few pages is still misplaced. Software engineering, as we shall see, is a very young discipline. Many of its monumental failures are very much in recent memory. On the other hand, no one is old enough to remember exactly how many bridges fell (at least the falling of the London bridge is canonized in nursery rhyme!) or how many trains tumbled (well, they still do), before the engineering behind these artefacts stabilized. Every engineering work is a trial-and-error game, as so brilliantly argued in books like [Petroski 1992], and software is no exception. To the prepared mind, failures are great learning aids, but to beginners, they are hardly inspiring.

Like any other human endeavour that is *alive*, software engineering is a work in progress. If I have the privilege of ushering bright, young minds (I am talking about you, the reader) into the field, I prefer to do so by outlining the challenges we face in building beautiful, flexible, and resilient software. Yet, at the same time underlining that you will be equal to those challenges in your lifetime with software engineering. This book is a journey to get you started with the best equipment we have now, so that you can fully utilize better equipment that comes to you in future. This chapter begins our journey.

1.2 DEFINITION OF SOFTWARE ENGINEERING

When asked to define his subject, one mathematician reportedly said, mathematics is what is done by mathematicians; and mathematicians are those who do mathematics [Hamming 1997]. This is surely a joke, and the humour perhaps lies in trying to define something in terms of itself. But this anecdote also highlights how difficult it is to define anything, even as established and important as mathematics. Defining software engineering poses more problems, at least quantitatively. First of all, when compared to 'mathematics', 'software engineering' is two words versus one. Moreover, both 'software' and 'engineering' are so-called *operative* words. There is no consensus on what 'engineering' means, and even less unanimity on what we mean by 'software'. Thus trying to make sense of software engineering by tying the definition of 'software' with that of 'engineering' is likely to create even more rancour. Instead of taking on such a task ourselves, let us see how others have tried to define software engineering.

- According to Boehm, software engineering involves the application of science and mathematics through which the facilities of computer equipment are made useful to human beings via computer programs, procedures, and associated documentation.

- Pfleeger identifies software engineering with the utilization of tools, techniques, procedures, and paradigms toward quality improvement of the software product.
- Naur and Randall see software engineering in terms of establishing and using sound engineering principles to obtain economically effective and reliable software that can work efficiently on real machines.
- According to Freeman and Von Staa, software engineering involves the organized application of methods, tools, and knowledge towards fulfilling stated technical, economic, and human goals for a software-intensive system. Interestingly, in recent literature [Booch 2006], 'software-intensive systems' is being increasingly used to denote what we customarily call 'software systems'. The new nomenclature highlights that to be successful, software has to successfully integrate within a larger framework of technological, commercial, and human concerns.
- Kacmar says simply applying engineering principles to designing and constructing computer software can be termed software engineering.
- To Schach, software engineering is the discipline that aims at producing fault-free software, to be delivered on time and within budget, which satisfies the user's needs.
- Whitmire describes software engineering as a 'slippery' term, and says for some it is something that can only be applied to a large project, while to others it is just a 'figment of collective imaginations'. He gives a *working definition* as, 'Software engineering is the science and art of designing and building, with economy and elegance, software systems and applications so they can fill the uses to which they may be subjected' [Whitmire 1997].

Now, what is the essence that ties these definitions together?

1.3 CHARACTERISTICS OF SOFTWARE

All the definitions in the previous section together make up our *current* understanding of software engineering, which may not necessarily be complete. Software engineering is very much a work in progress, due to its relative youth, as well as the very nature of software. We will consider these topics in more depth in Chapters 2 and 3. However, it is appropriate now to mention the set of software characteristics Brooks identified decades ago [Brooks 1995]; and whose depth and relevance we are still discovering.

- Software is inherently complex.
- Software must be made to conform to existing interfaces.
- Software is constantly subject to change.
- Software is invisible and unvisualizable.

> **Exhibit 1.1   What's in a Name?**
>
> Shakespeare, in the romantic classic *Romeo and Juliet*, has the hero say, 'What's in a name? That which we call a rose by any other name would smell as sweet'. This oft-quoted phrase is taken to mean that names do not matter, substance does. But for software engineering, in the beginning at least, names did matter.
>
> The phrase 'software engineering' was first used in a public discourse at a NATO Science Committee sponsored conference, held at Garmisch, Germany, from 7th to 11th October, 1968 [Bauer et al. 1968]. The conference was a visionary exercise, seeking as it did to bring together experts from the industry, academia, and user communities to chart out the course of software development for the future. Discussions were organized in the areas of Design, production, and service of software. The conference proceedings, now publicly available [Bauer et al. 1968] illuminate how much has changed with software engineering till date, with newer tools and technologies; as well as how little has changed, in terms of basic concerns and expectations. While deliberating on the 'nature' of software engineering, the importance of feedback was highlighted many times during the conference [Bauer et al. 1968]. This feedback aspect assumes much importance in the light of what we discuss later in this chapter.
>
> There have been many conclaves on software engineering ever since, but the 1968 NATO conference gave software engineering a name, in the most literal sense.

We will not get into the detailed discussion of each of the above characteristics at this time; let us wait till they unravel themselves as we get deeper into the book. We will remark in passing, however, that although the above may not capture *all* that is important about software, it certainly touches upon the essence of software as a unique artefact of human ingenuity and utility. The problems that software engineering addresses draw largely from these characteristics of software.

1.4 PROBLEMS CONFRONTED BY SOFTWARE ENGINEERING

Every engineering task starts off in response to some pressing problem. Civil or structural engineering served the need to have shelter; mechanical engineering addressed the need for locomotion; electrical engineering catered to growing energy demands; and chemical engineering unlocked the hidden potential of matter. What is, if any, the corresponding 'mission' for software engineering?

Software engineering confronts the problems of *change* and *complexity*.

1.4.1 Problem of Change

When a bridge, a house, or a car is built and given to us, we try to use it, love it, or hate it, continue using it, or move on to a new bridge, house, or car. When a software system is given to us, we try to use it, love it, or hate it, and want the same system to work the way we want it to. Typically, we do not know the way we want it to work, before we start using it.

The very nature of software—its *plasticity*—makes it amenable to a continuous cycle of change. It seems rather easy to accomplish. After all, tweaking one statement in a software program can radically alter the program's behaviour. But such tweaking—little by itself, but considerable in conjunction—can end up changing the intent of the program's Design in fundamental ways. It is absurd to expect a car to fly or float. But very often a software system built for one context is expected to function in drastically different contexts, with the same grace and efficiency. These expectations can be traced to our wide *cognitive gap* [Datta 2007] with the use of software. Decades and centuries of using cars and bridges respectively, and millennia of using houses has ingrained in our minds what cars, houses, and bridges can and cannot do. Accordingly, we tune our expectations as well as environmental factors to set the context for these systems to function. In comparison, the use of software amongst a large community of lay users has just begun. Our understanding of how and to what extent software can serve our needs is yet not complete. As a result, the problem of change for software comes primarily from changing user expectations, and also from changes in the environment—technological and social.

1.4.2 Problem of Complexity

Complexity is a complex word and there is no one definition to cover its ken; even reaching a definition is an onerous task [Nicolis and Prigogine 1989], [Waldrop 1992]. But we need to care about it in life as well as in software engineering as complexity arises out of simplicity, at times suddenly and surreptitiously. Think of a simple computer program of five lines of code. It is straightforward; by carefully reviewing each line, we can hope to have complete knowledge of the program's structure and behaviour. Now what if, a *loop* is introduced in the program—a simple construct that executes a set of statements repetitively, until a condition holds. The number of execution paths through the program has significantly increased now, and it has become far more difficult to know for sure what happens in each step when the program runs. (As we have illustrated in Chapter 17, for any non-trivial software system, an impractical amount of time and effort is needed to test each and every path of the program's execution.) This example is just a watered down instance of the *combinatorial* complexity software systems customarily face.

Then there are even more involved issues such as complexity of the problem domain, complexity in the interaction of the various forces—technological, commercial, political—that a software system has to balance to be successful.

We have made a case for software's complexity in Chapter 12 and will not go into the details here. However, one must note that a common feature of complex systems is that they are greater than the sum of their parts. Anyone who has done a class project to build a piece of software stretching across several files can appreciate the sense of this statement: A piece of software is made of individual files, but it delivers something that merely bunching the files together will not achieve. Now scale-up to a real world system—with hundreds, if not thousands of files; and thousands, if not millions of interfaces between them; perhaps simple by themselves, but certainly complex when functioning together. And this is just one, relatively less significant, facet of software complexity. Given that change and complexity are facts of life, what does software engineering do about them?

1.5 THE SOFTWARE ENGINEERING RESPONSE

The software engineering response to complexity and change comes in two parts: breaking down the problem into smaller, more manageable 'chunks' to confront complexity, and setting regular checkpoints during the process of building a software system to address the effects of change. The breaking down results in something we will call *workflows* and the checkpointing leads to *phases*; together they constitute the *software development life cycle* or the SDLC. The SDLC lies at the heart of software engineering and we take it up in right earnest later in the book (Chapters 4 and 14). We will now briefly discuss how the problems of complexity and change are addressed.

Workflows represent sets of activities starting from understanding what users want from a software system (Requirements), to translating the language of the problem into the language of the solution (Analysis), to expressing the solution constructs in the language of development (Design), to building the system using programming resources (Implementation), and finally, verifying whether the system matches the stated Requirements (Testing). Phases, on the other hand, are focused towards monitoring and managing change. During *Inception* we ask, what do the users want from the system? During *Elaboration*, we are interested in knowing if the system is feasible. Next comes the question: How do we build the system? This is the concern of *Construction*. Finally, during *Transition*, we enquire, how do we transfer the system from the developer domain to the user domain? In a particular development life cycle, we may not know the answers to these questions when we ask them. But based on our experience and understanding, we have an expectation of what the answers are likely to be. When expectations are not met, it serves as a reality check: A change, not budgeted for, must have occurred. This makes us aware of the need to find out what changed and what that change might affect.

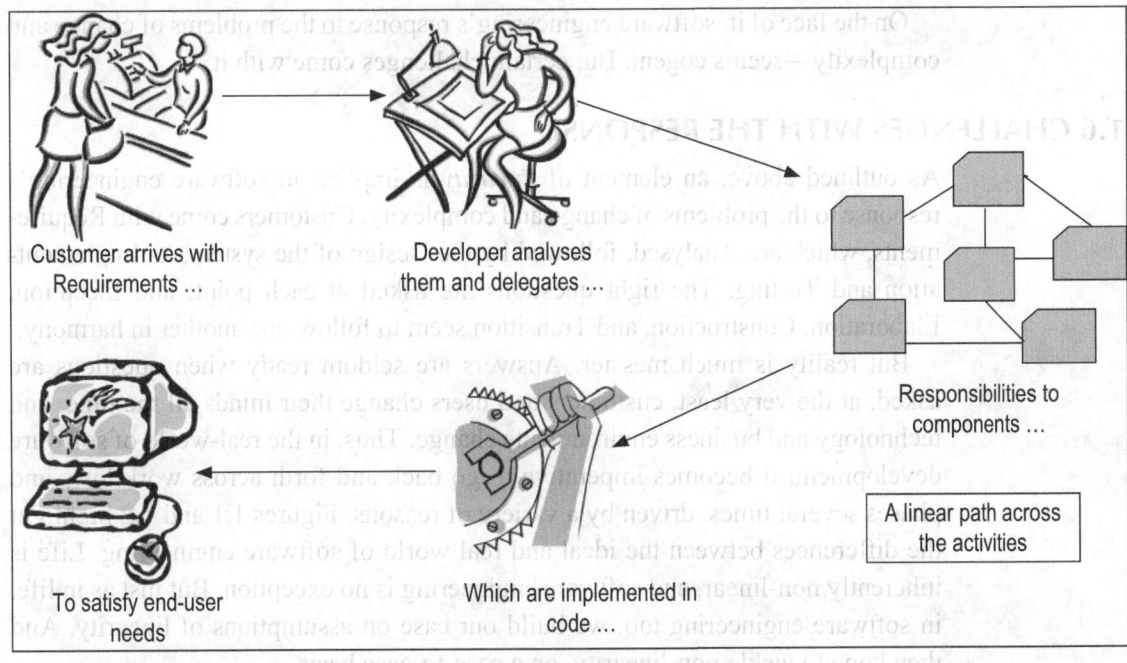

Customer arrives with
Requirements ...

Developer analyses
them and delegates ...

Responsibilities to
components ...

A linear path across
the activities

To satisfy end-user
needs

Which are implemented in
code ...

Figure 1.1 Software development, ideally

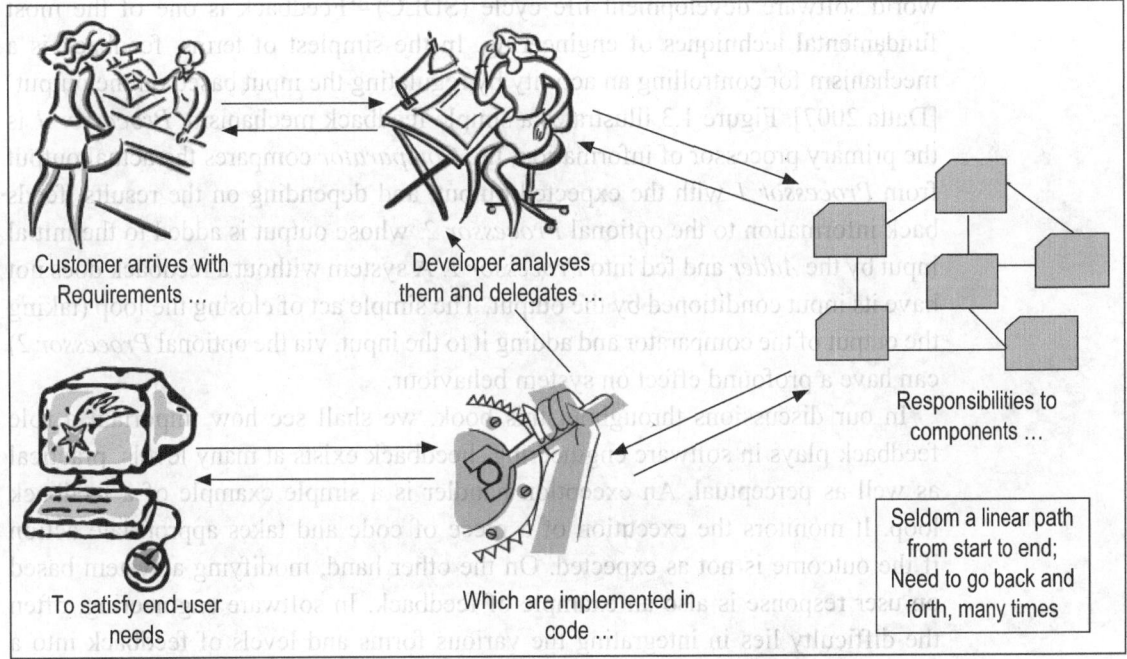

Customer arrives with
Requirements ...

Developer analyses
them and delegates ...

Responsibilities to
components ...

Seldom a linear path
from start to end;
Need to go back and
forth, many times

To satisfy end-user
needs

Which are implemented in
code ...

Figure 1.2 Software development in reality

On the face of it, software engineering's response to the problems of change and complexity—seems cogent. But certain challenges come with it.

1.6 CHALLENGES WITH THE RESPONSE

As outlined above, an element of *linearity* is implicit in software engineering's response to the problems of change and complexity. Customers come with Requirements, which are Analysed, followed by the Design of the system, its Implementation and Testing. The right questions are asked at each point; and Inception, Elaboration, Construction, and Transition seem to follow one another in harmony.

But reality is much messier. Answers are seldom ready when questions are asked; at the very least, customers and users change their minds all the time, and technology and business environments change. Thus, in the real-world of software development, it becomes imperative to go back and forth across workflows and phases several times, driven by a variety of reasons. Figures 1.1 and 1.2 highlight the differences between the ideal and real world of software engineering. Life is inherently non-linear, and software engineering is no exception. But just as in life, in software engineering too, we build our case on assumptions of linearity. And then hope to tackle non-linearity, on a case-to-case basis.

So the key challenge with software engineering's response boils down to being able to monitor, control, and utilize the many *feedback* paths that exist in the real-world software development life cycle (SDLC). 'Feedback is one of the most fundamental techniques of engineering. In the simplest of terms, feedback is a mechanism for controlling an activity by regulating the input based on the output' [Datta 2007]. Figure 1.3 illustrates a simple feedback mechanism. *Processor 1* is the primary processor of information; the *Comparator* compares the actual output from *Processor 1* with the expected output, and depending on the results, feeds back information to the optional *Processor 2,* whose output is added to the initial input by the *Adder* and fed into *Processor 1.* A system without a feedback does not have its input conditioned by the output. The simple act of closing the loop (taking the output of the comparator and adding it to the input, via the optional *Processor 2*) can have a profound effect on system behaviour.

In our discussions throughout this book, we shall see how important a role feedback plays in software engineering. Feedback exists at many levels, practical as well as perceptual. An exception handler is a simple example of a feedback loop. It monitors the execution of a piece of code and takes appropriate action if the outcome is not as expected. On the other hand, modifying a system based on user response is also an example of feedback. In software engineering, often the difficulty lies in integrating the various forms and levels of feedback into a consistent and repeatable development model. This is the central challenge with the software engineering's responses to the problems of change and complexity.

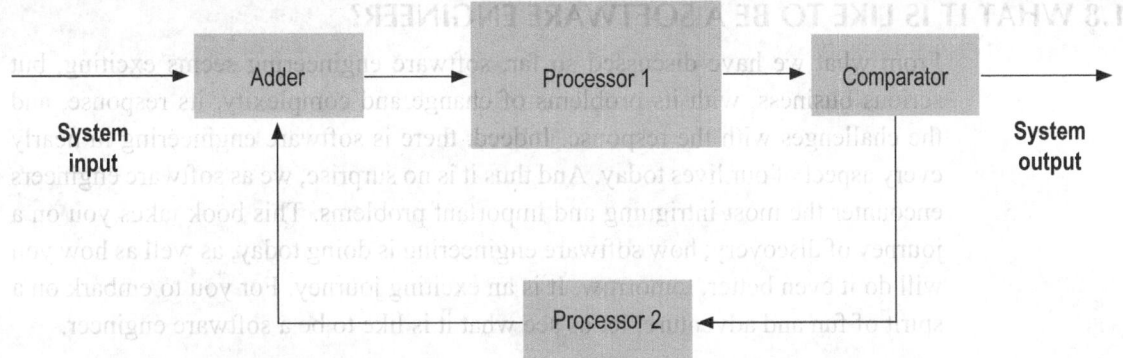

Figure 1.3 A simple feedback mechanism

1.7 GRAND CHALLENGE

A few years back, pioneering computer scientist, inventor of the *quicksort* sorting algorithm, and Turing award winner, C.A.R. 'Tony' Hoare, outlined a set of *grand challenges* for computing research [Hoare and Milner 2005]. According to Hoare, a typical grand challenge is like proving Fermat's last theorem (already accomplished), putting a man on moon (already accomplished), finding a cure for cancer in 10 years (not yet accomplished), and mapping the human genome (already accomplished). A grand challenge project typically lasts around 15 years, has world-wide participation, presents clear criteria for evaluating success, and offers a path-breaking advance in basic science and engineering.

One of the grand challenges Hoare identified for the next 15–20 years is 'Dependable Systems Evolution'. This aims to address the dependability of programs running in homes, offices, cars, planes, and rockets by developing tools and technologies to have the computer guarantee the integrity, safety, and correctness of its own programs. This guarantee should remain in place even as the programs evolve to deliver better service or meet new needs. The project is expected to illuminate the 'logical foundations of computer science and its application to software engineering' [Hoare and Milner 2005]. Note how the challenge to build dependable software systems involves *change* (better service, new needs) and *complexity* (multiple domains of operations, from homes to rockets): These are the same problems for software engineering we have discussed earlier in the chapter. The fact that someone of Hoare's erudition and experience identifies the problem of dependable software evolution as a 'grand challenge', points to how important it is for the world, and how difficult it is to solve.

Who will meet the grand challenge of Dependable Systems Evolution?

I am sure it will be you; the bright, young minds who study software engineering today and will research and practice it tomorrow.

1.8 WHAT IT IS LIKE TO BE A SOFTWARE ENGINEER?

From what we have discussed so far, software engineering seems exciting, but serious business, with its problems of change and complexity, its response, and the challenges with the response. Indeed, there is software engineering in nearly every aspect of our lives today. And thus it is no surprise, we as software engineers encounter the most intriguing and important problems. This book takes you on a journey of discovery; how software engineering is doing today, as well as how you will do it even better, tomorrow. It is an exciting journey. For you to embark on a spirit of fun and adventure, let us see what it is like to be a software engineer.

1.8.1 Knowing across Domains

As we discussed briefly earlier, and will discuss in detail in Chapter 3, software is unique amongst engineering artefacts in a number of ways. A bridge, a building, or a bicycle has some value per se. They help us cross a chasm, give shelter from the elements, or let us go from one place to another. On the other hand, a piece of software usually provides a *service* to an existing enterprise, helping get it done faster, better, and cheaper. Of course, additionally there are also truly game-changing uses of software. They let us do something that was never imagined before, like deciphering the human genome. But these lie in the realm of research for quite some time before getting integrated into the so-called mainstream of software engineering.

Over the course of a typical software engineering career, even when working on the usual customer-oriented projects, there is much scope for acquiring knowledge across a wide spectrum of application domains and industries. The domains of finance, travel, health care, entertainment, etc., are each different, with its own vocabulary, quirks, and challenges, as well as rewards. Some solutions work well for a particular domain, while wholly new ones have to be devised for others. Software engineers thus need to stay updated with working knowledge across various domains, in addition to refining their core skill of making software. This is not easy, but it provides for intellectual stimulation and variety not readily available in many other professions. It is enlightening to know of different industries, and their unique user expectations and functioning. Successful software engineers use this as an opportunity for hastening professional maturity.

1.8.2 Teaming across Cultures

Software engineering is very much a global enterprise now, and this trend will only grow in the future (see Chapter 21). This offers unique opportunities for interactions across cultures, even for those at the entry level. Indeed, to become a successful software engineer, it is becoming essential to acclimatize oneself quickly and easily to a variety of work environments. This includes an understanding of social, political, regional, and cultural sensitivities, in addition to technical and communication skills.

Being a team player is an important criterion for success in the engineering profession—after all, no serious engineering product comes out of the head or hand of a single individual. For software engineers, merely being a team player is not sufficient. The most successful software teams, the truly 'jelled' ones [Humphrey 1999] have the key ingredient of diversity. Most often, we need to work closely with people who are very different—in language, culture, background, and skill— yet united in a common professional purpose. Learning how to embrace and thrive in such environments is hardly something textbooks or class lectures can teach you. Your own attitude and temperament are your best—and often only—teachers.

1.8.3 Innovating across Technologies

As we mentioned before, software engineering is a young discipline; almost infant when compared to some of the conventional engineering disciplines. While this has its disadvantages—we are still groping for laws and first principles—it also makes our field fertile for innovation. With little discipline, focus, and imagination, every software engineer can innovate.

The burgeoning open source paradigm (see Chapter 22), has made it easy to interact with the software community at large. And this interaction fuels innovation. Ingredients for innovation are now available to every practicing software engineer as basic professional tools. Web access and efficient computing facilities, are necessary for software engineering innovation; but they are not sufficient. What remains is a key element: the willingness to think outside the box. Even amidst the grind of day-to-day work, with looming deadlines and delivery pressures, it is not impossible to think a little deeply on the most pressing issues at hand; why is a particular task taking so much time, how can performance of a component be improved, is there a general solution to a particular problem? Such 'lateral' thinking will alleviate the tension and ennui of everyday work. Also, sooner or later, it will lead to some innovation not only satisfying by itself but also offering valuable career boost. Hamming's description of how he developed his pioneering work on error-correcting codes, while doing his routine work is very inspiring [Hamming 1997] for today's software engineers looking to innovate. The facilities a software engineer has today even for routine work were unthinkable a few years ago. It is our onus—to society, to our profession, and most importantly, to ourselves—to utilize these facilities to their fullest.

SUMMARY AND TAKE-AWAYS

This chapter begins the book's journey of discovering what software engineering is, what its major challenges are, and how tomorrow's software engineers—the readers of this book—can stand up to those challenges and get way beyond them. Our discussion can be summarized as follows:

- There is no single universally accepted definition of software engineering; the essence lies in synthesizing the various definitions.
- Change and complexity are the two major problems confronting software engineering.
- To address the problem of complexity, software development is broken down into workflows, each of which addresses a specific concern in the development process.
- To address the problem of change, phases of software development monitors changes and their effects during the development process.
- Workflows and phases together constitute the software development life cycle; which lies at the heart of software engineering.
- The key challenge of software engineering is to be able to monitor, control, and utilize the many feedback paths that exist in real-world software development.
- Hoare has identified the evolution of dependable software systems, carrying with it the guarantee of acceptable behaviour across a wide variety of operating conditions, as one of the grand challenges of computing in the next 15–20 years.
- Knowing across domains, teaming across cultures, and innovating across technologies are the key elements of a software engineer's experience.

WHERE TO LOOK FOR MORE

Although software engineering is a young discipline, a body of informative and insightful writing has already been accumulated. The website http://tinyurl.com/100sebooks lists the so-called 'Top 100 Best Software Engineering Books, Ever'. While this may not be the definitive list—few of my own favourites are not featured—it does identify some very good books. Additionally, the author discusses the metrics he used in ranking the books, which may be generally helpful in choosing a good book.

EXERCISES

Review Questions

Review Questions test your understanding of the key concepts presented in this chapter.

1. Which of the following is not included in Whitmire's working definition of software engineering?
   (a) Economy
   (b) Use of a software system
   (c) Art
   (d) Performance

2. According to Brooks, which of the following is a characteristic of software?
   (a) Complexity
   (b) Changeability
   (c) Invisibility
   (d) All of the above

3. Software systems need to encounter the problem of change primarily because
   (a) users do not initially know what they want from software
   (b) user needs are complex
   (c) there is combinatorial complexity in software
   (d) of all of the above

4. Which of the following is not a concern associated with a workflow?
   (a) Testing
   (b) Feasibility study
   (c) Analysis
   (d) Implementation

5. Which of the following is a concern associated with a phase?
   (a) Testing
   (b) Feasibility study
   (c) Analysis
   (d) Implementation

Reflective Questions

Reflective Questions require you to think deeply about some of the ideas and come up with your own interpretations and answers.

1. Comment on the following statement in the light of the various definitions of software engineering: 'No matter how we define it, the most important component of software engineering is computer programming.'

2. Among the various definitions of software engineering given in this chapter, which one do you think comes closest to software engineering as you see it? Justify your answer.

3. Out of the four characteristics of software mentioned by Brooks (few decades ago), which one do you think is most relevant to software as it is perceived and used today, and which one the least? Support your choices with reasons.

4. Are invisibility and unvisualizability the same characteristic of software? If not, why? Explain with examples.

5. In this chapter, we have identified two major problems that software engineering needs to address. Can you correlate them with the characteristics of software Brooks identified?

6. Do you think the so-called cognitive gap mentioned in the context of software vis-à-vis other engineering disciplines is valid? Give reasons for your answer.

7. Change and complexity are the two major problems confronting software. Are these two related? Can the response to one serve the other?

8. How do you think workflows and phases are related? Are they aligned or orthogonal to one another?

9. Feedback is everywhere. Give an example of feedback and discuss the benefits it offers in that particular case and how the situation would have been without feedback.
10. How do you think we should approach the grand challenge of Dependable Systems Evolution? Is there a particular way that Hoare suggests? What do you think of his suggested path?
11. In this chapter, we have outlined some aspects of what it is like to be a software engineer. Which aspect attracts you most? Which one do you think is most boring?

REFERENCES

Bauer, F.L., Bolliet, L., and Helms, H.J. (1968), NATO Software Engineering Conference 1968, http://homepages.cs.ncl.ac.uk/brian.randell/NATO/nato1968.PDF, last accessed on Nov 8, 2009.

Booch, G. (2006), 'The Accidental Architecture', *IEEE Softw.*, 23(3): 9–11.

Brooks, F.P. (1995), *The Mythical Man-Month: Essays on Software Engineering,* 20th Anniversary Edition, Addison-Wesley.

Datta, S. (2007), *Metrics-Driven Enterprise Software Development: Effectively Meeting Evolving Business Needs*, J. Ross Publishing.

Hamming, R.R. (1997), *Art of Doing Science and Engineering: Learning to Learn*, CRC.

Hoare, T. and Milner, R. (2005), Grand challenges for computing research, *Comput. J.*, 48(1): 49–52.

Humphrey, W.S. (1999), *Introduction to the Team Software Process*, SEI Series in Software Engineering.

Nicolis, G. and Prigogine, I. (1989), *Exploring Complexity: An Introduction*, W.H. Freeman and Company.

Petroski, H. (1992), *To Engineer Is Human: The Role of Failure in Successful Design*, Vintage.

Saxe, J.G. (1850), 'The Blind Men and the Elephant', http://bygosh.com/Features/092001/blindmen.htm, last accessed on Nov 8, 2009.

Waldrop, M.M. (1992), *Complexity: The Emerging Science at the Edge of Order and Chaos*, Simon and Schuster.

Whitmire, S.A. (1997), *Object-Oriented Design Measurement*, Wiley Computer Pub.

Evolution of Software Engineering

Learning Objectives

In this chapter, we will discuss the evolutionary trends of software engineering. Specifically, we consider:

- Why knowing a little bit of software engineering history is helpful for a professional software engineer
- How software engineering has evolved across some of the major trends
- The important milestones in the journey of software engineering
- The so-called software crisis
- The developments that have collectively influenced the discipline of software engineering

---

2.1 MOTIVATION

This chapter is mainly about the history of software engineering. We will review some of the trends that have brought software engineering to where it is today and are likely to decide where it will be tomorrow. Why we need a chapter on history in a book on engineering is justified in the next section.

The story of evolution is always fascinating, be it of a biological species, an individual, or an entire discipline. Knowing this story endows us with a *perspective* that is essential in sensing which way the road goes next. Today's students are tomorrow's practitioners. It is imperative, especially for such a fast-changing area like software engineering, to equip students with the knowledge and skills that will help them most in the future. An understanding of evolutionary trends creates a foundation for such knowledge and skills.

Software engineering touches almost every aspect of our lives today. It wields critical influence on our needs and well-being. Surprisingly, something of such sweeping impact has evolved only over just a few decades. To measure the evolution of software engineering in time alone is not adequate. This is primarily due to software engineering's unique position (discussed in detail in Chapter 3).

As we shall see in this chapter, a number of interweaving trends, crisscrossing over time, have shaped the course of software engineering. Thus, while discussing these trends in the next few sections, we do not follow a chronological order. Later, we give a set of milestones by dates that mark the progression of the field.

In all that we paint in this chapter, details are not significant, but the broad strokes are. Unlike a typical history lesson, we need not pay special attention to dates, names, or places. So let us look forward to enjoy a good story and learn something from it.

2.2 NEED TO KNOW HISTORY

For as young a field as software engineering, history is not just something of the past. It is also very much what is unfolding in the present. As we shall see in the next few sections, many of the trends that have shaped software engineering and continue to do so are very much alive and amongst us. The rate at which computer technology moves is at once awe-inspiring and unnerving; we are left with fears of falling behind. Moore's law has reigned supreme over the last few decades: The number of transistors that can be economically placed on an integrated circuit has doubled every two years since the invention of the integrated circuit in 1958. Computing is becoming faster, cheaper, and more ubiquitous. When I was an undergraduate, this speed left me cold and grappling with the question: How do I know whether what I learn today will have any value tomorrow?

Addressing this question lies at the crux of learning a little bit of software engineering's history. A basic historical awareness can make us ready for change and what it means for individual careers, as well as the collective profession. Once again, the history we are talking about here is not one of dates, places, or names (and the annoying need to remember them). Neither is it exclusively a commentary of the past. In the threads of the story we are about to relate, details of the past may come, but they are not of primary importance. What is primarily important is the so-called 'big-picture', which tells us how and why one technology, or custom, or style, gives way to another, and how the new is always the harbinger of the newer.

The last chapter of this book (Chapter 23) reflects on a set of survival skills every software engineer needs to develop to meet their biggest professional challenge: obsolescence. I will not give away what that chapter says (you have to read the whole book to come to the last chapter; or perhaps you are like me, who reads the last, first), but those survival skills include a sense of history. Someone unaware of the past can hardly be expected to make reasonable projections about the future.

Hence, knowing a little bit of history is good for all, and we software engineers are no exception.

2.3 EVOLUTIONARY TRENDS

Over the next few subsections, we will explore some of the trends that have influenced software engineering and continue to do so. Nothing should be read into the order in which they are presented. The order is certainly not chronological, neither it is by importance. Also, even as we may remark both on the favourable or the unfavourable aspects of a particular trend, we do not intend to be judgemental about them. After all, we are talking about history here, which is the greatest judge of all. Besides, we are too close in time to many of these to know for sure whether they will turn out to be good or bad, in the long run. During the French Revolution's bicentennial in 1989, when asked about his opinion on the revolution's effect on human history, the Chinese patriarch Deng Xiaoping had reportedly said, 'it is too early to comment'! If bicentennial is too early, few decades is surely so.

2.3.1 Programming to Software Engineering

Many pioneering computer scientists, who have influenced software engineering to no little extent, like being called 'programmer'. Dijkstra's Turing lecture was titled 'The Humble Programmer' [Dijkstra, 1972], where he portrays with great humility, the great pride in his personal journey as a programmer. As Dijkstra says in the lecture, that when he started out, programming was not considered a profession enough to be mentioned in his marriage licence. He was entered as a 'theoretical physicist' instead, in deference to his PhD degree in physics, a subject which he had left for his love of programming. There are many other instances when the title of 'programmer' has been gladly clung to throughout a very distinguished career. Now, not even entry-level software engineers get designated as mere programmers. When did programming become software engineering? Or, has it?

As we shall see later in the book, programming plays a significant role in software development. This role has been redefined continuously with the evolution of software engineering. From a pre-eminently central position, programming has moved to be one of the concerns for the software developer. For a long time, software engineers have dreamt about the day when programs would write programs. In this chimera, we just feed in the Design to a *meta-program* and out comes the program we designed. This schema of development is widely followed in other engineering disciplines. Nowadays, a car is hardly built bolt-by-bolt by the human hand. An automated process (of course monitored and regulated by humans) transforms a Design into the physical artefact of the car. We are far from there in software engineering yet. Software engineers still have to program, building the system one line of a code at a time. In its functionality, programming seems close to the bolt-by-bolt building of a car, or the brick-by-brick masonry of civil engineering. But programming is also seen as an art; it has even been taken to be alike in some way to painting and poetry [Graham 2004], [Gabriel 2008].

This dichotomy represents programming's place in software development. Is it just the perfunctory act of implementing Design using constructs of the programming language, or is it the magic of doing smart and complex things through just a few cryptic instructions to the computer? In the former view, programming is the mechanical act of execution, hammering out lines of code as per a predefined master plan. This produces predictable results, barring inadvertent errors. In the latter view, programming is more about creativity, about the elegance and beauty of algorithms and language idioms. It may produce spectacular results at times, and not so good ones at others. Its outcome is much less predictable. Conventionally, predictability is a key aspect of engineering [Denning and Riehle 2009].

So, to make software development more an engineering, programming has to be seen essentially as an *executionary* step. Projecting this further, it is likely we may have those meta-programs in future to write programs for us, given the specifications. Many Integrated Development Environments or IDEs already generate skeletal code, and recent advances in Model Driven Development (MDD) have sought to tie in modelling with code generation more closely.

As later chapters of this book will illustrate, programs are only a part of an overall software system. So, even while giving programming its due in software development, there remain many key activities in producing a successful software system that are certainly not programming. Figure 2.1 shows how programming has been repositioned in the software development universe over the years. At the beginning, software development was perceived exclusively as programming. Then other concerns, such as Testing, began to be recognized. Later, programming became *one* of the concerns, not necessarily the biggest. Not too much should be read into the relative sizes, positions and orientations of the ellipses; the picture is illustrative only and not based on empirical data.

2.3.2 Hardware-Software: From Coupling to Congress

How do we differentiate between hardware and software? Well, hardware is *hard* (or at least tangible) to touch—all those circuits, transistors, and wires that lie in the guts of a computer. What about software? The name seems to suggest it should be soft to touch; but alas, only if we could touch it. We cannot because not only is software *not* amenable to touch, in Brooks' immortal characterization, it is also 'invisible' and 'unvisualizable' [Brooks 1995]. (For further discussions, see Chapter 1.) No wonder in the early days of computing, software was hardly recognized as something different from hardware then. Much of the instruction

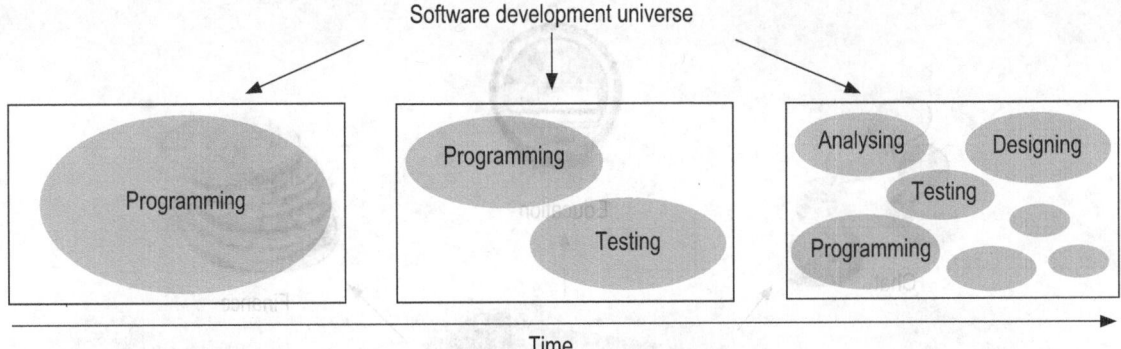

Figure 2.1 Repositioning of programming in the software development universe

needed to run a computer were hard-wired into the hardware. Additional instructions that came via software were specific to the hardware. When hardware was sold, software came with it; almost no one sold or bought software by itself.

Compare that with the present; probably the operating system is the only piece of software now that comes with the hardware we buy. Every other *application* software—those that let us do all those fun and useful things—is acquired and installed separately, often downloaded from the Web for free. Thus, the coupling between hardware and software has loosened considerably. Not completely however, as still there is software specifically designed to leverage some underlying hardware feature. There is also hardware that can support only a specific kind of software. But these tend to be confined largely to specialized areas such as scientific, high-performance and embedded computing.

Thus the erstwhile coupling can be said to have been replaced by something of a hardware-software congress. Different combinations of these two elements of the computer now allow a wide range of use. The beauty of the hardware-software congress comes largely from its uniqueness. As illustrated in Figure 2.2, a driver can make a car go fast, slow, reverse, forward, right or left. But the essence of the car's function, locomotion on land, cannot be changed. A car will neither fly nor float in its current configuration. Compare that with the range and diversity of uses we put our personal computers through. In the very least, we email, chat, do financial transactions, fetch information, and entertain ourselves. All of these are effectuated by *different* software running on the *same* hardware. If we get a different hardware tomorrow, chances are high these gratifications will continue unaffected. This is hardware–software congress in play; it has taken the coupling between these two elements to a new level.

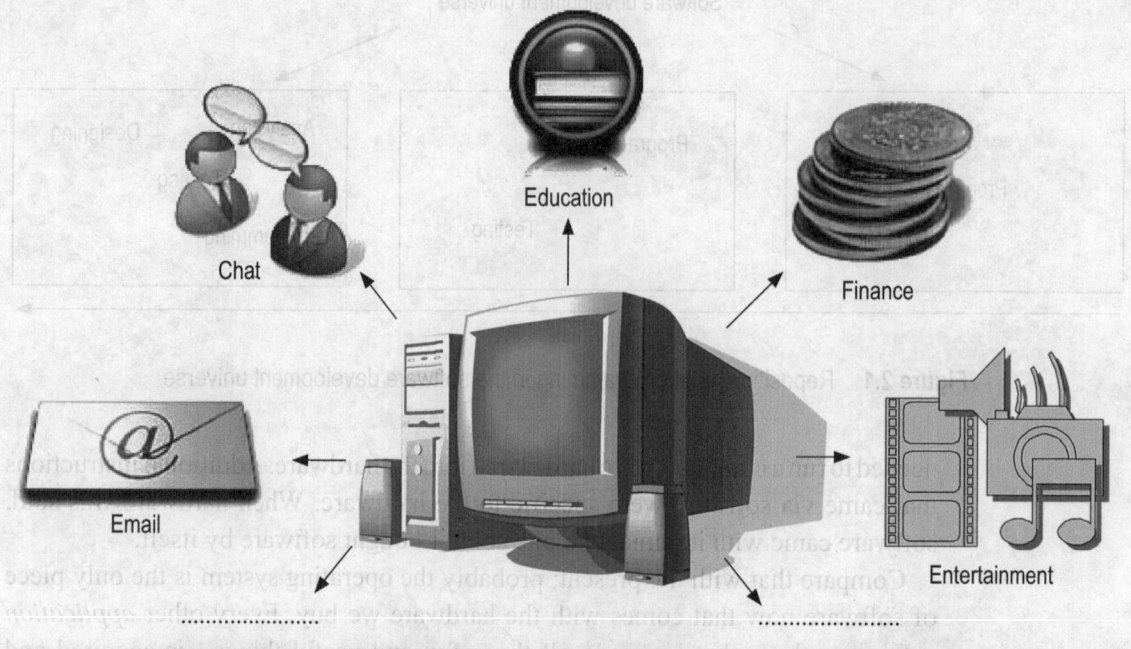

Figure 2.2 Beauty of hardware-software congress

2.3.3 Advent of High-Level Languages

I consider myself lucky to have had the experience of assembly language programming. And thankfully it was not in the real world, trying to satisfy paying customers, but in the cocoon of classrooms, when merely few exam marks were at stake. Assembly language programming is meticulous and intense. We communicate with the hardware using short, cryptic mnemonics; we have to keep track of all intermediate steps in calculations. Writing and running a program in assembly language to do even basic arithmetic made me break into a sweat. Yet that was exactly how 'real programmers' programmed in the early years of computing.

Programming after all, was seen to be not everybody's cup of tea.

Now, virtually anyone familiar with computers can pick up a book on basic Java for the first time, and write and run a 'Hello, World' program within few hours at the most. What has changed? Have humans become cleverer with computers, or have computers become cleverer with humans?

What has changed profoundly over the past few decades is how we communicate with computers. What has brought about this change is the vehicle of communication, that is, language. Very simply, the so-called high-level languages are programming languages whose syntax and idiom are closer to that of English. The evolution of high-level languages—as well as the case for calling them 'high-level'—has been discussed at length in Chapter 13, and we will not get into the details here. Let us instead try to get a flavour of what high-level languages allow us to do.

One of the main objectives of any language is to support *levels of abstraction* in communication. The word 'dog' is an abstraction for a certain type of animal. Whenever we hear, read, or say 'dog', we are sharing a common understanding of that type of animal. This understanding is not necessarily complete; it is said to be at a 'high-level'. In some cases, we may need more information about the dog; its colour, size, breed, etc. But the high-level abstraction of a dog denoted by 'dog' in English lets us communicate a commonly held conception of the animal. Now, what would be a low-level abstraction of a dog? One representation can be the 'genome, the dog's hereditary information encoded in DNA. It is certainly far more detailed and complete in describing the animal than the collection of the alphabets d-o-g. But using the DNA representation instead of the nifty little English word to refer to a dog in casual conversation is certainly more cumbersome. Note, however, that no sense of quality is stated or implied when we talk about high- and low-levels of abstraction. In this example, the low level abstraction of the DNA representation carries much more scientific information about a dog than the three-letter English word. High-level abstractions usually hide more details and make it easier for the human mind to broadly grasp an idea. But for someone who is only able to understand genomic information, the only way to talk about a 'dog' will be in terms of the DNA representation. Figure 2.3, shows these two abstractions of a dog. (The DNA representation is illustrative only and does not depict whole or part of the actual DNA of a dog.) For a detailed discussion of the place of abstraction in software development, see Chapter 12.

High-level programming languages can be thought to support 'dog'-like abstractions, whereas low-level ones can only communicate in terms of something like the DNA representations.

Very many of us have a working knowledge of dogs, but no knowledge of its genetic code, yet can go around life because English offers the 'dog' abstraction. Similarly, high-level programming languages let one be a programmer without

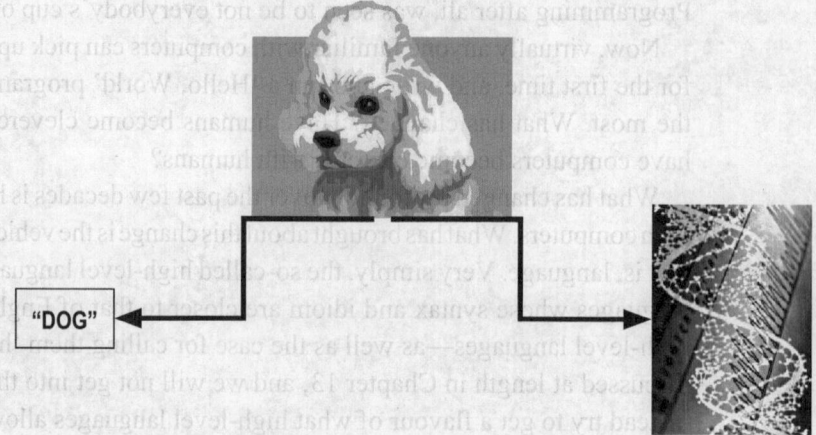

Figure 2.3 Abstractions of a dog

being able to communicate in machine or assembly language. So programming was no longer the domain of the nerds; 'programmer' was no longer an exclusive title. This caused an explosion in the number of programmers, which widened the scope of the software engineering profession.

2.3.4 Advent of the Personal Computer

With the advent of high-level language programming, programming became more plebeian. But a computer was still not everyone's cup of tea. In the 1950s and 60s, a computer was a considerable piece of equipment in price and size. Academic departments or large corporations owned them; it was unthinkable for individual users to have their own computers. Any computing pioneer will tell stories of how they 'timeshared' on terminals, very often giving the computer instructions on punched cards or magnetic tapes, and returning hours later to fetch the results of their program runs. When microprocessors—single electronic chips integrating large expanses of circuitry—began to be introduced in the early 1970s, a new breed of computers started emerging. The technology allowed computers to be made small and cheap enough to be owned by an individual, to serve his or her *personal* needs. Initially there was much skepticism as to the potential market for such a device. A senior executive of a very large computer company even publicly expressed doubt whether more than a handful of personal computers—or *PCs* as they were coming to be known—would ever be sold. The business decisions influenced by that line of thinking later came to be deeply regretted. Another very large software company was born through a contract almost accidentally won to develop operating systems for these PCs. It is estimated that more than one billion

PCs are in use in the world now. Certainly a far cry from that mere handful they were expected to sell in the early 1980s.

What we have been calling the PC till now is the so-called *desktop* computer. It gets its name from its usual position atop a desk at home or office. A sleeker and smaller mutant of the PC is the *laptop*, or the *notebook* computer. The notebook is now being upstaged by a still sleeker and still smaller cousin, the *netbook*, which is meant to be used primarily as a Web-enabled device. But the story does not stop here.

Computing is no longer confined to computers, personal or otherwise. The idea of *ubiquitous computing* involves spreading the functionality of computers to such devices as cellular phones and personal digital assistants. It is quite likely that any modern cellular phone has more computing power than all of the world's computers put together from just few decades ago. Computing will become even more widespread and transparent; it may not be long before clothes we wear will be PCs in their own right.

Computers have come from being isolated, redoubtable behemoths, to portable, friendly devices sitting on our desks or being carried around in hand bags. Concomitantly, there has been exponential growth in the amount, variety, scope, and power of the software that needs to support what we do with those 'personal' computing machines today. With the coming of the PC and its many subsequent variants, software engineers have become ever busier.

2.3.5 Global Software Development

As the Web's presence increased throughout the 1990s and then into the new millennium, software engineering became a truly global enterprise. This development has had far reaching implications on the very fabric of our profession, which we discuss at length in Chapter 21. It is customary now for the customers, users, developers and managers of a software project to be dispersed across continents, time zones, and cultures. This makes software development something of a unique enterprise; it is absurd to think of the customers, users, developers, and managers of a bridge building project to be so distributed, while the bridge is being commissioned, built, and used. In today's typical software projects, development proceeds round-the-clock, through the collaboration of offshore and onsite team members.

Global software development has deep economic as well as social implications. Perhaps for the first time in history, something close to the proverbial 'level-playing field' has been realized for a large-scale enterprise. Its impact on the dynamics of software engineering has also been profound. Additionally, global software development has put demands of cultural sensitivity, professional flexibility, and political awareness on the new breed of the global software engineer. To a large extent, this paradigm has re-written laws of international commerce, created new value-chains, and produced prosperity for many sections of the world. It has also brought new sets of challenges, such as transfer of jobs across geographies. India has so far been a pre-eminent hub of global software development. But we need to improve consciously and continuously to stay ahead in the field. Global software development will only increase in scope and diversity in the years to come.

2.3.6 Return of Open Source

We devote Chapter 22 entirely to the discussion of open source software development. Very simply, open source is about sharing the source code (as well as executables) of software developed by a particular group or individual for *free*, so that others can use or modify the code without paying any royalties. This seems rather detrimental to interests of the professional software engineer. After all, we are in the business of building software against remuneration. As we argue in Chapter 22, open source is no threat to the professional software engineer. But the main point about the *return* of open source is different, and that is what we discuss here.

Open source was nothing special in the initial era of computing. Companies sold hardware, and the software came with it, for free. Academics shared code just as they shared ideas. When software became a commercial commodity in its own right, large corporations made every effort to stifle the free flow of code. The coming of the Web in the early 1990s gave open source aficionados the fast, reliable, and cheap media they needed to connect and build exciting and useful software. And then give it away for free. This is how open source returned, after being in exile for many years.

Other than being an affirmation of intellectual freedom, the open source paradigm led to the re-examination of many fundamental principles of software development. Corporations large and small have come to recognize the power of open source and are keen to leverage it. The Linux operating system is one of the most visible demonstrations of how a very large, complex, and critical piece of software can be successfully built and continually supported by the open source community.

Open source has fundamentally changed software development, and the change is here to stay.

In Table 2.1, we summarize the essence and impact of each of the trends discussed above.

Table 2.1   Trends, essence, and impact

Trend	Essence	Impact
Programming to Software Engineering	Software development changes from being programming centric to recognizing other engineering concerns.	Enhanced awareness of the need for repeatable and consistent software production processes.
Hardware-Software: From Coupling to Congress	Software becomes less hardware specific and vice-versa.	Increased the demand for the development of general purpose software.
Advent of High-Level Languages	Programming languages are able to handle higher levels of abstraction.	Allowed software to better model real world concepts and processes, and implement more involved application logic.
Coming of the Personal Computer	Individuals can own portable computing devices.	Led to a great increase in the consumer base for software applications.
Global Software Development	Software development activities are distributed across continents and cultures.	Established a new paradigm of software development, with significant economic, social, and political implications.
The Return of Open Source	Supported by the burgeoning Web, the open source paradigm returns to mainstream software development.	Integrates a very large pool of motivated developers into the fold of software development.

2.4 MILESTONES IN SOFTWARE ENGINEERING

In the last section, we highlighted the trends that have influenced software engineering's evolution. These grew out of a series of milestones which are presented in Table 2.2 and Figure 2.4. The list of milestones in not exhaustive, we have selected the ones most relevant to the preceding discussion.

Table 2.2   Milestones in software engineering

Milestone	*Circa* Year
Design of Babbage's 'Difference Engine'	1849
Description of Turing Machine	1936
Unveiling of 'ENIAC': The first general-purpose electronic computer	1946
Invention of the first transistor	1947
Coming of the high-level programming languages	1957
Introduction of Simula: The first object-oriented programming language	1967
First use of 'software enginerring' in NATO-sponsored conference	1968
Coming of the ARPANET: Precursor to the Internet	1969
Invention of the first microprocessor	1973
Publication of Brooks' 'Mythical Man-Month'	1975
First 'PC' introduced in the market	1981
Coming of the World Wide Web	1992

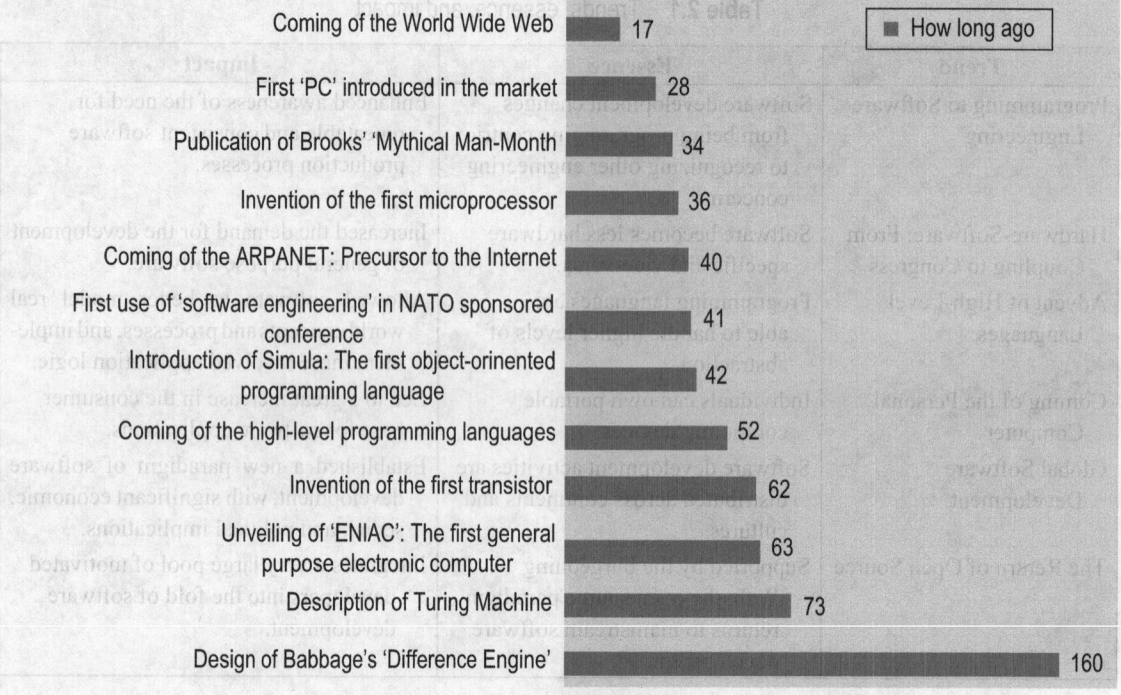

Figure 2.4   The number of years backwards from 2009 for the milestones listed in Table 2.2

2.5 TOWARDS A SLEW OF SILVER BULLETS

So far in this chapter, we have seen several important trends in software engineering's evolution. Which of these have had the maximum impact? Which have taken software engineering to the next level? In short, which has been a true *silver bullet*?

The allusion to silver bullets comes up often in software engineering. It draws from the celebrated 'no silver bullet' essay [Brooks 1995], which conjectures that there will be no new development in software engineering that brings in orders of magnitude improvement over past practices. The 'no silver bullet' prophecy has been debated endlessly, and it still remains a very polarizing issue. As each new trend erupts, there are claims and counter-claims of its potential as a silver bullet, followed by cynicism as it is inevitably found wanting (see Exhibit 2.1). Is this peculiarity—lack of silver bullets—unique to software? Do silver bullets pop up every now and then in other fields, and bring about dramatic improvements?

In medicine, the discovery of penicillin was indeed a great milestone. In electronics, the transistor brought about a generational shift. But to maximize their impact, each of these discoveries or inventions had to be followed up by many

Exhibit 2.1   Software Crisis

The public forum where 'software engineering' was first used—the NATO-sponsored conference of 1968—also coined the phrase 'software crisis' [Bauer et al. 1968]. This is ironical, as well as illuminating. Apparently, the need to have an engineering discipline of software arose out of a recognition of the crisis software was ensnared in.

Dijkstra, in his 1972 Turing Lecture, mentions how it was a 'blasphemy' to talk about a software crisis before the 1968 NATO conference. There it was admitted for the first time; there was a crisis, and something had to be done about it. The cause of the crisis is identified as: 'To put it quite bluntly: as long as there were no machines, programming was no problem at all; when we had a few weak computers, programming became a mild problem, and now we have gigantic computers, programming has become an equally gigantic problem.' [Dijkstra 1972].

Over the past few decades, 'software crisis' has been bandied about often, frequently as a casual (sometimes gloating) reference to the problems of software development. For example, an article provocatively titled *Software's Chronic Crisis* chronicles several egregious situations supposedly due to software failures. The author concludes with the statement, sweeping, and ominous, 'Until the lessons of computer science inculcate a desire not merely to build better things but also to build things better, the best we can expect is that software development will undergo a slow, and probably painful, industrial evolution.' [Gibbs 1994].

What are the symptoms of the so-called software crisis? They are known to almost every software customer, user and developer: budget and cost overruns, poor quality and performance, difficulties of maintenance and enhancement, even non-delivery of a promised system. These have happened and they continue to happen. Sometimes the software crisis is bracketed in a specific period of time, such as the 1960s, or the 1970s. The problem is that the crisis appears to come in every decade. Sometimes the crisis is said to be still persisting. Sometimes a bigger crisis is said to be looming ahead. One wonders how many structural, mechanical, electrical, or chemical crises were recorded for the respective engineering disciplines in their childhood and early youth.

Whether or not there *was* a software crisis is a contentious question, whether or not there *is* still one, is an even more contentious question. Instead of trying to settle them, let us hear what has been said about the advent of any transformational technology: 'First, there's a period of wild enthusiasm. ... Predictably, that fever passes, only to be replaced by utter disillusionment ... That passes, too, and the world finally gets down to the important work of taking the technology and integrating it into the structure and fabric of society and business' [Gerstner 2000].

Software is surely a transformational technology. Where does the software crisis fit into the cycle described above?

ancillary developments over time. A discipline moves forward through a slew of silver bullets, rather than single, unique ones.

All the software engineering trends discussed earlier have collaborated amongst and contributed to one another in taking the field forward. Software engineering of today is far from perfect. As we shall discuss in Chapter 3, we do not even have a set of basic principles like other engineering disciplines yet. But software engineering has come a long way to where it is today. It deeply influences how we do business, communicate with others, go from one place to another, entertain ourselves, and many other aspects of the very experience of living. Any particular trend may not have brought bliss overnight, and is unlikely to do so in the future. But over time, these trends together have shaped the discipline's journey. It is time we stopped searching for *one* silver bullet. The trends discussed in this chapter did not arise in isolation; they came out of interweaving threads of imagination and innovation.

Summary and Take-Aways

In this chapter we explored the evolution of software engineering. Our discussion can be summarized as:

- Software engineering has evolved over just a few decades. However, its evolution cannot be measured in time alone.
- With time, programming has lost its preeminent position in interrelated activities that make up software development. It has come to be seen as one amongst a set of software engineering concerns.
- The early trend of software being specific to hardware and vice versa has given way to significantly less hardware-software coupling.
- Advent of high-level languages enabled one to be a programmer without needing to program in machine or assembly languages. This caused an explosion in the number of programmers.
- The coming of the personal computer and its many variants has led to exponential growth in the amount, variety, scope, and power of the software applications that are run on these machines.
- The paradigm of global software development has distributed software development activities across economies and geographies.
- The return of the open source software development led to re-examination of many of the basic principles of software development and their use.
- 'Software crisis' is often used to denote difficulties of software development such as cost and time overruns.
- There may not be one silver bullet to solve all of software engineering problems. But the trends discussed in this chapter, in combination, have taken software engineering's journey forward.

WHERE TO LOOK FOR MORE

The following website has interesting resources on the history of software engineering:

- History of Computing: http://www.comphist.org/

EXERCISES

Review Questions

Review Questions test your understanding of the key concepts presented in this chapter.

1. Programming has been likened to
   - (a) painting
   - (b) translation of Design into code
   - (c) poetry
   - (d) all of the above
2. With time,
   - (a) software has become more coupled to hardware
   - (b) hardware has become more coupled to software
   - (c) a hardware-software congress has developed
   - (d) none of the above
3. Which of the following statements is correct?
   - (a) Assembly languages are at a higher level of abstraction than high-level languages.
   - (b) High-level languages are at a higher level of abstraction than assembly languages.
   - (c) Both are at the same level of abstraction.
   - (d) None of the above is correct.
5. Which of the following has had the most significant impact in establishing a global software development paradigm?
   - (a) The Internet
   - (b) The World Wide Web
   - (c) The microprocessor
   - (d) The transistor

6. Which of the following can be said to have directly hastened the rise of the personal computer?
   - (a) Invention of the vacuum diode
   - (b) Invention of the transistor
   - (c) Invention of the microprocessor
   - (d) All of the above

Reflective Questions

Reflective Questions require you to think more deeply about some of the ideas and come up with your own interpretations and answers.

1. 'To measure the evolution of software engineering in time alone is not adequate.' How would you justify this statement? If you cannot justify it, give three points in your defence.
2. While motivating the discussion of this chapter, we made the case for broad brush strokes being more important than details. Do you think this is relevant when we study the evolution of any field, or is this view unique to software engineering?
3. Why do you think Moore's law was mentioned in passing in this chapter? How do you think Moore's law is related to the evolution of software engineering?
4. Is there a dichotomy in programming's place in software development? Justify your answer.
5. What do you think is the most important

distinction between software and hardware other than their physical characteristics?

6. The concept of a dog denoted by the English word 'dog' versus its DNA representation has been presented as an analogy to the differences in abstraction offered by assembly language and high-level languages. What are the assumptions implicit in the analogy? When are those assumptions invalid?

7. Ubiquitous computing has many advantages, some of which can be inferred from the discussion of this chapter. What do you think are some of its disadvantages?

8. In your opinion, which other human enterprise is most similar to global software development? How does it differ from global software development?

9. 'This is how open source returned, after being in exile for many years.' What do you think is the most important factor in the putative return of the open source? Do you think open source can be sent to exile again?

10. With reference to Table 2.1, which of the trends do you think have had the most significant impact on the evolution of software engineering? Justify your answer.

11. Do you think the software crisis is still ongoing?

12. With reference to Figure 2.4, which recent development do you think qualifies as the major milestone after the coming of the Web?

REFERENCES

Bauer, F.L., Bolliet, L., and Helms, H.J. (1968), Nato software engineering conference 1968. http://homepages.cs.ncl.ac.uk/brian.randell/NATO/nato1968.PDF, last accessed on May 19, 2010.

Brooks, F.P. (1995), *The Mythical Man-Month: Essays on Software Engineering,* 20th Anniversary Edition, Addison-Wesley.

Denning, P.J. and Riehle, R.D. (2009), 'The Profession of It is Software Engineering?,' *Commun. ACM*, 52(3): 24–26.

Dijkstra, E.W. (1972), 'The Humble Programmer,' *Commun. ACM*, 15(10):859–866.

Gabriel, R.P. (2008), 'Designed as Designer,' In *OOPSLA '08: Proceedings of the 23rd ACM*

SIGPLAN conference on Object-oriented programming systems languages and applications, pp. 617–632, New York.

Gerstner, L. (2000), 'Lou Gerstner's E-business Expo Keynote Address,' http://www.informationweek.com/816/gerstner.htm, last accessed on May 19, 2010.

Gibbs, W.W. (1994), 'Software's Chronic Crisis,' http://www.cis.gsu.edu/~mmoore/CIS3300/handouts/SciAmSept1994.html, last accessed on May 19, 2010.

Graham, P. (2004), *Hackers and Painters: Big Ideas from the Computer Age*, OReilly Media, Inc.

Basic Ideas and First Principles

3.1 MOTIVATION

The standard approach to learn a subject is to start from the bottom. We learn about the basic ideas and first principles, in the hope that they will help us derive more involved concepts later. For example, we start off geometry with the axioms, graduate to the theorems, and then go on to solve riders using the axioms and theorems we learnt earlier. Such clear progression of ideas helps us learn a subject quickly and learn it well. Geometry has matured over millennia and the first principles seem to lead to the advanced ideas with silken ease. This is a chapter about the basic ideas and first principles of software engineering. Ideally, all else that is coming subsequently in this book should draw from what we present in this chapter. But the situation is not that *ideal* for software, and that is what this chapter is all about.

Every conventional engineering discipline draws its basic ideas and first principles from the underlying physical laws. By analogy, what are the laws that underpin software engineering? In general, how does software engineering compare with the other engineering disciplines in its approach? How do we characterize software and software engineering? These are the questions we consider in this chapter.

3.2 A WORD OF CAUTION

This chapter is titled Basic Ideas and First Principles. To set the right expectations, we should clarify now what this chapter can give, as well as what it cannot. Risking over-generalization, it may well be said that software engineering does not have its basic ideas and first principles. While this is too sweeping to be true, this is not entirely false either. As we shall soon see, software engineering is hardly able to offer ideas as fundamental or principles as primal as the axioms of geometry or the Ohm's law of electrical engineering. The starting point of software development is often a set of mere conjectures, as compared to the solidity of foundational principles of the other disciplines. In spite of this, we are building complex software systems that influence our lives in diverse and critical ways.

There is a definite tension between this current lack of completeness in the basic ideas and first principles and the need to build software to handle crucial responsibilities. The aim of this chapter is to explore that tension and highlight how software engineers can leverage it to their benefit, as well as avoid its pitfalls. We do not have all the answers we would like to have about the basic ideas and first principles of software engineering. But we do have a start towards creating that knowledge through our experience and intuition. And this is a start every software engineering career must be built on.

Thus the word of caution about this chapter boils down to: We may not find all the answers about software engineering's basic ideas and first principles here. But we will know the kind of questions we need to ask that will get us to the answers, eventually. After all, as we outlined in Chapter 2, software engineering's journey has just begun.

3.3 ARE THERE LAWS OF SOFTWARE ENGINEERING?

In other engineering disciplines, the basic ideas and first principles frequently come encapsulated in laws. Are there laws of software engineering?

Without giving a yes or no answer right away, let us see the layers of concerns in software development. With reference to Figure 3.1, Booch has identified the following levels in the *limits of software*: influence of politics, impact of economics, importance of organizations, problems of functionality, problems of Design, difficulty of distribution, challenges of algorithms, laws of software, and laws of physics [Booch 2005].

There is a hierarchy implicit in Figure 3.1. The figure can be read both ways; from top to bottom as well as from bottom to top. Starting with the laws of physics—the most definite in terms of clarity and representation—we end up with the influence of politics, the most ad hoc and fuzzy. On the other hand, politics represents a convergence of the human elements, whereas the laws of physics relate to the fundamental ways of the natural world. The so-called laws of software lie wedged between the laws of physics and the challenges of algorithms.

The laws of nature represent a discovery aspect, whereas algorithms are essentially inventions. Thus laws of software are seen to lie somewhere between discovery and invention, between what we have been given and what comes out of our skill and ingenuity. But what do we exactly mean, when we say laws of software?

At the top end of Figure 3.1 lie the laws of physics. For a software system, all information is ultimately carried by electrons, their behaviour governed by quantum mechanics. The medium in which these electrons operate are the integrated circuits of the computer's harware. Moore's law mandates that the number of transistors that can be placed inexpensively on an integrated circuit will grow exponentially, doubling approximately every two years, and it has held true from 1965 till date [Moore 2006]. On the other side of the laws of software is the world of algorithms. Turing's thesis (also called the Church-Turing thesis) relates to this world, it can be informally stated as: If an algorithm exists then there is an equivalent Turing

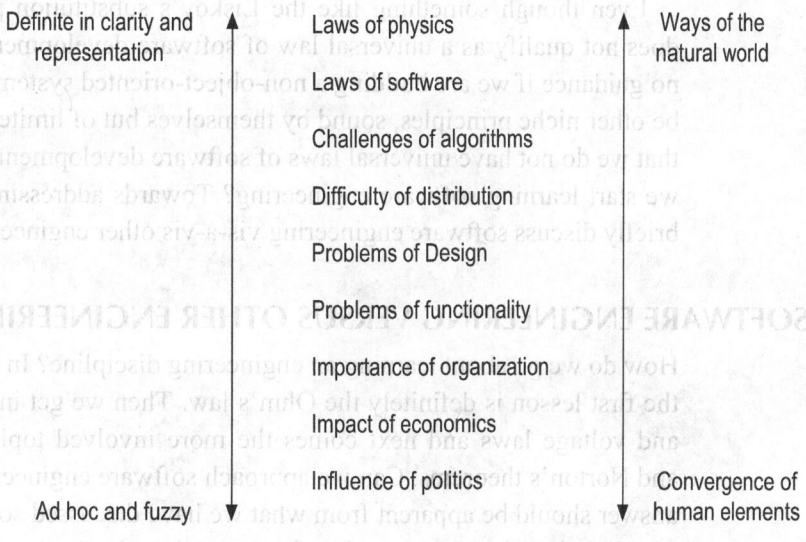

Figure 3.1   Limits of software (adapted from Booch 2005)

machine [Turing 1936]. (Very simply, a Turing machine is an abstract symbol-manipulating device that can simulate the process of any computation.)

Both Moore's law and Turing's thesis deeply influence software development. However, neither offers clear guidelines on specific concerns of software development; for example, how components should interact or how system performance can be enhanced.

Within particular paradigms of software development, we do have some specific principles, however. For example the Liskov's substitution principle defines the notion of *substitutability* in object-oriented programming [Liskov 1987]. (The object-oriented paradigm is discussed in Chapter 12.) It states that if *S* is a subtype of *T*, then objects of type *T* in a program may be replaced with objects of type *S* without altering any of the desirable properties of that program. This is basically saying that inherited classes should behave in a way such that they can be substituted for their ancestors without causing unexpected system behaviour. It is an important canon of object-oriented Design; we are reminded not to radically change behaviour of subclasses.

In Chapter 7 we will discuss the role of metrics in software development. There have been efforts at *empirically* deriving laws of program behaviour, Belady and Lehman (1976, 1979) proposed their laws of program evolution by scrutinizing a number of large systems. Recent studies on applying data mining techniques on software repositories have examined the validity of these laws. Belady and Lehman's work, though pioneering, does not give us guidelines that are universally applicable across software systems.

Even though something like the Liskov's substitution principle is helpful, it does not qualify as a universal law of software development. In fact, it can offer no guidance if we are building a non-object-oriented system. Similarly, there may be other niche principles, sound by themselves but of limited scope. The reality is that we do not have universal laws of software development yet. How then should we start learning software engineering? Towards addressing this question, let us briefly discuss software engineering vis-à-vis other engineering disciplines.

3.4 SOFTWARE ENGINEERING VERSUS OTHER ENGINEERING DISCIPLINES

How do we go about learning an engineering discipline? In electrical engineering, the first lesson is definitely the Ohm's law. Then we get into Kirchhoff's current and voltage laws and next comes the more involved topics around Thevenin's and Norton's theorems. Can we approach software engineering analogously? The answer should be apparent from what we have discussed so far; and the answer is no. How then do we approach software engineering?

Today, most software engineers learn programming first. Then, they are introduced to Design, and then into the wilderness of practice, occasionally via a smattering of processes and methodologies. This is a route very different from the laws that take us to the heart of other engineering disciplines. But then, is software engineering, an engineering? This is a question of much discussion, and much more contention. Let us get a feel of both sides of the debate.

Engineering can be defined as '... the application of scientific and mathematical principles to achieve the design, manufacture, and operation of efficient and economical structures, machines, processes, and systems' [Denning and Riehle 2009]. This definition broadly fits software engineering, but with a caveat software engineering also needs to play a significant role in managing the complexity of software itself. (A case for software's complexity has been made in Chapter 12.) However, to the end-user and customers, the most important criteria is not whether science and mathematics is applied to software engineering, but whether software engineering produces software systems that are usable, dependable, affordable, and flexible.

In general, the so-called 'engineering process' denotes a repeated cycle through Requirements, specifications, prototypes, and Testing. Several software development processes are in vogue (for more details, see Chapter 4). But none of them can deliver at a level of consistency (in terms of schedule, budget, quality, etc.) common in other engineering disciplines. Why is that? Following are some of the areas of divergence between the practice of software engineering and other engineering disciplines [Denning and Riehle 2009].

- *Predictable outcomes* Engineers are focused on ensuring there is lowest possible surprise in the production process so that its outcome can be accurately predicted. On the other hand, software engineers reared in the art of programming and/or computer science research often regard surprises and unpredictability of outcome as key motivators for software development.
- *Design metrics* Software engineers do not have corresponding metrics for allowable stresses, tolerances, performance ranges, structural complexity, failure probabilities, etc. These are routinely used in other engineering disciplines. Commonly used retrospective measures of software, such as lines of code or performance ranges, may present an oversimplified view of the problem domain. (These are, however, not the only software metrics, as discussed in Chapter 7.)
- *Failure tolerance* As Petroski has observed, failure plays a central and unifying role in all of engineering disciplines. The fundamental aim of any engineering effort is to prevent failure; and when it unfortunately occurs, to learn from it [Petroski 1992]. Software systems fail often and sometimes

critically. Yet software engineers do not have an established framework to document and share the lessons from such failures.

- *Separation of Design from Implementation* In other engineering disciplines, designers Design and handover the 'blueprints' to the Construction specialists. Software Design and Implementation together are too closely tied conceptually, to enable such a separation. The same software engineer usually has to do both, often simultaneously.

Thus it appears that a conventional engineering approach can both help, as well as hinder software development. Let us see how.

3.4.1 How an Engineering Approach to Software Helps

Engineering insists on consistency and precision in producing artefacts on time, within budget, and conforming to certain quality attributes. It has been reported (though not without controversy) that one in every three software projects fails to deliver working software, while another third delivers something that works but not the way users want it to. Even successful software projects need more time and money than was initially budgeted for, and large projects are often obsolete when they are delivered long after commissioning [Denning and Riehle 2009].

In the light of this situation, the typically engineering concerns of schedule, cost, and quality parameters are very relevant to software development. How this relevance can be translated to the activities of software development is far from straightforward though, as discussed at length in Chapter 17.

But, awareness of these engineering concerns is a key element that sets apart professional software development from mere programming.

3.4.2 How an Engineering Approach to Software Hinders

How do we build a bridge? Even if we may not have built a bridge ourselves, we have a fair idea of how one goes about it. Looking at it very simply, one analyses the chasm the bridge has to be built across, designs the bridge in terms of its components and how they fit together, and commits the design to construction. Then the drawings on paper are drawings no more, physical parts are made out of iron and brick and mortar and put together to make the bridge. Once the proverbial die has been cast, there is no going back.

On the other hand, how do we write? Not even the most consummate author can get a sentence right the first time, every time. Writing is essentially iterative. One keeps on refining what has already been written, hoping to take it closer and closer to some idealized level of expression. This is not unique to writing. Even painters and composers build their artefacts through countless iterations. Note how

Exhibit 3.1 Deriving from the First Principles

Towards the beginning of this chapter, we remarked that in any discipline, basic ideas and first principles lead to more advanced concepts. But how do we arrive at those basic ideas and first principles in the first place?

It is said that Man had been observing falling apples for thousands of years, but it took the genius of Newton to see gravitation in it. Recently there was an interesting news item titled 'Robot discovers laws of Newton in hours' (http://www.telegraphindia.com/1090403/jsp/frontpage/story_10767237.jsp), which claims a computer program 'has deduced Newton's laws of motion and laws of energy conservation by observing swinging pendulums without prior knowledge of physics.' For a serious discussion of the research behind the news, see the recent paper 'Distilling Free-Form Natural Laws from Experimented Data' by Michael Schmidt et al.; Science 324, 81(2009); DOI: 10.1126/science.1165893. Although the paper will surely be scrutinized further by the scientific community, its message sets us thinking on what it takes to deduce a fundamental principle. Do we need a large body of empirical data, a few insightful observations, or the genius of Newton?

different this is from the linearity of the engineering process; each return to something already written or painted or composed is different from the analyse-design-construct sequence of bridge building outlined earlier.

Small software systems with limited or no unknowns and negligible risk can be built the way bridges are. Indeed, the Waterfall model closely mimics this approach. But the iterative approach works much better for large-scale software development, due to the very nature of software and user expectations from a software system. (For details, see Chapter 4.)

Thus, a linearly defined engineering development cycle is often more of a hindrance than help, when blindly applied to software development.

3.5 CHARACTERIZING SOFTWARE AND SOFTWARE ENGINEERING

In the light of the preceding discussion, how then do we characterize software engineering?

To address this question, let us try and summarize some of the salient features of software and software engineering [Datta 2007]. We may have discussed a few of them before. They are repeated to highlight a general characterization of the discipline.

3.5.1 No Laws of Software Engineering, Yet

As we have observed, every other engineering is underpinned by natural laws, which serve as the foundation on which the engineers build their structures.

Software lacks these types of laws. Laws often are not formulated through mere cogitation; large collections of empirical data can play an important role in understanding and formalizing them. Similar efforts at collecting data for software engineering projects have just begun: Booch has commenced work on a handbook on software architecture, whose mission is to 'fill this void in software engineering by codifying the architecture of a large collection of interesting software-intensive systems, presenting them in a manner that exposes their essential patterns and that permits comparisons across domains and architectural styles' [Booch 2006]. At the handbook's website Booch quotes from Alexander's *The Nature of Order,* highlighting that Alexander and his colleagues 'made observations, looked to see what worked, studied it, tried to distill out the essentials, and wrote them down'. It may not be long before the first vestiges of software laws appear. But till then, we do not have laws to guide or constrain software development.

3.5.2 Development versus Production

DeMarco and Lister's classic book, *Peopleware,* has a chapter titled 'Make a Cheeseburger, Sell a Cheeseburger'. It begins with the credo, 'Development is inherently different from production' [DeMarco and Lister 1987]. Using the production of cheeseburger as a metaphor, the authors make interesting points about how making software is inherently different from assembly line production of any other engineering artefact.

DeMarco and Lister's points of view remain fresh even now. Although software making can benefit from certain aspects of the engineering approach, software is *developed* and not *produced*. What is this great divide between production and development? Production, in the conventional engineering context, can be thought to be the repetitive application of well-established rules and techniques, towards the manufacturing of a clearly defined product. The act of production, at its most sophisticated form, does not need human presence (humans may need to monitor or manage, though). Software cannot be produced in the same way as the rules and techniques for its development are not yet as universally established. But more importantly, the end product for software development is seldom well-defined at the beginning of development. This is primarily because people who want a software system developed often do not know what they exactly want from the system. This human element in software 'development' makes it so different from the 'production' of other engineering disciplines.

3.5.3 Plasticity of Software

A piece of software is never complete, in the sense a building or a car is. To significantly modify a building or a car, often the most sensible way is to make a new building or a new car. On the other hand software always seems so tantalizingly close to a quick tweak, which will make new and great things happen. This attribute is what may be called the *plasticity* of software. 'Plasticity' is used here in the sense of flexibility [Maier and Rechtin 2000].

It is true; changing one line of code in thousand lines can cause the behaviour of a software system to change dramatically, for better or for worse. But how much should we use such a facility? There is a tendency, driven by economics, indiscipline or just the spirit of adventure, to keep on changing code to make it do different things. It works spectacularly for a while and then fails just as spectacularly.

But code has to be worked on to make it better, and worked on in a streamlined and disciplined way. The idea of refactoring—practiced by many and crystallized by Fowler in his book *Refactoring: Improving the Design of Existing Code* and subsequent work [Fowler 2006]—is something essentially unique to software. (Refactoring is discussed in detail in Chapter 16.) It is unreasonable to think of refactoring a building, a bridge, or a car. The very need for software to be refactored points to the fact that its Design degenerates as the system is used and modified.

Software is perhaps the only engineering artefact that is customarily applied for purposes widely different from the intents of its original Design; and expected to deliver in ever-new circumstances at original or heightened levels of performance. The absurdity of a family car doubling as a Mars Rover is pretty obvious. But similar range of versatility is often demanded from software systems. 'Legacy' software written decades earlier is still in active service. They are successfully supporting vital business and other needs, way beyond the context and capacity for which they were built.

So, software development is a continuum rather than a one-off activity, and software never stops being built. This is one of the unique aspects of software and its engineering.

3.5.4 Macro- and Micro-states

Engineering usually deals with the macro-state of things. But conventional engineering also has a very good understanding of the micro-states that contribute towards a particular macro-state. Take for example Joule heating. It is something we studied in high school physics; the increase in temperature of a conductor when electric current passes through it. Many mechanical and electrical engineering processes are influenced by this phenomenon. Just as Joule heating can be perceived at the macro-level—the conductor getting hot—it can also be explained at the micro-level—the behaviour of molecules that lead to the rise in temperature. But similar connections between the macro- and the micro-levels in software are not yet as widely or deeply understood.

Kirwan gives an example which shows how the macro- and micro-states do not readily correspond in the software world [Kirwan 2000]. He talks about activating a computer, launching a word processing program, and on the blank screen, typing the word 'entropy' (any other word will do, just as well) and then deleting it. Before the word was typed and after it is deleted the screen is seen to be blank. Thus, evidently the macro-states are the same. But are the underlying micro-states, contributing to the same macro-state before and after the word was written and deleted, the same? Hardly so. Simply clicking 'undelete' or 'undo edit' or 'ctrl-Z' may bring 'entropy' back on the screen. So the information was somewhere in what is loosely called the computer's 'memory' even when it was not on the screen. One of the conclusions the author draws from the observation is: 'In principle, all macro-states are observable or distinguishable; that is, each can be assigned a unique name. Micro-states may either be distinguishable or indistinguishable.' Software engineers still have a shallow understanding of the gamut of micro-states that affect the macro-state of a working program. We know for sure that everything is ultimately stored and transported via 0's and 1's in a computer, but how that relates to utility we derive from using a computer is not easily apparent.

Lack of even reasonably complete understanding of the macro- and micro-states is a unique characteristic of software.

3.5.5 Importance of the Human Aspects

In *The Psychology of Computer Programming*, Weinberg lays down the book's purpose as 'to trigger the beginning of a new field of study: computer programming as a human activity, or, in short, the psychology of computer programming' [Weinberg 1971]. Several decades have passed since these visionary words. In the meantime, computer programming has graduated to software engineering. Yet the human aspects of building software systems remain hardly understood (or worse, misunderstood) and ignored.

One of the central themes of this 'human activity' viewpoint is communication. Sometimes communication is confused with public relations, or even marketing. For a software engineer, communication plays a deeper and more fundamental role. When a building is commissioned for office, hospital, school, residential, or any other purposes, the users have a clear idea of what the finished product needs to be, how they will use it, even how they should not use it. This understanding is almost subliminal, coming as it does from the fact that buildings have been in use for centuries, if not millenniums. So, the cognitive gap between what is wanted and what is engineered is manageably narrow. But it has been only a few decades since software systems are being used by people other than those who build them. So not knowing what they want is a common characteristic amongst software users. Accordingly, a software engineer's first task is to listen. To listen and to hear what is being said, but additionally, á la the solitary reaper, to hear what is not being said. Without being a good communicator, a software engineer can seldom be good at his or her job. No other engineering needs this degree of human interaction. We have discussed the human aspects of software development in detail in Chapter 9.

3.5.6 Concept of Co-evolution

In the 2000 Turing lecture, titled 'The Design of Design', Brooks brought to light how making buildings differs from building software [Brooks 2000]. For a building, customers pay for a design phase, have the design approved, and contract its implementation; or a builder pays for the design phase and sells its implementations. According to Brookes, software Design has to let go of the 'spurious' assumption that 'function and performance are what matters about software'. He goes on to mention other attributes that matter: reliability, changeability, structure, and testability. Brooks advocates the co-evolution perspective: A conjoint evolution of the problem and solution spaces, the former coming more to light as the latter is progressively explored. Co-evolution leads to an understanding of the solution influencing understanding of the problem and vice versa. We may note how different this view is from the conventional engineering approach discussed earlier in this chapter. The phenomenon of co-evolution deeply influences many aspects of software development. (See Chapter 9 for a deeper discussion of co-evolution.)

3.6 TYING THE THREADS TOGETHER

In this chapter, we have been viewing software engineering through a variety of lenses. Is there a final word, putting into perspective all that we have said and heard so far?

A question commonly asked is whether software development in sum should be viewed as an art or a science. Perhaps a better question would be whether software development is a craft or an engineering. As we have seen in this chapter, software can be *crafted* like an essay or a painting, going back and forth iteratively, improving over refinements. We have also seen how it needs to be *engineered* to satisfy the time and cost constraints, and deliver consistent quality. This craft-engineering dichotomy lies at the heart of software development. There is an element of craft in every engineering. It is more apparent at the beginning, and becomes subtle as the field matures. Till it assumes a reasonable maturity in comparison with other engineering disciplines, the craft-engineering dichotomy will remain significant in software development. As software engineers, we have to live with this dichotomy, leverage its advantages, and work around its disadvantages.

For a software system that is in use, it is important to recognize another nearly universal phenomenon. It is referred to in many different ways: continual development, moving target problem, etc. Gabriel interprets the idea of *piecemeal growth* as introduced by Alexander, in the software context, as: 'Piecemeal growth is the process of Design and Implementation in which software is embellished, modified, reduced, enlarged, and improved through a process of repair rather than of replacement' [Gabriel 1998]. In other words, we try to change an existing software system to fit our changing needs, instead of replacing it with a new system. This captures the very essence of software development (see Chapter 18).

On one hand, a software system is in a constant flux due to changing user needs, changing operating environments, changing technology, and changing business concerns. While on the other, the flux is due to the mutation of the software system itself—'piecemeal growth'—in response to these changes. This flux represents the biggest challenge towards building robust, resilient, and beautiful software systems. Accordingly addressing the craft-engineering dichotomy and piecemeal growth are central themes of software engineering.

In the following chapters, we will explore these themes in depth.

SUMMARY AND TAKE-AWAYS

In this chapter, we explored the basic ideas and first principles of software engineering. Our discussion can be summarized as follows:

- Unlike other engineering branches, software engineering does not relate directly to natural laws. The laws of software are yet to be discovered.
- Due to the relative youth of the discipline, we do not yet have a full grasp on the basic ideas and first principles of software engineering.
- There is a definite tension between this lack of completeness in the basic ideas and first principles and the need to build software to handle crucial responsibilities.
- Moore's law and Turing's thesis influence software development but they are not laws of software development.
- Liskov's substitution principle offers valuable guidance for object-oriented Design, but is not a general software Design principle.
- Software engineering differs from other engineering disciplines in the perception of predictable outcomes, Design metrics, failure tolerance, and separation of Design from Implementation.
- The typically engineering concerns of schedule, cost, and quality parameters are very relevant to software development, also.

- Due to the very nature of software, a linearly defined engineering development cycle is often more of a hindrance than help.
- Software development is inherently different from the production of other industrial artefacts.
- The plasticity of software refers to the flexibility of its Design and use.
- Unlike other engineering disciplines, the relationships between the macro- and micro- states are not always clear for software.
- Due to the cognitive gap inherent in the use of software, human aspects have major importance in software engineering.
- The concept of co-evolution—simultaneous evolution of the problem and solution spaces—is a key characteristic of software.
- This craft-engineering dichotomy lies at the heart of software engineering.
- Piecemeal growth, or the fact that we try to change an existing software system to fit our changing needs, instead of replacing it with a new system, captures an essence of software development.
- A software system is in a constant flux due to changing user needs, changing operating environments, changing technology, and changing business concerns.

WHERE TO LOOK FOR MORE

- The Software Engineering Institute's website has links to a number of interesting resources: http://www.sei.cmu.edu/

═══ EXERCISES ═══

Review Questions

Review Questions test your understanding of the key concepts presented in this chapter.

1. Moore's law relates to
   (a) algorithms
   (b) the number of transistors on an integrated circuit
   (c) the interaction of software components
   (d) none of the above
2. Turing's thesis is about
   (a) algorithms
   (b) the number of transistors on an integrated circuit

   (c) the interaction of software components
   (d) none of the above
3. Which of the following is not a part of the so-called engineering process?
   (a) Specifications
   (b) Prototypes
   (c) Testing
   (d) None of the above
4. Piecemeal growth relates to
   (a) replacing a software system
   (b) repairing a software system
   (c) constructing a software system
   (d) retiring a software system

Reflective Questions

Reflective Questions require you to think more deeply about some of the ideas and come up with your own interpretations and answers.

1. In the day-to-day work of a software engineer, which of the elements depicted in Figure 3.1 do you think has the most influence and which the least? Justify your answer.

2. We have reflected how laws are usually discovered and algorithms invented. Can we invent laws, or discover algorithms? Justify your answer with examples.

3. Can the Liskov's substitution principle be extended to a universal law of software development? Will there be any assumptions?

4. A definition of engineering has been suggested in this chapter. What would the corresponding definition of software engineering be?

5. In which aspects do you think software engineering and other engineering disciplines come closest in similarity, and in which are they farthest apart?

6. Do you think a more useful approach to large-scale software development is iterative rather than linear? If not, why?

7. Do you agree with the statement 'development is inherently different from production', in the software context? Justify your answer. Is the statement valid for any artefact other than software?

8. Is co-evolution unique to software? Can you think of some other area where it is relevant?

9. In your opinion, is software engineering, as it stands today, more of a craft than an engineering?

10. Why do you think piecemeal growth is related to the essence of software development?

REFERENCES

Belady, L. and M.M. Lehman (1976). 'A model of large program development, IBM', *IBM Systems Journal*, 15(3): 225

Booch, G. (2005). *The complexity of programming models*. AOSD '05, Chicago, US, March.

Brooks, F.P. (2000), 'The Design of Design', http://weatherhead.case.edu/requirements/Brooks_Plenary_0604.ppt, last accessed on May 19, 2010.

Datta, S. (2007), *Metrics-Driven Enterprise Software Development: Effectively Meeting Evolving Business Needs*, J. Ross Publishing.

DeMarco, T. and Lister, T. (1987), *Peopleware: Productive Projects and Teams*, Dorset House Pub. Co.

Denning, P.J. and Riehle, R.D. (2009), 'The Profession of IT: Is Software Engineering Engineering?', *Commun. ACM*, 52(3): 24–26.

Fowler, M. (2006), Refactoring home page, http://www.refactoring.com/, last accessed on May 19, 2010.

Gabriel, R.P. (1998), *Patterns of Software: Tales from the Software Community*, Oxford University Press, US.

Kirwan, A. (2000), *Mother Nature's Two Laws: Ringmasters for Circus Earth–Lessons on Entropy, Energy, Critical Thinking and the Practice of Science*, World Scientific Publishing Company.

Liskov, B. (1987), 'Keynote Address – Data Abstraction and Hierarchy', In *OOPSLA '87:*

Addendum to the proceedings on Object-oriented programming systems, languages and applications (Addendum), ACM, pp. 17–34, New York, US.

Maier, M.W. and Rechtin, E. (2000), *The Art of Systems Architecting*, 2nd ed, CRC Press.

Moore, G.E. (2006), 'Cramming More Components onto Integrated Circuits, reprinted from electronics', vol. 38, number 8, April 19, 1965, pp.114 ff. *Solid-State Circuits Newsletter, IEEE*, 20(3): 33–35.

Petroski, H. (1992), *To Engineer Is Human: The Role of Failure in Successful Design*, Vintage.

Turing, A. (1936), 'On Computable Numbers, With an Application to The Entscheidungs Problem', http://web.comlab.ox.ac.uk/oucl/research/areas/ieg/e-library/sources/tp2-ie.pdf, last accessed on May 19, 2010.

Weinberg, G.M. (1971), *The Psychology of Computer Programming*, Van Nostrand Reinhold.

PART II

PLANNING AND MANAGING SOFTWARE DEVELOPMENT

Software Development Methodologies

Learning Objectives

As any engineering discipline matures, methodologies are devised to guide the repetitive and large-scale production of artefacts. In this chapter, we discuss the following topics:

- The role of methodologies in software engineering
- Important software engineering methodologies
- The relevance of different methodologies in different development scenarios

4.1 MOTIVATION

> "If you want to build a ship, don't drum up the men to gather wood, divide the work and give orders. Instead, teach them to yearn for the vast and endless sea."
>
> —Antoine de Saint-Exupery

If the words above were as apt for software development as they are inspiring, software engineering would not need any methodologies. But the reality is somewhat different. Software engineers operate under many situations, constraints, and pressures; they also have a variety of backgrounds and skill sets. A methodology is one way of ensuring that in spite of such diversity, an end product of consistent quality can be repetitively delivered.

Over software engineering's journey in the past few decades, many methodologies have come up and many have fallen by the wayside. That there has been so much trial and error points to the fact that deciding on a sure-shot way of developing quality software is not easy. Methodologies operate in a flux; they need to be continuously refined if they are to stay current. Often, one methodology begets another. This chapter narrates the story of software engineering methodologies.

In the following sections, we build a case why methodologies are needed in the first place and follow up with a discussion of the software development life cycle. We then discuss how algorithm, process, and methodology differ from one another and why we need to understand the difference. Next, we highlight two different development philosophies that underpin almost all the major software development methodologies. A review of methodologies is presented subsequently, followed by a discussion on the relevance of methodologies in real-life scenarios.

4.2 A METHOD TO THE MADNESS

To be able to appreciate the place of methods and methodology in the context of software development, we need to clarify our understanding of these terms. In colloquial use, *method* is defined as 1. a way of doing something, 2. orderliness of thought or behaviour (http://www.askoxford.com/concise_oed/method?view=uk). The first definition tallies with our notion of a guided sequence of steps towards a particular end. For example, in middle school, all of us learnt a (rather messy!) method of dividing one algebraic polynomial by another. The method did not guarantee the correct result whenever we applied it—we had to guard against computational errors—it was nonetheless an assured procedure for arriving at the quotient and remainder when one polynomial divided another. The second definition of 'method' reflects on something of the opposite of disorder in thought and action. It is precisely this meaning we use in the title of this section, to suggest an element of regularity in the seemingly chaotic realm of software development. Both the meanings of 'method' blend together when the word is used in the software engineering context—we seek to use methods to have a *method* for the production of quality software. Note how we used 'method' in the plural the first time it was used in the last sentence. The plurality of methods brings us to methodology. Methodology is defined as a system of methods used in a particular field (http://www.askoxford.com/concise_oed/methodology?view=uk).

Let us go back to the algebraic example. In school algebra, we do not learn just the method for polynomial 'long' division, but along with it, a slew of other sleek or stodgy methods for manipulating algebraic expressions. Algebra, as we know it in middle school, can be thought of as a collection of these various methods— it is more a methodology for manipulating symbols and their relationships. A methodology offers a collection of procedures for performing a set of related activities in a *consistent* and *repeatable* manner. As we shall see later in this chapter, consistency and repeatability are key characteristics of any useful methodology.

These ensure that no matter how widely varied the inputs are, the output from applying the methodology will lie within an expected range. Software engineering methodologies are a collection of procedures to guide the activities of the software development life cycle (SDLC). Before we go deeper into the methodologies, it is imperative that we discuss the SDLC.

4.3 SOFTWARE DEVELOPMENT LIFE CYCLE

In general, whenever we refer to life cycles, we are talking about a progression of stages in the existence of an entity, usually with a clearly defined start and end. We see the human life cycle play out in our lives—birth, childhood, adulthood, old age, and death. These are just some of the broad epochs or events; one can always look at finer details—gestation precedes birth, adolescence comes between childhood and adulthood, and each culture has its own conjectures about what succeeds death. We also observe that every stage of any life cycle, in spite of being clearly distinct from others, may not have clear start or finish lines. True, in the human life cycle, adulthood officially begins at the age of eighteen. But this is more a legal specification than a physical or psychological one. For all practical purposes, growing up is a gradual phenomenon and we often cannot pinpoint a date and time when we ceased to be a child and became an adult.

The software development life cycle (SDLC) can be thought of as a series of circumstances a software system goes through from the time it is *conceived* by someone who needs the software to deliver some value, to the time it is designated to be of no further practical use (although it may still retain historical or academic value). Simply put, SDLC is the story of how a software system is born, how it matures, and how it dies. However, none of the states of being born, maturing, and dying, can be pinned to specific points of time in the software's lifetime. Strictly speaking, there should be a difference between a software system's life cycle and its development life cycle. The former is a commentary on the stages in the system's life, while the latter is a view on the activities that must be performed to *build* the system. But this differentiation is seldom made, not out of negligence, but due to a peculiarity of the software medium, as discussed in preceding chapters.

For any other engineering product—from a bolt to an automobile—once the product has left the factory and is in the hands of the end-user, it follows its own path from use (or misuse!) to obsolescence or breakdown. A particular bolt or an automobile is never expected to be radically improved upon by its developers even as it is in use. New and improved versions of the same product may come out or the product may be repaired, but a given bolt will never be expected to be enhanced to serve also as a bottle opener, or an automobile sought to be turned into an airplane.

But such ambitious enhancements, driven by changing user needs, are routinely expected from software systems. We discuss these concerns in detail in Chapters 17 and 18. Software development never really ceases for the software system to live a life of its own; a continuum of changing Requirements and concomitant development envelope a piece of software's existence. The orientation of the software development life cycle vis-à-vis the software life cycle is illustrated in Figure 4.1.

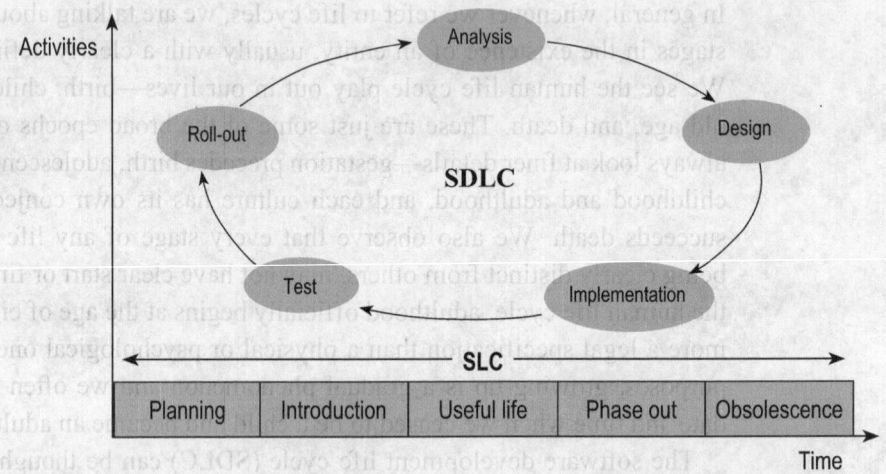

Figure 4.1 Software development life cycle (SDLC) versus software life cycle (SLC)

As mentioned earlier, the SDLC is a collection of activities that facilitate the building of software systems. The activities include specifying and understanding Requirements, Analysing them, Designing the system's components and their interaction, Implementing the Design, Testing the Implementation, and handing over the system to the end-users. Even as the software development process does not end here—software needs to be continuously modified and maintained—these are the main activities that go into building a software system. They are often referred to as *workflows* to highlight the *flowing* nature of the work from one activity to another. The other dimension of the development life cycle consists of the *phases* of Inception, Elaboration, Construction, and Transition, which identify the various epochs in a software development project. We shall discuss these in more detail in Chapters 11 to 16. Let us for the moment regard the SDLC as something of a black box, and see how different methodologies suggest different ways of going about it. Before wading into the various—often contradictory—ways of building software, we will briefly discuss three important concepts.

4.4 ALGORITHM, PROCESS, AND METHODOLOGY

The terms *algorithm*, *process*, and *methodology* are often used in similar contexts with vaguely similar connotations, which is very misleading. Let us clarify what they mean. The IEEE Standard 610.12-1990 defines *algorithm* as: 1. A finite set of well-defined rules for the solution of a problem in a finite number of steps; for example, a complete specification of a sequence of arithmetic operations for evaluating sine *x* to a given precision, 2. Any sequence of operations for performing a specific task. The same IEEE Standard defines *process* as: 1. A sequence of steps performed for a given purpose; for example, the software development process, 2. An executable unit managed by an operating system scheduler, 3. To perform operations on data. No definition for *methodology* is given in the aforementioned IEEE Standard, but we may revisit the meaning we have given earlier: Methodology is a collection of methods.

In the software development scenario, *algorithms* most closely relate to computer programs, *processes* to coordinated tasks that call for the involvement of people (usually in addition to computer programs), and *methodologies* extend across computer programs, people, and organizations, as illustrated in Figure 4.2. So, algorithm, process, and methodology respectively embrace the computational, behavioural, and organizational aspects of software development. In this book, unless otherwise clarified, we will use each term in its particular sense. We may note, however, that in some cases the distinction between methodology and process is somewhat blurred, as in the Unified Software Development Process [Jacobson et al. 1999] or the Rational Unified Process [Krutchen 2004] (see Chapter 5). Strictly speaking, the former is more of a methodology while the latter is more of a process, even as both call themselves process.

In a software engineering career in the industry, as one moves from being a programmer, to an analyst, a designer, a technical lead, an architect, a project manager, and so on, one's primary concern shifts from algorithms (how program segments do their tasks), to processes (how people interact to understand what programs need to be written), to methodologies (how organizations align themselves to deliver best value to their customers).

Before we go into the discussion of different methodologies, let us identify some of the underlying philosophies that lead to the differences.

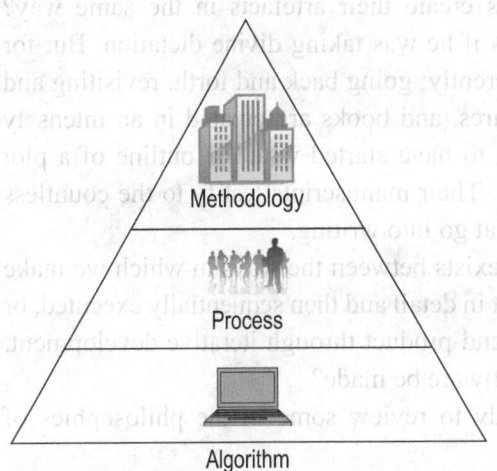

Figure 4.2 Algorithm–process–methodology

4.5 DIFFERENT DEVELOPMENT PHILOSOPHIES

Methodologies are very simply recipes for making software. What are the different ways in which we make things? Though we touched upon this topic in Chapter 3, a bit of repetition is due.

To build a bridge, we must first plan in detail how long and wide the bridge needs to be, whether it will be pontoon, suspension, or cantilever (or any other style), how much material and resources we need, etc. Only when every detail has been worked out to its most minute quanta can the actual building begin. Once the building starts—the proverbial die has been cast—it must progress and end according to the plan. Mid-way through the bridge building, neither of those who have commissioned the bridge or those who are building it, can afford to have a change of mind or heart to have the bridge built higher or lower or built to a style different than the original plan. If they do, such a bridge will most likely never get built; if it manages to get built, it is likely to be a very dangerous bridge. In this context, we may recount a joke many of us may have come across in college canteens: Two friends get a contract to build a tunnel under the English Channel. They start by each of them beginning to dig, one from the English side, the other from the French. When asked how they think they can ensure both the excavations will meet to give one continuous tunnel, the reply was—if we meet by chance, it is good; if not even better—you will get two tunnels for the price of one! The joke's humour draws from the two friends' ignorance of a basic engineering credo; you must plan before you build. All conventional engineering products get built more or less in a *sequential* succession of steps that have been decided and documented before any 'action' starts.

Do painters, composers, or writers create their artefacts in the same way? Mozart was said to have composed as if he was taking divine dictation. But for other mortals, art gets made very differently; going back and forth, revisiting and revising countless times. Music, pictures, and books are created in an intensely *iterative* way. Often, authors are said to have started with the outline of a plot but end up with a very different story. Their manuscripts testify to the countless deletions, additions, and corrections that go into writing.

So, it is apparent that a dichotomy exists between the ways in which we make things—either every step is thought out in detail and then sequentially executed, or a broad idea gets crystallized into an end-product through iterative development. The question is: Which way should software be made?

With this background, we are ready to review some of the philosophies of software development.

4.5.1 Sequential Development

The sequential school prescribes that software should be built the way bridges are, in a linear sequence of activities driven by detailed pre-planning. As we shall see when we discuss specific methodologies, this works well for certain scenarios, and does not work at all for significantly large number of others. The implicit assumption in sequential development is that there are no major unknowns in the development life cycle, all factors that can potentially derail the project can be seen upfront and accounted for. Linearity is the key theme of sequential development, with the additional rider of unidirectionality. In sequential development, software building proceeds from start to finish without retracing steps. The rationale for this approach comes from seeing software as an engineering artefact, it is thus sought to be built via the time-tested engineering way.

Sequential development is highly concerned about making detailed initial plans and following the plans with due diligence. To address this concern, copious documentation needs to be generated. This has led to sequential software development being sometimes called *document-driven development*.

4.5.2 Iterative Development

Iterative development for software recognizes the *plasticity* of software, an idea that was discussed in Chapter 3. As we mentioned earlier, users expect a software system to continually change and meet their expectations. Thus software development ends up being more a continuum than means to a clearly defined end. To be able to continually change, being bound to a steadfast initial plan is of no help. It is best to build in bits, going back and forth in understanding new or changed Requirements and implementing them. Revisiting past deeds to undo and improve upon them is a basic tenet of iterative development. This sounds very expedient, but what ensures iterative development will converge to a coherent system? This is a challenge any methodology based on iterative development must address.

Iterative software development is sometimes referred to as *code-driven development* due to the fact that the most important artefact is code—fully functional parts of the final system that real users can test and give feedback on. Code-driven development, vis-à-vis document-driven development, is more focused on producing working software than paper trails, but runs the risk of descending easily into random modification of code, to meet immediate goals.

In the next section, we look at specific methodologies that are based on these contrasting philosophies of sequential and iterative development.

4.6 BRIEF REVIEW OF SOFTWARE DEVELOPMENT METHODOLOGIES

In the following subsections, we discuss some of the important software development methodologies. The ones discussed here either have pedagogical, historic, or practical importance—sometimes all of these. Software engineering is an evolving discipline and it is likely that new methodologies are being tried out even as you read this book. There is no one methodology that fits all situations. In the following discussion, we will use 'model' and 'methodology' interchangeably.

4.6.1 Code-a-Bit-Test-a-Bit

This one should be familiar to all who have written and run a computer program. In fact it should be so familiar that we may be surprised at it being set up on the pedestal of a 'methodology'. Code-a-bit-test-a-bit (CABTAB) is often the way many of us learn to program. Frequent and consistent Testing is one of the surest signs of a good programmer. What is interesting and sometimes disturbing is that in many organizations, no matter what methodology is initially adopted with much fanfare, the pressure of deadlines brings down software development to CABTAB. While it is useful for individual programmers working on a small piece of software, CABTAB has some serious difficulties with scaling, that is, when many developers start working on a large system with many components. Even then, it is important to recognize the utility of CABTAB to deliver small programs of limited scope. CABTAB is sometimes referred to as the build-and-fix model [Schach 2005].

4.6.2 Waterfall

Its name captures the essence of the Waterfall model. As we know, almost intuitively, water flows downhill, and never the other way (unless we expend energy to pump it up). The unidirectionality of falling water is something of a universal truth. The Waterfall model mandates that software development also be unidirectional—the activities of Requirement Specification, Analysis, Design, Implementation, and Testing follow one after the other in a linear sequence. The biggest and most implicit assumption the Waterfall methodology makes is that Requirements are *frozen* before Analysis can start, Analysis must end before Design commences and so forth. Each activity is assumed to have a clear beginning and an end, with the artefacts and experience derived from one activity driving the subsequent ones. Waterfall is very much like the way bridges and other conventional engineering products are made. Software systems can also be successfully made through the Waterfall model, but only when they conform to Waterfall's assumptions. Figure 4.3 shows the Waterfall model—a succession of steps one after the other as the

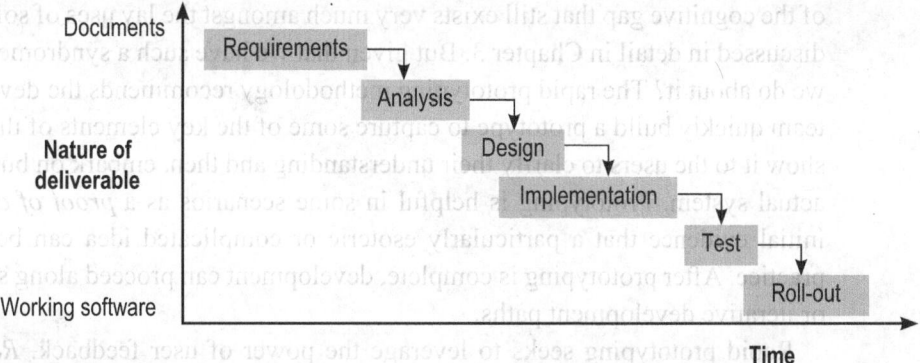

Figure 4.3 The Waterfall model of software development

development cascades down. It may be pointed out that there are no return paths (feedback loops) in the Waterfall model. Once you are done with one activity you move on to the next, and never come back to it. Life without feedback would be very difficult in all but the most trivial situations. This is one of the main reasons why a purely Waterfall approach so rarely succeeds in large and complex systems. Another point to note is that the vertical axis in Figure 4.3 represents the shifting nature of deliverables: At the beginning there are only documents and towards the end there is working software. Waterfall's main significance lies in the fact that it is the first streamlined model of software development, and even today it is effective for small and relatively risk-free projects. For an interesting trivia on the background of Waterfall, see Exhibit 4.1.

4.6.3 Rapid Prototyping

A *prototype* is defined as "first or preliminary form from which other forms are developed or copied", or "a typical example of something" (http://www. askoxford.com/concise_oed/prototype?view=uk). In the rapid prototyping model, a prototype of the final software system to be developed is first built. The prototype needs to implement a representative part of the whole system. What qualifies as a representative part? There is no clear answer to this question, but to be able to address it, we need to see how an initial prototype is useful.

A common situation with users of software systems is that they know what they want (and very often, what they do not) *only* when they see (and are able to work with) the software system. This is sometimes referred to as the IKIWISI (I know it when I see it) syndrome. This is not so much of a syndrome as the consequence of the cognitive gap that still exists very much amongst the lay uses of software; as discussed in detail in Chapter 3. But given that we have such a syndrome, what do we do about it? The rapid prototyping methodology recommends the development team quickly build a prototype to capture some of the key elements of the system, show it to the users to clarify their understanding and then, embark on building the actual system. Prototyping is helpful in some scenarios as a *proof of concept*— initial evidence that a particularly esoteric or complicated idea can be put into practice. After prototyping is complete, development can proceed along sequential or iterative development paths.

Rapid prototyping seeks to leverage the power of user feedback. *Rapidity* is indeed a key idea behind the utility of the model: The prototype must be developed in a 'quick and dirty' way. The development team needs to remember that the prototype is *not* the system, neither should the former be allowed to morph or grow into the latter. The most useful prototype is the one that is thrown away after it has served its purpose. Brooks' words in a different context—'Plan to throw one away; you will, anyhow' [Brooks 1995]—echo the spirit of rapid prototyping. Figure 4.4 illustrates the concept of rapid prototyping.

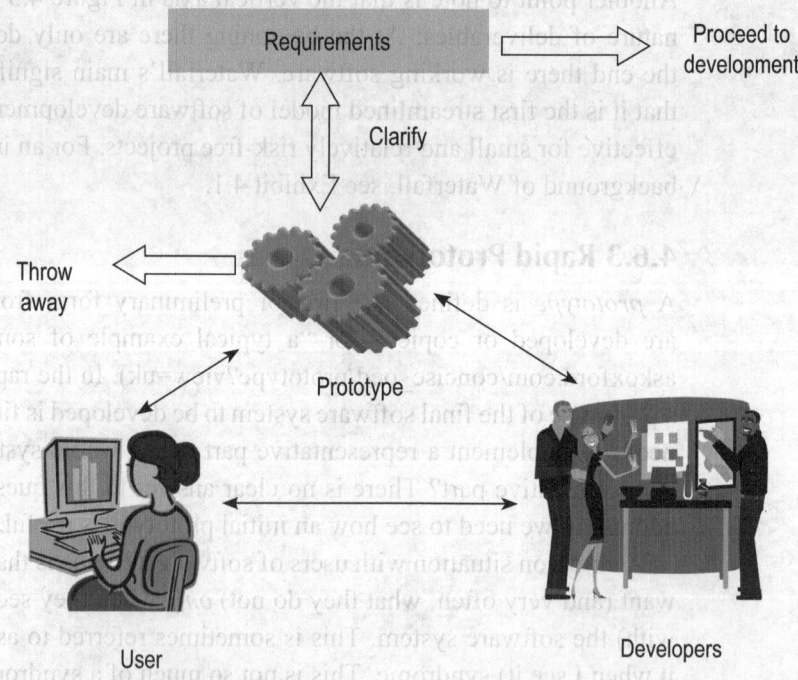

Figure 4.4 The spirit of rapid prototyping

Rapid Prototyping Techniques

There are different types of rapid prototyping such as throwaway prototyping, i.e., building a prototype that will certainly be discarded; evolutionary prototyping, i.e., building a robust prototype that will eventually become the heart of the system; incremental prototyping, i.e., incrementally building a succession of prototypes that will finally come together to form the system; extreme prototyping, i.e., building a Web-based system in a sequence of phases, starting with screen mockups and ending in implementing the full back-end functionality.

Some widely used rapid prototyping techniques are dynamic systems development method, operational prototyping, evolutionary systems development, evolutionary rapid development, etc. In their own ways, each of these techniques seeks to maximize the advantages of prototyping while minimizing its disadvantages.

User Interface Prototyping

An important objective of prototyping is to understand how users interact with the system being built. User interface prototyping is thus an area of much interest within rapid prototyping. User interface prototypes serve as Analysis and Design artefacts as well as planks for exploring and refining the usability of the system. According to Ambler, some of the successful approaches towards user interface prototyping are working with real users, understanding the business objectives of the system whose interface prototype is being built, being consistent, avoiding Implementation decisions as long as possible, clearly explaining the scope of the prototype to the users, and prototyping only what can be actually built [Ambler 2010].

4.6.4 Iterative and Incremental Development

Iterative and incremental development (IID) seeks to take the best out of all the methodologies discussed so far, and succeeds to a large extent in doing so. Iterative and incremental development is not a radical new idea; it has been around under other aliases for a long time [Fowler 2005].

An iterative approach towards building things both appeals to and militates against our subconscious notions of how engineering 'things' should be built. As we said earlier, for painters, composers, writers, and film makers, iteration underpins the creative process. Engineers (civil, mechanical, electrical, etc.), on the other hand, hardly build in this vein. It is absurd to start constructing a bridge, road or a home, build a bit, have users test whatever is built, and use their feedback to modify or build further [Datta 2007]. On the surface, an iterative approach seems closest to CABTAB. As mentioned earlier, CABTAB is well-suited to small and simple systems, but can easily descend into chaos when applied to anything that is neither small nor trivial. The idea of *incrementation* brings in the vital element of control that makes iteration such a powerful way of building software.

Incrementation makes sure the development activities undertaken iteratively are actually converging; that is, after each iterative cycle there is a *tangible* and *testable* addition to the body of functionality the software system delivers to the user. Being tangible ensures the end-user can perceive an incremental change in the system's behaviour over a past datum (as opposed mere managerial glib talk—'you can't see it, but the underlying Design is pretty neat now; that's what we did in this iteration.') Being testable is qualifying in a set of user-defined tests (as opposed to self-Testing by the developers, who are instinctively easy on the part of the system most likely to break). Often the idea of a *release* is closely associated with incrementation. To release a part of the system is to deploy it in the actual production environment where real users (or a subset of real users as in the case of beta releases) are free to test as hard as they can and report all problems. The combination of iteration and incrementation embodies a simple, powerful, and controllable strategy for developing large-scale software systems. Iterative and incremental development works best in situations where rigid upfront planning is impossible due to a variety of factors [Datta 2007].

What goes on inside an iteration; and what actually is an increment? We need to address these key questions to appreciate why IID is such a powerful and widely used development paradigm now. With reference to Figure 4.5, we see that IID in fact prescribes the Waterfall activities be performed in each iteration. In a rather naive but illuminating sense, IID may be thought of as Waterfall contained in a 'for' loop [Datta 2007]. Then each iteration of IID is a mini-Waterfall, where we go over all the motions, guided by user feedback after each release. The edge of IID comes from its clever use of feedback; users get to see and test parts of the final system as early as possible and voice their opinions on what they like and what they do not. The finesse in executing IID lies in filtering the feedback for the signs of what needs to be done differently and using it to change course if necessary. For large and complex projects, IID remains by far the most suitable development methodology; but applying it successfully in practice is far from easy. IID can easily degenerate into uncoordinated patchwork, unless managed with experience and discretion. Every process that uses the IID framework has to devise a set of checks and balances against such a contingency.

Although the Waterfall model seems to epitomize sequential development, the spirit of iterative development was recognized when Waterfall was first proposed. This is explored in Exhibit 4.1.

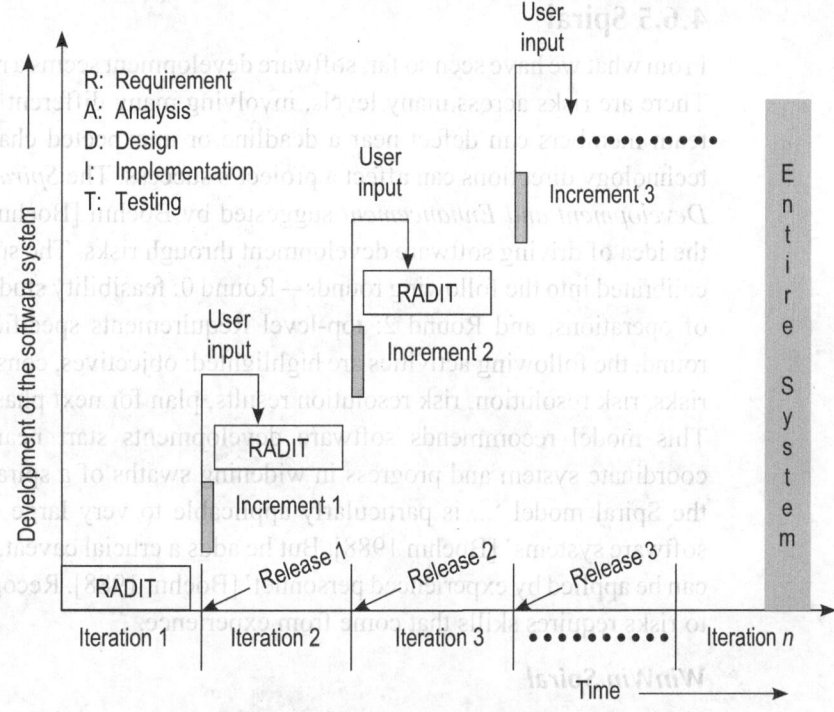

R: Requirement
A: Analysis
D: Design
I: Implementation
T: Testing

User input

User input

User input

User input

Increment 3

Increment 2

Increment 1

RADIT

RADIT

RADIT

RADIT

Release 1

Release 2

Release 3

Development of the software system

Entire System

Iteration 1   Iteration 2   Iteration 3   •••••••••   Iteration *n*

Time

Figure 4.5   Iterative and incremental development

Exhibit 4.1   Is Waterfall Iterative?

In general perception, Waterfall appears to be strongly antithetical to the iterative development philosophy. After all, Waterfall prescribes software be built through a sequential progression of activities, whereas IID recommends frequent revisits and retracing of paths. Surprisingly, Royce's article 'Managing the Development of Large Software Systems' (originally published in 1970)—regarded as the first exposition of the Waterfall model—says, 'If the computer program in question is being developed for the first time, arrange matters so that the version finally delivered to the customer for operational deployment is actually the *second* version insofar as critical design/

operations areas are concerned' (italics added) [Royce 1987]. So Waterfall, at its very source, is talking about repeating the development life cycle a second time, which is close to the iterative philosophy.

It helps to regard both Waterfall and IID as being subtly complementary. We have seen, what goes on inside an iteration is quite similar to the Waterfall activities. The acts of specifying, Analysing, Designing, Building, and Testing a software system must be done in that order. But it is also imperative that there be scope to return and refine based on user feedback. So, the original spirit of the Waterfall model, is not wholly different from the theme of IID!

4.6.5 Spiral

From what we have seen so far, software development seems a rather *risky* business. There are risks across many levels, involving many different parameters. Critical team members can defect near a deadline or unexpected changes in business or technology directions can affect a project's success. The *Spiral Model of Software Development and Enhancement* suggested by Boehm [Boehm 1988] is based on the idea of driving software development through risks. The spiral model has been calibrated into the following rounds—Round 0: feasibility study, Round 1: concept of operations, and Round 2: top-level Requirements specification. Within each round, the following activities are highlighted: objectives, constraints, alternatives, risks, risk resolution, risk resolution results, plan for next phase and commitment. This model recommends software developments start near the origin of the coordinate system and progress in widening swaths of a spiral. As Boehm notes, the Spiral model '... is particularly applicable to very large complex, ambitious software systems' [Boehm 1988]. But he adds a crucial caveat, '... the spiral model can be applied by experienced personnel' [Boehm 1988]. Recognizing and reacting to risks requires skills that come from experience.

WinWin Spiral

Although primarily of historical interest now, we may briefly discuss the WinWin Spiral methodology in this context.

The WinWin Spiral model of development was proposed by Boehm to address some of the difficulties of the conventional spiral model (discussed earlier). A key theme of the WinWin Spiral model is its involvement of *all* stakeholders from the Inception of the development process, with a view to reaching a "win-win" situation for all. With this objective, the following three additional activities are introduced at the beginning of each spiral cycle:

- Identifying the main stakeholders of the system being developed.
- Identifying the win condition for each stakeholder.
- Negotiating "win-win reconciliations of the stakeholders' win conditions" [Boehm et al. 1998].

The WinWin Spiral model also uses a set of three process milestones or "anchor points" to help connect the completion of spiral cycles with the decision points of the organization. These are: life cycle objectives (LCO), life cycle architecture (LCA), and initial operational capability (IOC). LCO commits to establishing a sound business case, LCA commits to establishing the candidate architecture and using it to understand all major risks, and IOC commits to delivering the working software. In a lighter vein, it has been suggested that LCA is similar to the commitment you make in getting married (just as LCO is like getting engaged and IOC like having your first child). It should be noted that recognizing LCO,

LCA, and IOC are not unique to the WinWin Spiral model, they have been used in other development methodologies also, as we shall see later.

4.6.6 Extreme Programming and Agile Processes

So far we have discussed *discrete* methodologies, in the sense that each represented a particular way of software building. Now we will turn to a slew of the latest software development methodologies. This is less of a single methodology and more of a collection of related ones. Usually, these have grown out of the collaboration of groups of practitioners who found the existing ways inadequate.

Extreme programming (XP) tries to take the best of the existing methodologies, while leaving aside their worst. In some of its features, the novelty is certainly 'extreme'; while in others, it is merely a small, calculated departure from the norm. 'Rather than planning, analysing, and designing for the far-flung future, XP programmers do all of these activities—a little at a time—throughout development' [Beck 1999]. The XP major practices, called the 'circle of life' [Newkirk 2002] such as Planning game, Small releases, Metaphor, Simple Design, Tests, Refactoring, Pair programming, Continuous integration, Collective ownership, On-site customer, 40-hour weeks, Open workspace, Just rules, etc. are unconventional prescriptions on how software in-the-large can be built. (The maverick names do enhance the aura of unconventionality.) For example, *Pair programming* mandates the actual writing of software be done by *two* programmers working together; *On-site customer* is having a customer representative co-located with the development team; *40-hour weeks* means, well, exactly what it says—no overtime for members of the development team. XP is described as '... a deliberate and disciplined approach to software development' and the metaphor of a jigsaw-puzzle invoked, underscoring the fact that individual pieces of XP 'make no sense', unless combined together to give the total picture (http://www.extremeprogramming.org/what.html). XP has gained significant traction in the past few years as one of the Agile Processes.

The word 'agility' indicates quickness of action or understanding. Agility for a software development methodology is its facility to absorb change and its effects. Change in what? Change principally in Requirements and its ripple effect downstream. Every methodology seeks to better understand, evaluate, and manage change that inevitably occur as software is designed, built, and used. It is a software engineer's fact of life that Requirements will undergo change, customers will change their minds, their perception of the utility of the software being built will change, the environment in which the software operates will change and so will the technology with which the software is being built. As Fowler says, 'In building business software Requirements changes are the norm, the question is what we do about it' [Fowler 2005]. The most important aspect of a software process is its ability to coordinate and control the effects of such a change. The word 'agility', though co-opted only recently by software methodologists, reflects a lasting objective of software processes: the capacity of adapting to, and delivering in spite of, change [Datta 2007]. The *Agile Manifesto* lists the following principles of agile software development: setting customer satisfaction as the highest priority; welcoming changing Requirements even late in the development process; delivering working software frequently; encouraging developers and customers to work closely throughout development; building projects around motivated individuals, promoting sustainable development, recognizing the value of self-organizing teams, etc. (http://agilemanifesto.org/). These are lofty virtues, and the effectiveness in applying them in recent times has led to the increasing popularity of Agile Processes. Agile Processes are thus a family of methodologies that represent some of the recent thinking in software development and point to the most promising of future trends.

In the preceding sections, we have discussed some of the important software development methodologies. Every organization has its own consideration in adopting a methodology for a particular project. In the Exhibit 4.2, we reflect on methodologies and organizational cultures.

Table 4.1 summarizes our discussion on software development methodologies.

Exhibit 4.2   Different Ways of Functioning

The preference of one software development methodology over another is often closely related to organizational cultures. Organizational best practices—the body of in-house wisdom accumulated over years of doing and learning, something which outlasts individuals—usually play a major role in deciding which methodology will be encouraged. However, it is important to realize that successful organizations become successful not because they merely follow a certain methodology, but since they follow it successfully. McConnell makes this point tellingly in his book *Professional Software Development: Shorter Schedules, Higher Quality Products, More Successful Projects, Enhanced Careers* [McConnell 2003]. He derides the 'slavish devotion to process for process's sake'. Just as generating lots of documents and holding frequent meetings are the symptoms of process adherence in some successful companies, just mimicking these symptoms will hardly ever make an organization use processes effectively. On the other hand, some organizations seem to be doing so well because they have done away with stifling processes altogether: They call their software engineers 'techies' (or even more flatteringly—'geeks'), encourage unconventional attire in office, and allow employees to create and manage their own work schedules. But just by letting employees turn up for the day at 3:30 PM with a geeky hairstyle may not necessarily ensure quality software and customer satisfaction. So, living up to the *spirit* of a process is much more important than merely following it to the *letter*.

Table 4.1 Comparison of methodologies

Name	Pros	Cons	Relevance
Code-a-bit-test-a-bit (CABTAB)	Quick and easy.	Can degenerate into random and ad hoc modification of code.	Very small systems, with few developers.
Waterfall	Mandates detailed up-front planning and documentation, creates enough paper-trails for all stakeholders to know exactly what the system *should* do.	May only end up generating documents, project resources may be significantly expended in planning and documentation only.	Systems with limited risk and low levels of uncertainty and ambiguity, and very clearly defined deliverables.
Rapid prototyping	Offers valuable opportunity for establishing a proof-of-concept of the most difficult features of the system through a prototype, helps understand user needs better.	Prototype may end up being the final system, too much time and effort may be spent in developing the prototype, what to prototype is difficult to decide early in the project.	Systems which have clearly identifiable features that need prototyping where users are unsure of specific functionalities that prototyping will help clarify.
Iterative and incremental development	Helps meet real needs of real users through continuous feedback and modification, suitable for systems that have significant unknown parameters.	Difficult to decide what goes into an iteration, and to ensure increments converge into a cohesive system.	Systems across a wide range of size and complexity.
Spiral	Helps evaluate risk at every stage of development and address them accordingly.	Very often risks are not known upfront and only evolve as the project moves forward.	Systems with well-documented and well-understood risks.
Extreme programming and Agile Processes	Takes the more useful features from existing methodologies.	Sometimes the prescriptions are somewhat 'extreme', like having a customer representative sit with developers, all the time.	Project teams with experienced and skilled personnel.

4.7 PEOPLE AND PROCESSES

As Figure 4.2 depicts, methodologies are supported by processes. Why have processes in the first place? This is one of those questions which have many answers; we explore some of them in Chapter 5. For the time being, we can justify the need for processes as insulators against the variability of people. Ideally, a process should be able to consistently produce a deliverable, irrespective of who executes it. But in reality, a process' utility is strongly influenced by the people who run it.

This effect is very pronounced in the Agile Processes. Take for example, *pair programming*. This agile tenet recommends two programmers work on the same piece of code at the same time. If the pair consists of two bad programmers, the effect can be disastrous, if there is one good and one bad programmer, hopefully things will even out; and two good programmers can lead to great things, or get in one another's way. No matter how good a process is, reliance on good people will never entirely go away. In a freewheeling conversation between Boehm and DeMarco, the former questions the latter's assertion that 'premium people' are essential for Agile Processes to succeed. Boehm complains that premium people sounds too much like Nietzsche's supermen and Huxley's Alphas, implying that reliance on premium people may not be sustainable in the long run [DeMarco and Boehm 2002].

People need good processes to give their best. But a process is only as good as the people who work with it.

SUMMARY AND TAKE-AWAYS

In this chapter we discussed the context and importance of methodology in software development, introduced the idea of the software development life cycle (SDLC), and briefly reviewed some of the important methodologies. Here are the salient points:

- The SDLC is the sequence of activities covering Requirements, Analysis, Design, Implementation, and Testing, over which a software system is developed.
- Software development philosophies can be broadly classified into two groups: sequential and iterative. The former recommends development proceed in a series of steps sequentially from start to finish, while the latter encourages development be moderated through feedback and refinement.
- Code-a-bit-test-a-bit (CABTAB) is hardly recognized as a formal methodology, although it is widely used in programming. It is unsuitable for anything other than very small systems of limited scope.
- The Waterfall model uses the metaphor of falling water to underline the sequential nature of software development. It is suited to projects of limited uncertainly and risks.
- Rapid prototyping recommends the building of a prototype to clarify Requirements and system scope. The prototype, however, should never become the final system.
- In iterative and incremental development, the software system is built through a series of time-boxed development cycles—iterations—leading to tangible and testable additions to the overall system functionality—increments.

This is an expedient model for building systems with initial ambiguity of scope and changing Requirements.

- Spiral model is risk driven; it drives software development through recognizing and addressing risks, at every stage.

WHERE TO LOOK FOR MORE

- 'Extreme Programming: A Gentle Introduction': http://www.extremeprogramming.org/
- 'The New Methodology by Martin Fowler': http://www.martinfowler.com/articles/newMethodology.html

EXERCISES

Review Questions

Review Questions test your understanding of the key concepts presented in this chapter.

1. In the software engineering context, method and methodology
   (a) mean the same thing
   (b) mean different things
   (c) may mean the same or different things
   (d) are not relevant in the software engineering context

2. The software development life cycle is included within the software life cycle.
   (a) This is a true statement due to the unique characteristics of software.
   (b) This is a false statement since software is similar to all other engineering products.
   (c) This may be true or false depending on the specific software system under consideration.
   (d) Software development life cycle is the same as the software life cycle.

3. Two major software development philosophies are
   (a) Waterfall and spiral
   (b) code-a-bit-test-a-bit and rapid prototyping

   (c) sequential and iterative
   (d) none of the above

4. In the Waterfall model,
   (a) there is one feedback path
   (b) there are no feedback paths
   (c) there are multiple feedback paths
   (d) there is none of the above

5. The initial prototype developed in the rapid prototyping model
   (a) should be modified to become the final product
   (b) should be released for initial Testing
   (c) should be used for fine tuning performance
   (d) should be thrown away

6. The power of iterative and incremental development arises from
   (a) its similarity to the artistic process
   (b) its basis on the sequential development philosophy
   (c) its popularity
   (d) continual user involvement in Testing and feedback.

7. The key objective of the spiral model of software development is
   (a) to control the spiraling cost of development

(b) to use the metaphor of a spiral to understand Requirements better

(c) to drive software development via risks

(d) to facilitate better Testing of software

8. In the iterative and incremental model, increments are

   (a) salary hikes for team members

   (b) additional Requirements

   (c) tangible and testable addition to the body of functionality the software system delivers to the user

   (d) may be any of the above

9. Waterfall model is suitable for

   (a) systems with frequently changing Requirements

   (b) large systems with significant ambiguity in initial Requirements

   (c) systems with limited risk and low levels of uncertainty

   (d) none of the above

10. Agility, for a software development methodology, is

   (a) its power to support distributed development

   (b) the flexibility of working hours of project team members

(c) both (a) and (b)

(d) its facility for absorbing change and its effects.

Reflective Questions

Reflective Questions require you to think more deeply about some of the ideas and come up with your own interpretations and answers.

1. In his 2000 Turing lecture 'The Design of Design', Brooks said, 'The Waterfall Model is Dead Wrong'. Do you agree or disagree with this observation? Write three points in favour of the view you take.

2. We have highlighted how earliest exposition of the Waterfall model recognized the need for iterations. Why do you think Waterfall lost the so-called spirit of iterative development in the subsequent years of its use?

3. You are asked to build a website for your class, which will have static information (name, home town, educational background, etc.) about students. Which methodology will you choose? What are the factors that will influence your choice?

REFERENCES

Ambler, S.W. (2010), 'User interface prototyping tips and techniques,' http://www.ambysoft.com/essays/userInterfacePrototyping.html, last accessed on Feb. 7, 2010.

Beck, K. (1999), 'Embracing Change with Extreme Programming,' *Computer*, 32(10): 70–77.

Boehm, B.W. (1988), 'A Spiral Model of Software Development and Enhancement,' *Computer*, 21(5): 61–72.

Boehm, B., Egyed, A., Kwan, J., Port, D., Shah, A., and Madachy, R. (1998). 'Using the winwin spiral model: a case study,' *Computer*, 31(7): 33–44.

Brooks, F.P. (1995), *The Mythical Man-Month: Essays on Software Engineering*, 20th Anniversary Edition, Addison-Wesley.

Brooks, F.P. (2000), 'The Design of Design,' http://www.siggraph.org/s2000/conference/turing/index.html, last accessed on May 19, 2010.

Datta, S. (2007), *Metrics-Driven Enterprise Software Development: Effectively Meeting Evolving Business Needs*, J. Ross Publishing.

DeMarco, T. and Boehm, B. (2002), The agile methods fray, *Computer*, 35(6): 90–92.

Fowler, M. (2005), 'The New Methodology,' http://www.martinfowler.com/articles/newMethodology.html, last accessed on May 19, 2010.

Jacobson, I., Booch, G., and Rumbaugh, J. (1999), *The Unified Software Development Process*, Addison-Wesley.

Krutchen, P. (2004), *The Rational Unified Process: An Introduction,* Third Edition, Addison-Wesley.

McConnell, S. (2003), *Professional Software Development: Shorter Schedules, Higher Quality Products, More Successful Projects, Enhanced Careers*, Addison-Wesley Professional.

Newkirk, J. (2002), 'Introduction to agile processes and extreme programming,' in *ICSE '02: Proceedings of the 24th International Conference on Software Engineering*, pp. 695–696, New York, NY, USA. ACM Press.

Royce, W.W. (1987), 'Managing the development of large software systems: concepts and techniques,' In *ICSE '87: Proceedings of the 9th international conference on Software Engineering*, pp. 328–338, Los Alamitos, CA, USA. IEEE Computer Society Press.

Schach, S. (2005), *Object-Oriented and Classical Software Development,* Sixth Edition, McGraw-Hill International Edition.

Place of Process in Software Development

5.1 MOTIVATION

Process is a word every software engineer hears often; right from when he or she is a novice till the time it takes to become an expert. It should come as no surprise that such a common word has many meanings in many contexts, each different, and some quite contradictory. Every textbook on software engineering has a pet definition of a process, some have several. Instead of wading into those definitions, let us see if we can first appreciate a process intuitively.

A process is always for something that has been done in the past and needs to be repeated in the future. For something that is being done for the first time— for example, the moon mission of 1969—one cannot expect to have a process. But for something more routine, like flying a passenger jet across the Atlantic Ocean, which is being done hundreds of times each day, there do exist processes.

These allow pilots and cabin crews—individuals without any unique qualities, but with the preparation, discipline, and discernment to understand and follow a pre-defined set of guidelines, and go beyond them when necessary—to complete trans-Atlantic flights safely and on schedule almost always. Flying an airplane across the Atlantic Ocean or even attending to the needs of hundreds of passengers ten kilometres above the Earth is not the easiest of tasks. Yet it is done with precision and grace, over and over again. If there is a single word answer to how that is possible, the answer is 'process'. Many of us would not need to fly an airplane over the Atlantic ocean, but we will need to understand and work with processes.

The beauty of an effective process lies in making a complicated task simple. Something that involves much intermediate thought, many decision points, quite a bit of individual judgement, becomes a relatively straightforward succession of steps that anyone with adequate training can perform, when successfully encoded in a process. A process is built on the wisdom of what has worked in the past, precluding the folly of what has not. Processes in any discipline of engineering record and transmit what Booch has called 'tribal memory'—the invaluable collection of tips and tricks that have evolved over countless trials and as many errors along the discipline's journey.

In software engineering, processes have matured considerably in the past few decades. This chapter will present some basic ideas surrounding processes and illu-strate them through examples of some established processes. We will also discuss the scope and necessity of process improvement. Processes, like all other useful things of life, are useful in situations where they are relevant; and not useful, even detrimental, in situations where they are not relevant. A rare and valuable skill is to be able to detect when a process is relevant and when it is not. Or, at a deeper level, the dilemma is to decided what needs to be addressed through a process, and what should be left to the more ad hoc instruments of originality and serendipity. As Figure 5.1 shows there is always a tension between discipline and originality; while doing a task, how much should one adhere to past precedent or the dictum from some authority, and how much should one think outside the proverbial box? An effective process helps practioners balance this tension.

In the next section, we see what a process is in the software engineering context.

5.2 WHAT IS A PROCESS?

It is difficult to find one, universally accepted definition of a process even within the software engineering context. A process may be said to be a set of coordinated, pre-defined activities recommended to team(s) of practitioners(s) with the intention of fulfilling an objective [Datta and Vaishnavi 2004]. (A team may consist of a single practitioner.) Humphrey describes a process as 'the sequence of steps required to

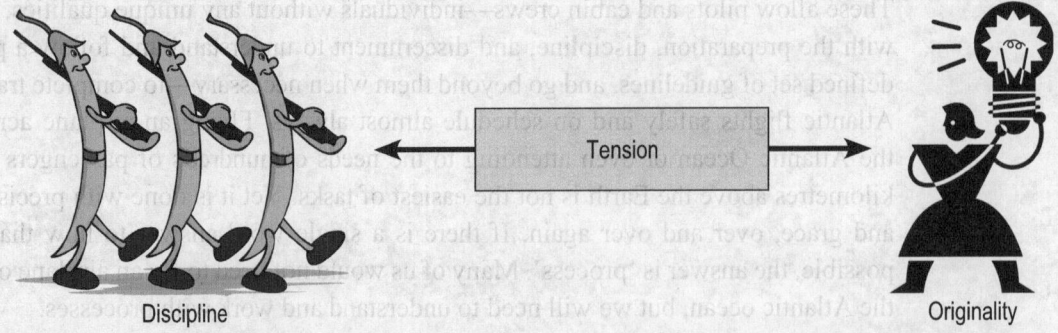

Figure 5.1   Successful processes need to balance the tension between discipline and originality

do a job, whether that process guides a medical procedure, a military operation, or the development or maintenance of software' and clarifies the *software process* as something that 'establishes the technical and management framework for applying methods, tools, and people to the software task' [Humphrey 2005]. A caveat is due here: Although 'software process' is often referred to in a sense of totality, there is no *one* single software process subsuming all of software development. There are very many processes—large and small, formal and informal, widely practiced or of limited currency—that go into the making of a software system and the running of a software project. Although we will look into some specific examples in this chapter, our aim is to broadly outline the place of processes in software development.

5.3 PROCESSES AND SOFTWARE ENGINEERING

Recognizing the risk of overgeneralization, it can be said that engineering is about the repetitive production of artefacts with predictable and consistent quality, on an adequate scale and within manageable cost. Software engineering seeks to fulfill this charter for the development of software systems. Every engineering discipline needs processes, but processes have a special role in software engineering. Why?

As we have remarked in earlier chapters, we are yet to discover laws that directly guide the Design or interaction of software components. There are very few, if any, cut-and-dried first principles to return to, whenever we need direction. Processes thus become very important in a software engineer's daily work; they give him or her the broad contours of how to address a particular situation. In the absence of laws, processes encapsulate past learning into reusable nuggets of insights for the future. As they are primarily based on intuition and experience, processes can never command the infallibility of natural laws. A good process is one that is amenable to change, a great process inspires newer processes and goes into oblivion when its time is up.

Processes as heuristics are very important in software development. But there is also the danger of a blinkered over-dependence on processes, which wrecks projects and organizations. So far we have been talking about processes in abstraction. We will talk about some of the established processes in software engineering; but before that we need to understand the scope of processes.

5.4 FROM MICRO TO MACRO

From what we have said so far, it is clear that a process is a set of prescribed steps towards performing some pre-defined task. The question that comes next is, who is the prescription meant for? This is a very crucial question.

Whenever an individual is given a task, he or she has two related but divergent needs. On one hand, the individual needs some guidance to get started, some directions to keep moving, and some indications to judge completion. On the other hand, no one likes to be told exactly what to do in excruciating detail; there needs to be space for originality and creativity, as motivators for a task to be done well. Addressing this dichotomy is the central challenge for devising successful processes. Processes are what make average people produce adequate results, competent people produce superior results, and talented people produce outstanding results.

In software engineering, there are processes for nearly everything. Rather than getting mired in the merits and demerits of individual processes, we will briefly discuss a representative process each from three broad areas—guiding an individual practitioner, guiding a software development team, and guiding the development of a software system. As depicted in Figure 5.2, the discussion is meant to give an impression on the swath of software processes, from the micro (personal level) to the macro (project or system level)—from trees to a forest.

5.5 PERSONAL SOFTWARE PROCESS

The Personal Software Process (PSP) is positioned as a self-improvement process for individual software engineers to help 'control, manage, and improve' the way they work as a '... structured framework of forms, guidelines, and procedures for developing software' [Humphrey 2005]. The PSP is *not* a set of specific guidelines to do a particular work. It provides practitioners with data and analytical techniques to choose technologies and methods that are most effective, and make routine activities more 'predictable and efficient'.

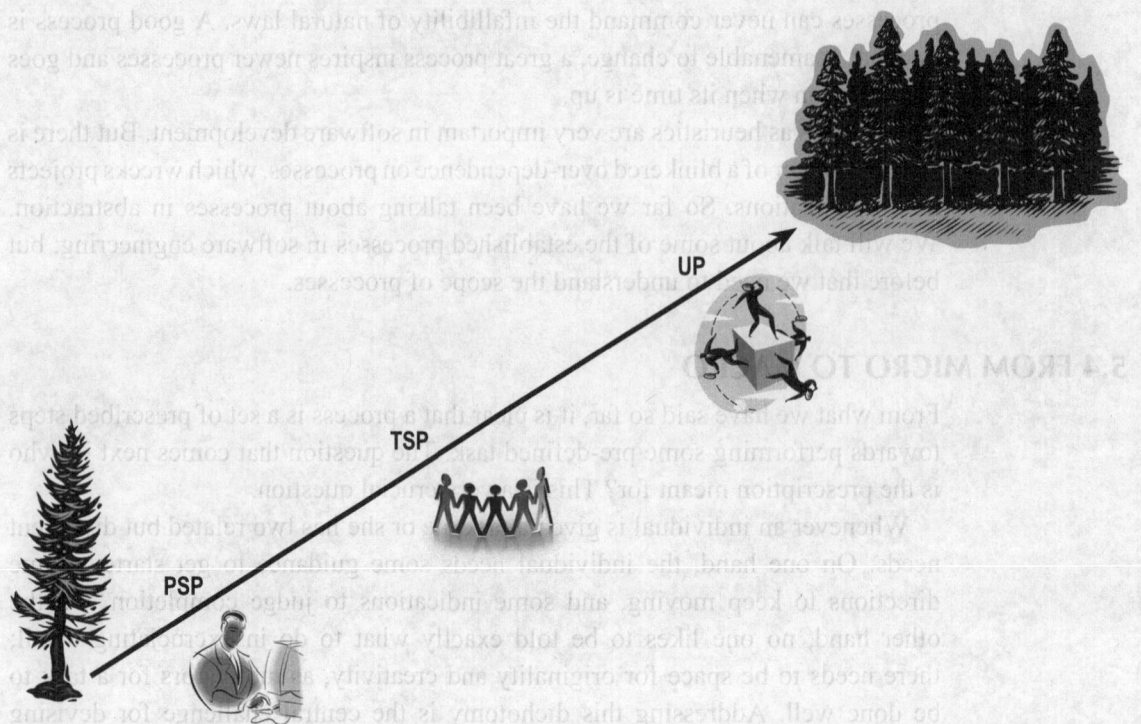

Figure 5.2   The swath of software processes: Personal Software Process (PSP), Team Software Process (TSP), and Unified Process (UP)

The PSP starts from the most basic task for a software engineer—writing a computer program. Across a series of processes—PSP0, PSP1, and PSP2—PSP illustrates how historical data can facilitate the estimation of program size and development time as well as give valuable inputs into quality improvement.

Humphrey identifies the following as *process elements* [Humphrey 2005]:

Scripts: To describe how a process is 'enacted' and to point to standards, forms, guidelines, and measures whenever necessary.

Forms: To provide a placeholder for recording data.

Standards: To guide activities and give a basis for verification of the quality of product and process.

Process Improvement: To facilitate the improvement of an existing process through a process improvement proposal (PIP).

Defining a process needs attention to a number of specific concerns: the assessment of needs and priorities, definition of process objectives, goals, and quality criteria, characterizing current and target processes, putting into place a process development strategy, defining and validating an initial process, and enhancing the process [Humphrey 2005]. Two points are worthy of special emphasis here. Firstly, when thinking about a new process, the current process also has to be taken into account. Often there may not be any clearly defined current processes; things are being done in an ad hoc and haphazard way. Still, the knowledge of how things are being done now can give vital inputs into the making of a new process. Secondly, the *incremental* nature of process development needs to be recognized. A process has to be continually scrutinized for compliance and delinquency, and improved upon by retiring some tasks and introducing new ones.

When and where can the PSP be used to make a difference? Humphrey mentions the following areas: development challenges, negotiation of commitments, controlling projects, ensuring product quality, and sustaining effective teamwork [Humphrey 2005].

The individual practitioner is an atomic unit of software development. The PSP seeks to guide an individual's behaviour by instilling discipline, purpose, and objectivity at the personal level. But as we discuss in Chapter 9, teams play a vital role in the production of large-scale, industrial strength software systems. What guides team synergy in the software development context?

5.6 TEAM SOFTWARE PROCESS

When software projects fail, the reasons are seldom technical. As DeMarco has said, most often projects fail due to human interaction problems, or teamwork issues [DeMarco and Lister 1987]. The timeless classic *Peopleware* conjures the vision of a 'jelled team', whose members together produce results that are qualitatively different from the sum of what their individual efforts could have produced; and perhaps most importantly, the "jelled" team members derive enjoyment from their work together [DeMarco and Lister 1987]. How do we collect a bunch of diverse individuals, and expect them to jel into a team? Or rephrasing the question, is there a process that can turn a group of diverse individuals into a jelled team?

The Team Software Process (TSP) is an attempt to address this question. 'It is an industrial process for teams of as many as twenty engineers who develop or enhance large-scale software-intensive systems' [Humphrey 1999]. Using TSP, individuals who have practiced PSP on a personal level can come together to form effective teams. TSP is based on the following principles:

- Team members learn best when a defined process is followed which gives rapid feedback. The TSP scripts and forms offer 'a defined, measured, and repeatable framework for teams', enabling teams to deliver products over a number of short development cycles and evaluate results after each cycle.

Exhibit 5.1 The Promise of a Process

An issue worth examining in passing is what does any software development process, promise? Is it a *potion* that the prince (development team) needs to fell the monster (project vicissitudes) and win the princess (customer satisfaction)? Or is it a *panacea* that delivers a terminally ill project to eternal health? Or is it a *prescription*, which, like every prescription a doctor writes, is given on the assumptions: 1. The patient must take the medications as advised. 2. The patient must not indulge in activities which, though not explicitly proscribed, will clearly prevent him or her from getting better (e.g., climbing a mountain while being treated for a sprained ankle). A prescription gives directions on regaining and maintaining health. But there is a lot it does not say, gaps which are expected to be filled by the experience and common-sense of the patient or the caregivers. Similarly, software development methodologies contain prescriptions for building systems that offer best returns on their investment. To be general and widely applicable they need to have many ellipses and innuendos. Indeed, in reading between these lines lie the challenge of building software.

- Productive teamwork comes out of the confluence of clear goals, supporting work environment, and consummate coaching and leadership. Out of these, TSP provides the supporting work environment.

Assuming that we have disciplined individuals and jelled teams, does that automatically ensure that we will develop effective software systems? The answer is: not necessarily. There needs to go into the mix a way for successful software making. The Unified Software Development Process, or the Unified Process for short, or UP for shorter, is one such recipe. (See Exhibit 5.1 for what a process does and does not promise.)

5.7 UNIFIED SOFTWARE DEVELOPMENT PROCESS

In the words of the proponents of the Unified Software Development Process (UP), '... the Unified Process is more than a single process; it is a generic process framework that can be specialized for a very large class of software systems, for different application areas, different types of organizations, different project sizes' [Jacobson et al. 1999]. The key characteristics of the UP are highlighted as being use-case driven, architecture-centric, iterative, and incremental [Jacobson et al. 1999]. It needs to be underlined that the UP, more than being just a process, is also a process *framework*. It is a sort of template to create processes, given a broad objective. One such *instantiation* of the UP is the RUP or the Rational Unified Process [Krutchen 2004].

Among the UP's foundational principles of being use-case driven, architecture centric, and iterative and incremental, the first two relate closely with how software is designed, while the latter concerns the way software is built. The UP is a matrix of *phases* and *workflows* as we discuss in detail in Chapter 14.

The workflows correspond to the activities that are readily associated with software development—Requirements, Analysis, Design, Implementation, and Test. The phases Inception, Elaboration, Construction, and Transition, embody the monitoring and control mechanism. The four major milestones, life cycle objectives (LCO), life cycle architecture (LCA), initial operational capability (IOC), and product release (PR) calibrate the outcome from the phases' in their respective order. However, end-users can and should see results independent of the completion of phases. The system grows incrementally over iterations. Each increment leads to tangible and testable results. As the project moves forward, the iteration window can be thought to move across the spectrum of the phases. It may be noticed, in a clear departure from the Waterfall way, each phase of this model has components of all workflows (that is every type of development activity, from understanding Requirements to Testing). The extent to which a workflow is addressed in a phase varies from one iteration window to another.

As we remarked earlier, the Rational Unified Process (RUP) is process created out of the template of the UP. RUP has been described as a '... a process product. It is developed and maintained by Rational Software and integrated with its suite of software development tools' [Krutchen 2004]. RUP seeks to incorporate many of the best practices for software development and make them applicable to a wide range of development scenarios such as [Krutchen 2004]:

- Iterative software development
- Requirements management
- Use of component-based architecture
- Visual modelling of software
- Continuous verification of software quality
- Controlling of change in software systems

As we remarked earlier, good processes always have scope for improvement–we examine this idea in detail in the next section. Table 5.1 gives a snapshot of the process described in this chapter—PSP, TSP, and UP in terms of scope, utility, and open issues.

Table 5.1   Personal Software Process (PSP), Team Software Process (TSP), and Unified Process (UP)—a snapshot

	Scope	Utility	Open issues
PSP	Guiding the individual practitioner's work.	Helps individuals organize and improve their work.	If a practitioner is involved in software development activities other than programming, for example, Design, how helpful will the PSP be?
TSP	Facilitating the combination of individuals into effective teams.	Helps the fulfillment of collective tasks.	What is the level of preparation needed for individual practitioners to function together effectively?
UP	An end-to-end process template for building software systems.	Helps perform the workflows and phases of software development in a consistent and repeatable way.	How relevant is the UP for maintenance or enhancement projects?

5.8 TOWARDS PROCESS IMPROVEMENT AND PROCESS MAKING

Our discussion so far may have imparted the impression that processes are always for the good, their power largely lies in enabling ordinary people achieve extraordinary results. While this is true, there is a different perspective on the place of processes in software development. Following processes is advisable, but blind, unthinking adherence to processes is often a bane. Experience endows us with the wisdom to know when to follow a process, when to go beyond it, or when to create a new process. Processes are never cast in stone, a good process is always a work in progress.

Whenever we work with a process, we should always be on the lookout for how the process can be improved. What do we mean by improving a process?

Improvement in a process can happen at many levels. We can tweak the individual tasks that make up a process; we can redefine the process objectives, or refine levels of practitioner participation in a process. Whichever way we address process improvement, it should be borne in mind that every process necessarily confronts a moving target problem. The goal posts keep shifting, and to keep up with the change, process parameters need to change too. Processes are useful artifices to get any repetitive task done with a consistent level of quality and performance. There is nothing sacrosanct about them, processes can and should be mended or modified to make way for newer and more useful processes.

CASE STUDY

Like all interesting topics, software engineering theory presents unique challenges when it comes to putting them into practice. In some of the following chapters, we will present case studies to illustrate theory in the context of practice. The case studies give glimpses into the real world of software development—the world every practicing software engineer has to inhabit. Let us get introduced to our main "character".

After being a hard-core technical resource for several years, Preeti finds herself in a new role. She has to work with several project teams, ensuring each team fulfills some tasks that are part of a larger process. Defining the overall objective, as well as the individual tasks of the sub-process was rather easy for Preeti, and explaining them to the teams and getting assurances for adherence was not that difficult either. But the fun started later, after a few process cycles had run.

Preeti found that every practitioner and every team made what appeared to be very sincere efforts at completing the process tasks. But with deadlines, customer satisfaction issues, and other day-to-day pressures, process tasks were the first to be put on a backburner. Coaxing or cajoling were of no use, as everyone agreed process tasks needed to be done, but simply did not have the time to do them.

Preeti changed tack, and devised a set of metrics. The metrics on one hand gave a quantitative reflection of the extent to which a team was following the process tasks, while on the other, they identified the relative level of how much each process tasks was being performed. Devising the metrics and putting them to use did not automatically increase compliance, but after few cycles of metrics calculation and discussion, a clear picture started emerging on how teams as well as the process tasks were faring. Preeti's experience leads us to an interesting observation: Processes work best when some measurement of their compliance is in place.

SUMMARY AND TAKE-AWAYS

In this chapter we have outlined the place of processes in software development. We discussed how a software process is defined, and highlighted the salient aspects of the Personal Software Process, Team Software Process, and Unified Software Development Process. Processes are important in software engineering, but they are not cast in stone—they can and should be modified and improved upon as and when necessary.

WHERE TO LOOK FOR MORE

• Software Engineering Institute: www.sei.cmu.edu

EXERCISES

Review Questions

Review Questions test your understanding of the key concepts presented in this chapter.

1. The discovery of a new medicine is an activity that must be guided by a process.
   (a) True
   (b) False
   (c) True or false depending on the nature of the new medicine
   (d) True or false depending on the nature of the process

2. Compared to other engineering disciplines, the use of processes in software development is
   (a) widespread
   (b) moderate
   (c) negligible
   (d) cannot be ascertained

3. Processes are most helpful for
   (a) repetitive tasks
   (b) one-off tasks
   (c) tasks requiring high degree of originality
   (d) tasks that can only be done by a large team of people

4. The importance of processes in software engineering is tied to the
   (a) need for heuristics
   (b) the 'softness' of software
   (c) the relative youth of software engineering
   (d) none of the above

5. In the context of the processes discussed, we have seen that processes can guide
   (a) individual practitioners
   (b) teams of practitioners
   (c) projects across an organization
   (d) all of the above

6. The Personal Software Process (PSP) is a
   (a) piece of software
   (b) structured framework of forms, guidelines, and procedures
   (c) the most recent software process
   (d) none of the above

7. Team Software Process (TSP) is
   (a) an academic exercise to teach students about processes
   (b) an industrial process for teams
   (c) a team for building software processes
   (d) none of the above

8. The Unified Process
   (a) unifies all software engineering processes
   (b) unifies the major software engineering methodologies
   (c) is a generic process framework
   (d) fosters harmony among different software Design techniques

9. Which of the following statements is valid regarding the relationship between the Unified Process and the Rational Unified Process?
   (a) The former is instantiated from the latter.
   (b) The latter is instantiated from the former.
   (c) The latter is a rationalized version of the former.
   (d) The two are unrelated.

10. Processes in software development
    (a) should be followed with unquestioned loyalty
    (b) stifle creativity and originality
    (c) should be broken down into smaller sub-processes
    (d) should have scope for continual improvement

Reflective Questions

Reflective Questions require you to think more deeply about some of the ideas and come up with your own interpretations and answers.

1. Each process can be thought to comprise of a sequence of atomic activities – tasks. Identify the tasks of a process for making a cup of tea. Would the tasks change if the process is scaled up from making one cup of tea to one thousand cups of tea?
2. Do you think the Unified Process can be applied to any activity other than software development? If yes, give examples. If no, justify your answer.
3. 'The Personal Software Process (PSP) is more beneficial to practitioners if they learn it as a part of their engineering education, rather than if they learn it on the job.' Write two points in favour and two against this statement.
4. You are the leader of a five-member development team. The team members are being negligent about tracking and managing change requests (that is, requests from end-users for changes in the system's functionality, after the system has been released). In the context of the discussion in the Case Study section, devise a process to manage change requests and a set of metrics to measure the process.
5. Comment on the open issues of PSP, TSP, and UP highlighted in Table 5.1.

Programming Problems

Programming Examples require you to analyse or write a program or a program segment to understand a specific problem.

1. Write a Java program to check whether a given integer is prime or composite. Test the program with a random set of inputs and record all the test outcomes. Write down the steps you performed while writing and Testing the program. Can these steps collectively be called a process?   Justify your answer.

REFERENCES

Datta, S. (2007), *Metrics-Driven Enterprise Soft-ware Development: Effectively Meeting Evolving Business Needs*. J. Ross Publishing.

Datta, S. and Vaishnavi, N. (2004), Process compliance and process improvement algorithm, IT Process Model knowledge network, ICM AssetWeb of IBM. Recognized as an Intellectual Capital (IC) and recommended for patent filing evaluation.

DeMarco, T. and Lister, T. (1987), *Peopleware: Productive Projects and Teams*, Dorset House Pub. Co.

Humphrey, W. S. (1999), *Introduction to the Team Software Process*, SEI Series in Software Engineering.

Humphrey, W. S. (2005), *PSP: A Self-Improvement Process for Software Engineers*, Addison-Wesley.

Jacobson, I., Booch, G., and Rumbaugh, J. (1999), *The Unified Software Development Process*, Addison-Wesley.

Krutchen, P. (2004), *The Rational Unified Process: An Introduction*, Third Edition, Addison-Wesley.

6

Software Estimation

6.1 MOTIVATION

Estimation plays a major role in our lives. We are required to project into the future on a variety of situations; and based on such projections, regulate our resources and actions. We usually rely on experience, intuition, gut-feelings, and hunches to get us through our day-to-day estimation needs. For example, to estimate when we will return home from work on a given day, we may consider among other factors, how long it usually takes for the commute (historical data), when we are likely to leave the place of work on that day (specific information), and whether any particular traffic condition is likely to affect the trip (individual judgement). If we have moved to a new city or a new job, and this is the first time we are making the trip from office to home, we may also consider how long a similar trip has taken for us or for someone else we know (analogy). As we shall see in this chapter, all of these factors—historical data, specific information, individual judgement, analogy—strongly influence software estimation. Software development, like any other engineering activity, needs the commitment of time, effort, and money on a large scale. For those who need to make such commitments, some idea of 'how much' is needed upfront. Software estimation helps us reach an *a priori* understanding of how much time, effort, or money an initiative may need. The following sections highlight some of the basic principles and techniques of software estimation.

6.2 WHAT IS ESTIMATION?

Estimate is defined as: 1. an approximate calculation; 2. a written statement indicating the likely price that will be charged for specified work; 3. a judgement or appraisal; (http://www.askoxford.com/concise_oed/estimate?view=uk). Software estimates involve calculations, judgement and appraisal, and may result in promises based on factors such as size, effort, and time. Estimation is the process by which estimates are arrived at.

'The project will require 70 person-months of effort.' One hears statements like this all the time in software projects and on the face of it, this statement seems reasonable. But does the statement mean it can be said with 100% surety that the project will require 70 person-months of effort? Given the uncertainties inherent in every real-life scenario, this seems too audacious a claim. So if the level of surety is not 100%, what is it? McConnell warns that in all such 'single-point' estimates, the probability of the statement being true is often not 1 (absolute certainty) [McConnell 2006]. What the probability is has to be explicitly specified and understood. This can be expressed in a number of ways: The estimator is 80% confident that the project will need 70 person-months or the project will at the most need 90 person-months and at the least need 60 person-months. Both these statements illuminate the vital aspect of how much the estimate is likely to *vary* from the claim of 70 person-months. So every estimate has an element of probability attached to it, either implicitly or explicitly. If the probability is *not* mentioned, it should be clarified, but *never* assumed to be 1. Figure 6.1 illustrates this idea through an analogy; it is no use merely saying an archer can hit a target 200 metres away. The key issue is the probability of getting a hit; how many hits out of how many shots?

Ideally, we would want to be 100% sure about all estimates. But the costs involved in trying to go anywhere near that target are impractically high. We explore why that is so in the next section.

6.3 SCIENCE AND ART OF SOFTWARE ESTIMATION

Weather forecasting is an estimation system that is used extensively in our day-to-day lives. Its reasonable accuracy allows us to plan our activities beforehand. Predicting

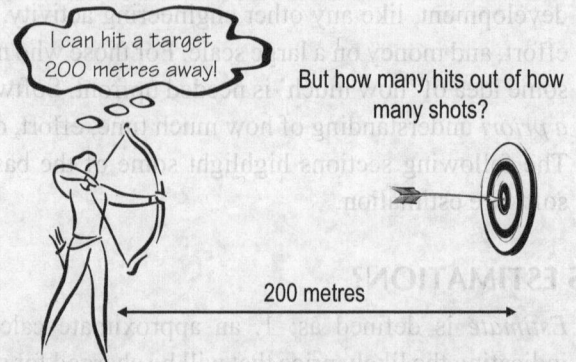

Figure 6.1   Single-point estimates always need to be clarified

the weather is however is no trivial task—the climatic influences are varied and complex, and the daily forecasts come from rigorous theoretical foundations as well as very intense computations. The science of weather forecasting depends very heavily on mathematical models as well as processing of large volumes of data. Similarly, the *science* of software estimation models the software development scenario through rigorous techniques and uses involved number-crunching, and seeks to produce estimates that are between ±10% to ±5%.It is an area of active research. On the other hand, software development organizations are interested in heuristics, rules of thumb, and simple, intuitive formulas that would prevent software estimates from being off the mark by 100%, and would enthusiastically try out techniques that reduce estimation error by 25% or less. The collection of such procedures can be called the *art* of software estimation [McConnell 2006]. The so-called science and art of estimation are complementary and one can freely draw from the other. However, it needs to be underscored that complicated analyses may not necessarily give commensurately accurate estimates, and often simple rules of thumb very well serve the estimation needs of real-life software projects.

While fully recognizing the importance of the science of estimation, we will focus more on the art of estimation in this chapter, to give you an idea of the most widely used estimation techniques in the industry.

6.4 IMPORTANCE OF ESTIMATION IN SOFTWARE DEVELOPMENT

Although it is obvious that software estimation plays a significant role in successful software development (so much so that it merits an entire chapter in this book!), it is helpful to highlight *why* and *how* it is important. The importance of estimation can be seen at the following three levels, which we will call *getting the work*, *getting the work done*, and *getting the work done well*.

6.4.1 Getting the Work

The software industry is highly competitive. Competing companies submit their own proposals on how much money and how much time they require to deliver a project, and the customer organization takes a decision on whom to award the development contract to. The basis for such proposals is *estimation*. The estimate that drives the proposal is crucial; under-estimation may lead to enticingly low time and cost projections which cannot be met during actual development, and over-estimation may take away the competitive edge. To be able to bid for, secure, and deliver a software development project contract, estimation is a key skill. The criticality of estimation is likely to increase in future, as global software development (see Chapter 21) raises demands of higher quality at lower cost.

6.4.2 Getting the Work Done

As every good software engineer knows, getting the work is often far easier than getting the work done. A sound estimate gives a framework for comparing current status of work with the planned status, thus identifying the need for corrective action, if any. Effectively estimating size, time, and effort also allows for efficient resource planning and team building, as well as determining the scope and timing of releases. Without reliable estimates, software development practically becomes blind flying, thriving merely on hope and depending primarily on heroics to complete the project on time and within budget. A typical software development team also has to interface with many other groups and stakeholders, such as business partners, Testing teams, marketing personnel, higher management. Estimates are the basic idioms of communicating and coordinating with these entities to get the work done.

6.4.3 Getting the Work Done Well

But getting the work done is only a part of the story. To be able to sustain itself in the industry and make profits, a software organization has to ensure it gets the work done well, which is to deliver *quality* software consistently, on time, and within budget. Good estimation is a necessary condition for producing good quality software. Empirical data has established [Jones 1993] that unreasonable schedule and deadline pressure contribute to the so-called quick-and-dirty development, leading to somehow meeting the dates but producing bug-infested, fragile software that dissatisfies users and earns a bad name for the organization. As has been so evocatively discussed in Yourdon's *Death March* [Yourdon 2003], software projects with impossible deadlines are rampant in the industry. In many such cases awry estimations are to be blamed (although it is not uncommon for death-march projects to be launched in cold blood to suit some vested interest). So, for members of a development team to be in a physical and mental state to develop quality software, the first criterion is to have a reasonable schedule, which can only be ensured through sound estimation. Estimation also affects software quality by helping unearth risks early. If the estimate for time is calculated as eight months, and a project stakeholder tries to *negotiate* to have it done in four months—without foregoing any functionality—a red flag should certainly go up. In such a situation, without the backing of a reliable estimate, it is easy to end up making commitments that can not be met. Good estimation helps developers hold their ground in such situations. Thus, getting the work done *well* needs the solid foundations of estimation.

6.5 WHY IS GOOD ESTIMATION SO DIFFICULT?

In our discussion so far, we have mentioned 'good estimation' several times. How do we identify good estimation? A number of different criteria have been suggested for identifying *good* estimates. According to Capers Jones, estimates with accuracy ± 10% are good, but it is only attainable with effective project

Exhibit 6.1   Expert Estimation

In the next few sections, we will introduce some of the techniques of making software estimates. As we shall see, expert judgement is often very helpful in successful estimation. What is so special about expert judgement? Experts are guided to a conclusion by a variety of factors; a process hardly replicable or even comprehensible by lay people. A very illuminating example of estimation by expert judgement can be found in an anecdote involving the Italian physicist Enrico Fermi (1901–1954). Fermi was closely engaged in the development and initial Testing of the technology for a nuclear bomb, carried out by the United States on 16 July, 1945 in the deserts of New Mexico. Fermi was almost 15 kilometres away when the bomb detonated and he started dropping small bits of paper immediately after seeing the flash. When strong winds from the explosion arrived in moments, the paper bits were no longer falling directly beneath his hand, but fell on the ground some distance away. After measuring their displacemnt he announced an estimate of the energy released by the bomb. The accuracy of Fermi's estimate was later confirmed by sophisticated instruments. Fermi had arrived at his estimate by reading from a table he had prepared earlier which correlated displacement of paper bits to energy released from the explosion (http://www.anl.gov/Media_Center/logos20-1/fermi01.htm) This anecdote brings to light two very important factors of estimation by expert judgement: Experts must have a deep understanding of the estimation problem (Fermi's thorough knowledge of the science behind the explosion) and they often use tools to quickly get their numbers (Fermi's pre-formulated table linking displacement with energy).

control [Jones 2007]. Conte et al. call estimates good if they are in the range of 25% of actual results 75% of the number of times [Conte et al. 1986]. In general, as organizations progress to higher levels in the Capability Maturity Model (CMM) (refer to Chapter 17), predictability of project outcomes ascend to higher degrees. There have been reports of organizations where 98% projects have been delivered within time and budget [Pitterman 2000], [Putnam and Myers 1997].

It should be evident by now that arriving at good estimates for software systems is difficult business. McConnell has identified the following factors that contribute to estimation inaccuracy [McConnell 2006]:

- Inaccurate information about the project
- Inaccurate information about the level of maturity and preparedness of the organization
- Too much uncertainty in the project and instability of Requirements
- Inaccuracies in the estimation process itself

Figure 6.2 highlights these factors. Out of these, all except the third can be seen upfront and factored into consideration. But project uncertainty and instability

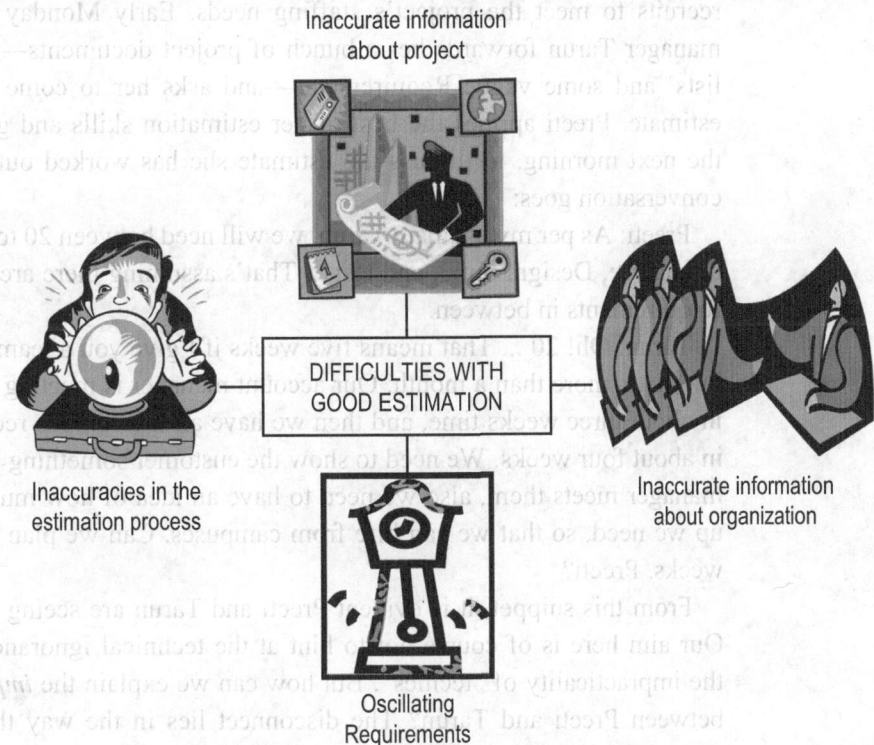

Figure 6.2   Factors influencing the quality of estimates

in Requirements is endemic to enterprise software development (as discussed in detail in Chapter 20). Thus estimation often becomes a moving target problem, with the object of estimation changing even as it is being estimated.

Before discussing standard estimation techniques, we will study the crucial interrelations between *estimate*, *plan* and *target* through the following case study. See Exhibit 6.1 for an anecdote on expert estimation.

CASE STUDY

Software estimation does not exist in a vacuum; estimates are not made merely to take pride in perfecting the art of prediction, or for the fun of foretelling. Decisions depend on estimates, and upon those decisions lives and livelihoods depend. Before we get into the technicalities of estimation, it will help to clarify common confusions that surround the estimation process as well as what comes out of it. It is best to clarify such matters by example.

Our protagonist Preeti has recently moved to a new project. The project is just about starting, but it can potentially grow to be very large, stretching over many months and needing many people. Preeti's organization may need to hire new recruits to meet the project's staffing needs. Early Monday morning, Preeti's manager Tarun forwards her a bunch of project documents—mainly user 'wish lists' and some vague Requirements—and asks her to come up with an effort estimate. Preeti applies the best of her estimation skills and gets back to Tarun the next morning, to discuss the estimate she has worked out. Here is how the conversation goes:

Preeti: As per my estimate Tarun, we will need between 20 to 24 person-weeks, to analyse, Design, build, and test ... That's assuming there are no changes in the Requirements in between.

Tarun: Oh! 20 ... That means five weeks if I give you a team of four people,... but that's more than a month. Our account manager is meeting with the customer in about three weeks time, and then we have all the campus recruitments starting in about four weeks. We need to show the customer something before the account manager meets them, also we need to have an idea of how much resource ramp-up we need, so that we can hire from campuses. Can we plan to deliver in three weeks, Preeti?

From this snippet, it is evident Preeti and Tarun are seeing things differently. Our aim here is of course not to hint at the technical ignorance of managers or the impracticality of "techies". But how can we explain the *impedance mismatch* between Preeti and Tarun? The disconnect lies in the way the words *estimate* and *plan* are being used. In general conversation, we often end up using them interchangeably. But it is very dangerous to confuse them in software development.

As McConnell says so succinctly, 'Estimation should be treated as an unbiased, analytical process; planning should be treated as a biased, goal-seeking process' [McConnell 2006]. As a manager, Tarun's charter is to plan and meet goals; as an engineer Preeti's work is to deliver the best technical solution. Estimation can and does play an important role in planning, in fine tuning budget and staffing levels, in deciding what should be implemented in an iteration, etc. However, estimation is *not* planning and vice-versa. They should be approached and executed with different mindsets, and it is advisable that different people do it.

Let us return to our story. Tarun is now talking to Sam, the customer product manager. This is what goes on between them:

Tarun: Good morning, Sam! We worked out the numbers, and I think we can have the first release out four weeks from now.

Sam: Four weeks! Hmm ...that's a bit on the high side. We need to give something to business by three weeks max. Your techies can meet that target, can't they? We really need to have business start playing with the system as soon as we can.

No estimates or plans are being talked about now; so what is the keyword? *Target*. Sam is least concerned about analytical estimation or passionate planning; he has to have his target met. (Once again, we are not reflecting here on how unreasonable Sam—and customers in general—may sound.) So how do estimates translate to targets?

The only way Preeti and her team can deliver something that was estimated in the *best case* at 20 person-weeks of effort in three weeks (assuming a four member team) is by reducing the scope of the deliverable; that is, by choosing some functionality to implement now, and leaving out others. So whenever someone talks about doing something by a given date, it is a target, neither an estimate nor a plan. If the target has to be met, estimates and plans cannot be tweaked; the target has to be made manageable to be met with the resources at hand.

Estimate, plan, and target are closely related in software development, and tend to get intertwined very easily. Each of them should be seen clearly, and addressed separately.

6.6 SOME STANDARD ESTIMATION TECHNIQUES

From the discussion so far, it is clear that estimation involves foretelling the future. We tend to think the future can be foretold in two ways—*perceptional* and *computational*. The common image of perceptional estimation is going to a know-all entity, asking questions, and getting (the right) answers. Computational estimation, on the other hand, conjures the image of a machine guzzling in information and spewing out insights. In software estimation, there is no purely

Exhibit 6.2   Cone of Uncertainty

McConnell introduces the idea of a *cone of uncertainty* [McConnell 2006] to highlight the relation between the accuracy of an estimate and the stage it was made in the development life cycle. The best way to understand the cone of uncertainty is to think about a one-day cricket match. If we make an estimate in the first over as to how many runs the batting team will make, the chances of an accurate answer are much less than that from the estimate we will make in the 45th over. The increased accuracy comes from the decreased uncertainty between the first and 45th overs—many factors like the weather and pitch conditions or the "forms" of the top batsmen are now known. Something similar happens in software development also; the cone of uncertainty encompasses all that is unknown about the project, being widest at the beginning and narrowing towards the end. But, as McConnell points out, the cone does not narrow by itself. It has to be narrowed by conscious and active project control, through uncertainties being systematically laid to rest as the project progresses. So the cone of uncertainty indicates that estimates made earlier in the software development life cycle are more prone to inaccuracy than those made later.

perceptional or purely computational way; best results are found when perception meets computation. This most preferable 'handshake' of perception and computation is shown in Figure 6.3.

McConnell identifies *count*, *compute,* and *judge* as the three cornerstones of software estimation [McConnell 2006]. The best way to estimate is to count whenever we can, compute based on what we have counted, and apply judgement

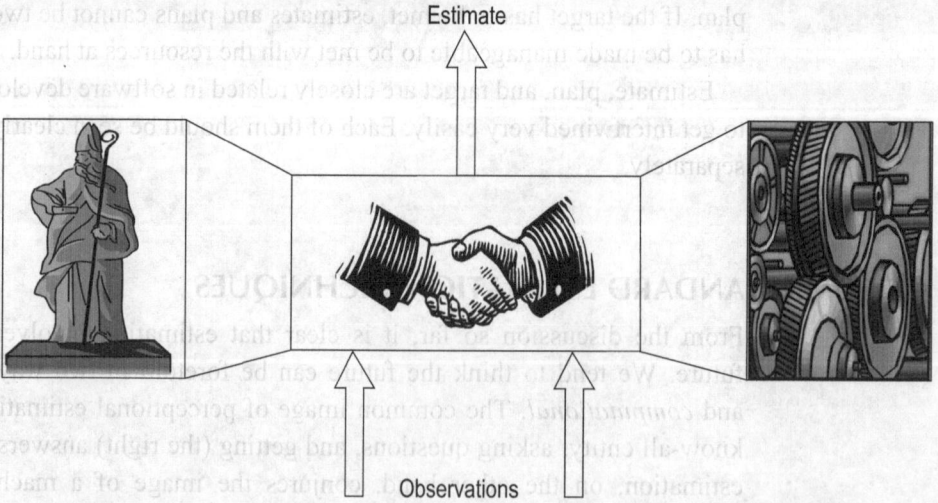

Estimate

Observations

Figure 6.3   Perception and computation must combine to give successful estimates

when necessary. This is much easier said than done. Historical data plays a very vital role in any estimation, and software is no exception. Historical data can be calibrated to produce a model that is useful for making estimates. Diligent collection of historical data and useful calibration can help answer questions like—"if component x is expected to be around y lines of code in size, what is the estimated effort in its development?" Any estimation, no matter which technique it has been arrived through, is helped by historical data. Recording and maintaining a personal repository of data is a helpful habit for any software engineer.

Now we will study some standard procedures for estimation: *estimation by judgement, estimation by comparison*, and *estimation by correlation*. There is no one exclusive way to estimate all these procedures complement one another. See Exhibit 6.2 for a key concept related to estimation.

6.6.1 Estimation by Judgement

Evidently, the easiest and quickest way to get an estimate is to ask an expert. But the problem is to find the appropriate expert. All of us like to think ourselves as experts in something or the other. But being an expert in software Design may not necessarily translate to expertise in software estimation. However, expert judgement can help in making useful estimates; we will now discuss two ways expert judgement can be utilized. As a general rule, it is always good to look for experts from within the group who will be involved in the activity whose output is being estimated. This will ensure experts are in the best position to estimate and are also sufficiently invested in the estimation process.

6.6.1.1 Estimation from individual experts

Estimation from individual experts can come in two flavours—*free-form* and *structured*. Free-form expert estimation involves one expert pulling out an estimate from the blue. Magical as this may sound, it is rarely a source of even reasonably accurate estimate. Better results are obtained when expert judgement is structured. Following are the steps in structured expert judgement, based on [McConnell 2006], and program evaluation and review technique (PERT) [Putnam and Myers 1991]:

1. Break down the entity to be estimated in as small *grains* as possible. For example, if we want to estimate how much time it is going to take to develop a system, the system should be broken down into individual Requirements, or use cases, and the time taken to develop each estimate.

2. For each Requirement or use case, the *Best Case, Most Likely Case*, and *Worst Case* estimated from expert judgement.

3. Calculate the *Expected Case* based on the formula
$ExpectedCase = [BestCase + (4 * MostLikelyCase) + WorstCase]/6$

4. If the Most Likely Cases tend to be unduly optimistic, the above formula can be corrected as

$$ExpectedCase=[BestCase+(3*MostLikelyCase)+2*WorstCase]/6$$

The above procedure serves as a good starting point for individual expert-based estimates. However, for every expert there is scope for enhancing his or her expertise. A good way to do it is to compare estimates with reality and calculate the Magnitude of Relative Error (MRE) [Conte et al. 1986] using the formula:

$$MRE=[ActualResult-EstimatedResult]/ActualResult$$

Collection and review of the MRE data will help judge the Best Case, Worst Case, and Most Likely Case with greater precision and lead to better estimates.

Referring to the adage 'several heads are better than one', sometimes groups of experts are expected to make better estimates than individuals. On this note, we next discuss the *Delphi* technique and its variation, the *Wideband Delphi*.

6.6.1.2 Estimation from groups of experts

The seemingly logical way to have something done in a group is to call a meeting or form a committee. Such congregations often have the potential to tap into the collective wisdom of the group members, and more often, the potential goes unfulfilled. Design by committee is usually a deprecatory phrase, as underlined by the maxim—'a camel is a horse designed by committee.' Usually in meetings, some attendees have original ideas, some attendees have original counter ideas, and still some make their participation conditional to whether their ideas are accepted. The Delphi and the Wideband Delphi techniques try to ensure useful output comes out of meetings of experts, inspite of these usual meeting dynamics.

Originally devised by Rand Corporation in the 1940s to forecast technological developments, the Delphi technique requires experts to make their own estimates and meet as long as needed to reach a consensus on a single estimate [Boehm 1981]. Subsequent studies have shown that results from applying the Delphi techniques are in no way better than those reached through mere meetings. This is because the so-called experts' consensus mandated by the Delphi technique is often influenced by politics or vocal strength. Thus, Boehm et al. extended the technique into what came to be known as the Wideband Delphi.

Figure 6.4 highlights the steps of the Wideband Delphi technique. McConnell reports that he found Wideband Delphi reduced estimation errors by 40% on average when compared to estimates arrived by taking the mean of the estimates of group members. He also recommends Wideband Delphi be used in estimating work in areas with high levels of novelty and asserts it is most helpful in estimating single, clearly defined entities. It is not very useful for detailed estimates of tasks [McConnell 2006].

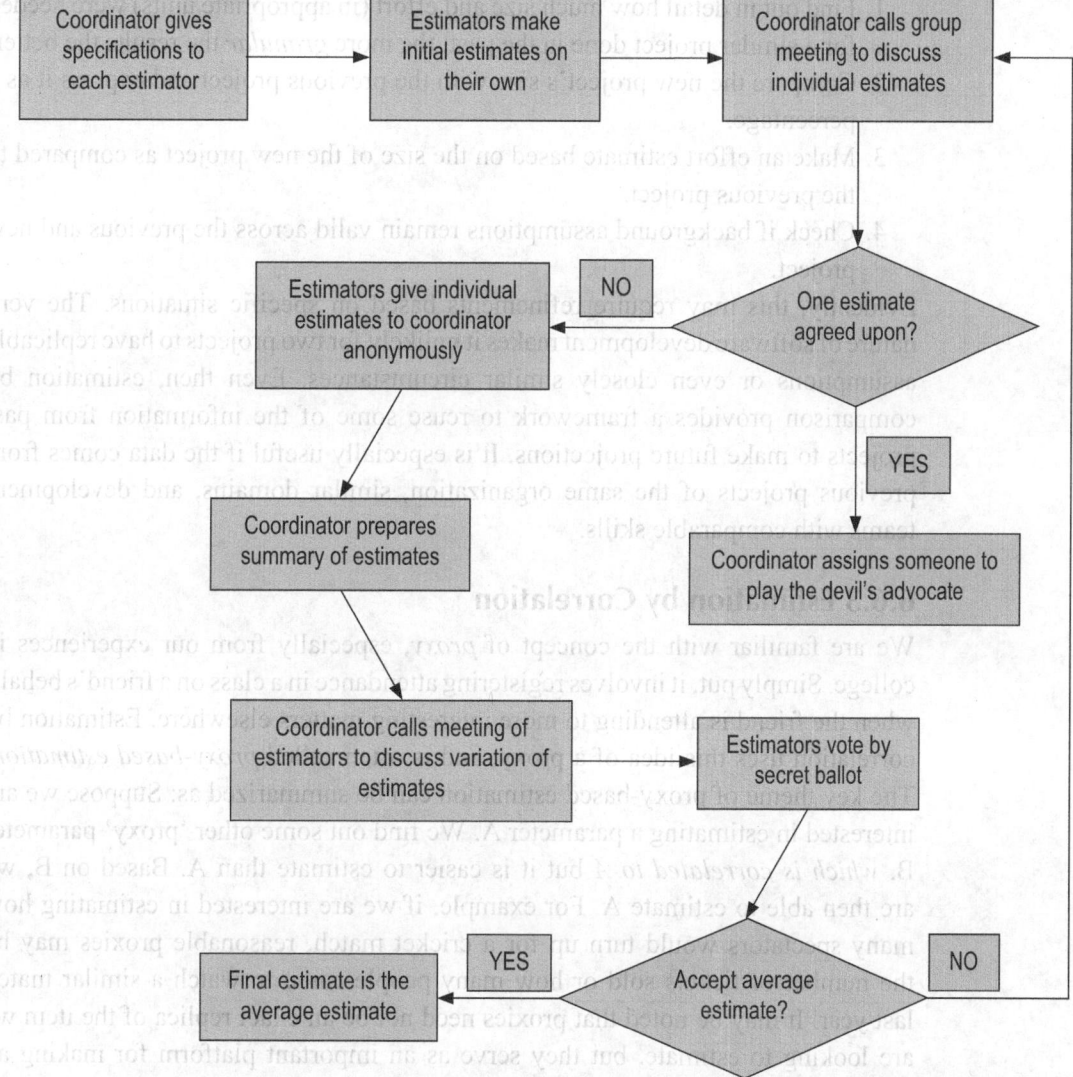

Figure 6.4   Flow chart representing the Wideband Delphi technique (steps taken from McConnell (2006))

6.6.2 Estimation by Comparison

Whenever confronted with a new problem or situation, we seek similarities in what we have already seen or known. Finding something similar helps us predicate the new situation in terms of the old, and reuse some of the experience and intuition we already have. Estimation by comparison, sometimes called estimation by analogy, builds on this idea. McConnell lists the following steps in estimation by analogy for a new project [McConnell 2006]:

1. Find out in detail how much size and effort (in appropriate units) were needed for a similar project done in the past; the more *granular* the results the better.
2. Compare the new project's size with the previous project and express it as a percentage.
3. Make an effort estimate based on the size of the new project as compared to the previous project.
4. Check if background assumptions remain valid across the previous and new project.

Evidently, this may require refinements based on specific situations. The very nature of software development makes it unlikely for two projects to have replicable assumptions or even closely similar circumstances. Even then, estimation by comparison provides a framework to reuse some of the information from past projects to make future projections. It is especially useful if the data comes from previous projects of the same organization, similar domains, and development teams with comparable skills.

6.6.3 Estimation by Correlation

We are familiar with the concept of *proxy*, especially from our experiences in college. Simply put, it involves registering attendance in a class on a friend's behalf, when the friend is attending to more interesting matters elsewhere. Estimation by correlation uses this idea of a proxy, and is often called *proxy-based estimation*. The key theme of proxy-based estimation can be summarized as: Suppose we are interested in estimating a parameter A. We find out some other 'proxy' parameter B, *which is correlated to A* but it is easier to estimate than A. Based on B, we are then able to estimate A. For example, if we are interested in estimating how many spectators would turn up for a cricket match, reasonable proxies may be the number of tickets sold or how many people came to watch a similar match last year. It may be noted that proxies need not be an exact replica of the item we are looking to estimate, but they serve as an important platform for making an educated guess. Converting the count from the proxy to the actual estimate can be done based on historical data.

We will now discuss some of the techniques for estimation by correlation.

6.6.3.1 Estimation by Fuzzy Logic

Putnam and Myers (1991) and Humphrey (1995) have suggested a fuzzy logic-based approach in estimating a project's size in lines of code. Although it is very difficult to come up with specific numbers on how large or small the Implementation of a particular use case or Requirement will be, it is relatively easy to classify their size in terms of Very Small, Small, Medium, Large, and Very Large. In Table 6.1, Average LOC per Requirement is based on the organization's historical data and

Table 6.1   Estimation by fuzzy logic

Requirement size	Average LOC per Requirement	Number of Requirements	Estimated LOC
Very Small	104	13	1352
Small	231	21	4851
Medium	465	5	2325
Large	940	32	30080
Very Large	1890	3	5670
Total		74	44278

the Estimated LOC is calculated by multiplying number of Requirements with the Average LOC per Requirement. Evidently, a crucial issue in this approach is the differentiation between the categories of size. There should at least be a factor of 2 between the categories while some authors recommend a factor of 4 [Putnam and Myers 1991].

6.6.3.2 Estimation by Story Points

Similar to fuzzy logic based estimation, but with some unique features is estimation by *story points* [Cohn 2005]. In this technique, the development team goes over the set of stories (use cases, or Requirements, or lists of features) that need to be implemented, and annotates each story with a numerical score, which is meant to be an indication of size. The numerical scores are assigned on a predefined scale; a widely used scale being the powers of 2 as points on the scale—1, 2, 4, 8, 16 ... For example, let us have a project with 50 story points, with the points assigned as, Story 1 = 4, Story 2 = 2, Story 3 = 1 ...Let there be 50 stories in total and the total points add up to 180. Now the development team plans an iteration to deliver say 11 story points. At the end of the iteration, it is found that delivering those 11 story points has taken 6 person-weeks of effort and 2 calendar weeks. So as the initial calibration, we can say effort = 11 story points / 6 person-weeks = 1.83 story points / person-week and schedule = 11 story points / 2 calendar weeks = 5.5 story points / calendar week. Based on this calibration, the delivery of the entire project of 50 story points is estimated at 50 / 1.83 = 27.32 person weeks and 50 / 5.5 = 9.09 calendar weeks. As the project progresses through iterations, the initial calibration is further refined.

The story point approach is very suitable for iterative development; the shorter the first iteration is, the earlier we can have the initial calibration data. The value of this approach comes from the fact that it does not rely on historical data (although historical data is helpful) for the initial calibration and also since the development team can assign scores to each story on a consistent and unbiased rating scale.

6.6.3.3 *Estimation by Standard Components*

When a project team builds a number of software systems that are similar in terms of their architecture, Design, language of Implementation, etc. they can use the *standard components* approach for estimation [McConnell 2006]. Say for example, an online financial application is being developed, which has Web pages with dynamic content, middle-tier business objects, and database access components. If the development team had earlier developed a system of the same scope, it is likely they will have information on the typical sizes of these three types of components. So past sizes can help them estimate the sizes of different standard components for the new system being built. The crucial point about this approach is of course how to decide on the 'standard' nature of the components; two software projects are seldom closely similar. Thus the standard components approach has to be used with discretion; it can yield reasonable results only in the early stages of the project.

After introducing some generic estimation approaches in the preceding sections, we will now discuss some specific techniques for estimating size, effort, and time.

6.7 ESTIMATING SIZE

The easiest way to estimate size is to count the number of lines of code (LOC). The aptitude of the LOC measure has been discussed at length in Chapter 7. We will now discuss an alternative measure of size—*function points*. A function point is a *derived* measure of a software application's size that is useful for estimation earlier on in the project's life cycle [Albrecht 1979]. The International Function Point Users Group (IFPUG) (http://www.ifpug.org/) maintains the standards for measuring and analysing function points.

Function point calculation is a two-step process: In the first round, the number of each of *user inputs*, *user outputs*, *user queries*, *files*, and *interface files* are multiplied by a complexity factor (high, medium, and low) and the sum of these numbers gives the *unadjusted function point* (UFP). The function point (FP) is then calculated from the UFP by using the formula,

$$FP = UFP * [0.65 + 0.01 * \Sigma(F_i)]$$

F_i, (*i* ranging from 1 to 14) are *complexity adjustment values* based on answers to the following 14 questions on the range of 0 (not important or applicable) to 5 (essential) [Pressman 2000].

1. Is reliable backup and recovery required?
2. Are data communications needed?
3. Are there functions for distributed processing?
4. Is performance a critical factor?
5. Will the system need to work in an existing, heavily utilized operational environment?

Exhibit 6.3 COCOMO

Boehm's pioneering book *Software Engineering Economics* [Boehm 1981] introduced *COCOMO* (COnstrcutive COst MOdel) for software cost estimation. COCOMO is more of a set of estimation techniques than a single technique and it has been subsequently enhanced into COCOCMO II (http://sunset.usc.edu/research/COCOMOII/). COCOMO and COCOMO II are among the most widely used software estimation methodologies. COCOMO II recommends the following models for the different stages of software development:

- *Application composition model*—Relevant for the early stages of software development.

- *Early Design stage model*—To be used when initial Requirements have been agreed upon, and the software architecture devised.

- *Post-architecture stage model*—To be used during the time the software system is being actually built.

As input to the model, COCOMO needs sizing information in terms of object points, function points, or lines of source code. COCOMO arrives at its estimation through an involved procedure using complexity weights and rates of productivity that are determined based on experience from past projects.

---

6. Is on line data entry required?
7. Does the online data entry need input transactions over multiple screens or applications?
8. Do the master files need to be updated online?
9. Is there complexity in the inputs, outputs, files, or inquiries?
10. Is there complexity in the internal processing logic?
11. Does the code need to be designed for reusability?
12. Does the Design need to include conversion and installation?
13. Does the system need to be designed for multiple installations at different organizations?
14. Does the system need to support ease of use and user initiated changes?

Evidently, there is significant subjectivity in answering these questions, and also in deciding on the level of complexity of the user inputs, etc. Also, few of the above questions have lost some of their relevance in the Web-based software systems of today. Still, function points remain an important approach for measuring size.

6.8 ESTIMATING EFFORT

Once we have the size estimated in function points, we can use the method developed by the International Software Benchmarking Standards Group (ISBSG) to estimate effort based on the criteria of application size (in function points—FP), the type of development environment, and the maximum team size (MT)

[ISBSG 2004]. Given below are the equations that give the effort estimates is staff months (SM), assuming 132 *project focused* hours per staff month, which excludes all activities of the staff not related to the project.

- General project: $SM = 0.512 * FP^{0.392} * MT^{0.791}$
- Mainframe project: $SM = 0.685 * FP^{0.507} * MT^{0.7464}$
- Mid-range project: $SM = 0.472 * FP^{0.375} * MT^{0.882}$
- Desktop project: $SM = 0.157 * FP^{0.591} * MT^{0.810}$
- Third generation language project: $SM = 0.425 * FP^{0.488} * MT^{0.697}$
- Fourth generation language project: $SM = 0.317 * FP^{0.472} * MT^{0.784}$
- Enhancement project: $SM = 0.669 * FP^{0.338} * MT^{0.758}$
- New development project: $SM = 0.520 * FP^{0.385} * MT^{0.866}$

It should be noted that the estimates from the ISBSG method are directly dependent on the team size. So, smaller teams will lead to smaller estimates and vice-versa. This can be intuitively justified by the fact that larger teams necessarily mean more inter-team communication overheads which may affect individual work output [McConnell 2006].

6.9 ESTIMATING TIME

Once we have an idea of size and effort, we are somewhat well-placed in terms of a project's estimation. However, estimating time or schedule remains another important concern. The *Basic Schedule Equation* gives us the schedule in months (SIM) in terms of the staff months (SM), which can be calculated using the ISBSG method:

$$SIM = 3.0 * SM^{1/3}$$

The coefficient 3.0 is sometimes replaced by values from 2.0 to 4.0, but estimation experts generally agree on the SIM being proportional to the cube root of the SM [McConnell 2006]. The Basic Schedule Equation has been feted as one of the most widely replicated results in software engineering [Boehm 1981]. See Exhibit 6.3 for an outline of cost estimation and Exhibit 6.4 for algorithmic models for estimation.

6.10 ESTIMATION AND EXPERIENCE

Over the last few sections, we have discussed many of the widely used techniques of software estimation. The utility of these techniques are enhanced when complemented by the experience of those who apply it. Many a time, the choice of a technique for a particular situation can also be helped by prior knowledge. So, as one uses estimation techniques more, and learns from the successes and failures, one becomes better at estimating. This statement seems something of a platitude,

---

Exhibit 6.4   Algorithmic Models for Estimation

Earlier in the chapter, we have reflected on the science of software estimation. Science of software estimation uses many of the so-called algorithmic models. They use regression Analysis to empirically derive some of the key parameters. The most generic algorithmic model can be expressed in the form $E = A + B * S^c$ [Laird and Brennan 2006], where E = effort, S = size (in either lines of code or function points), and A, B, C are empirically determined constants. There are many variants of this general equation suited to specific project data, such as the Watson-Felix:

$E = 5.2 * (KLOC)^{0.91}$

or, Bailey-Basili:

$E = 5.5 + 0.73 * (KLOC)^{1.16}$. Similarly there are equations which use function points instead of lines of code, such as Albrect-Gaffney:

$E = 13.39 + 0.0545 * FP$.

As Laird and Brennan points out, many of these models were derived from data points of limited scope, so they have to used with 'extreme caution' [Laird and Brennan 2006]. Sometimes, if we have sufficient data at hand, the parameters A, B, and C can be determined for the current project, instead of having to rely on pre-determined values.

---

as it is so universally true: You get better at doing something as you do more of it! But there is something peculiar about software development which makes software estimation so dependent on experience. As we discuss in Chapter 18, a software system, from the time it is conceived to the time it is completed, is in a state of continuous flux. From changing user needs, to new technological trends, to turnover of project personnel, *what* was estimated is often difficult to match *what* one gets when the system turns into reality. There are no formulas to negate or normalize many of the influences seen and unseen that influence a software project. Only experience imparts a *sense* of these factors and helps make informed decisions about the estimation process and its outcomes. Also, as illustrated in Figure 6.5, software estimation is as much about techniques as it is about skills to strike a balance between the (often conflicting) needs of developers, managers, and customers.

SUMMARY AND TAKE-AWAYS

In this chapter, we have introduced what software estimation is, reflected upon the science and art of software estimation, underlined the importance of estimation and the factors that make good estimation so difficult. We have also discussed the following standard estimation techniques:

- Estimation by judgement
- Estimation by comparison
- Estimation by correlation

Figure 6.5 Trinity of software estimation: balancing different stakeholder needs

We have outlined specific approaches to estimating size, effort, and time, such as function points, ISBSG method, and the Basic Schedule Equation. We have also highlighted the crucial role experience plays in effective software estimation.

WHERE TO LOOK FOR MORE

- Software Benchmarking Standards ISBSG Homepage (www.isbsg.org) is a helpful starting point for reviewing historical data.

- CSE Center for Software Engineering—COCOMO (http://sunset.usc.edu/research/COCOMOII/) is a definitive guide to the many useful estimation techniques.

WORKED-OUT EXAMPLES

1. We have an expert who has estimated the time taken to implement a Requirement as the following (in suitable units): Best Case = 4, Worst Case = 11, Most Likely Case = 5. What is the Expected Case without and with considering the Most Likely Case as being unduly optimistic.

   Answer: Using the equations from section 6.6.1.1, the Expected Case = (4 + (4 * 5) + 11) / 6 = 5.8. Compared with the Best and Worst Cases, the Most Likely Case seems unduly optimistic; so the estimation can be corrected as: Expected Case = (4 + (3 * 5) + (2 * 11)) / 6 = 6.8.

2. For a project of size 290 function points, which uses FORTRAN as the development language, and a team size of ten, what is the estimated effort in staff months? Are there any assumption?

   Answer: Using the fourth equation from Section 6.8, we have $SM = 0.157 * 290^{0.591} * 10^{0.810} = 28.9$ staff months. The assumption is 132 project focused hours per staff month.

3. Use the information of the previous example and the Basic Schedule Equation to calculate the Schedule in Months for the project.

   Answer: Using the Basic Schedule Equation and the information from the previous example, we have $SIM = 3.0 * 28.9^{1/3} = 8.22$ schedule in months.

EXERCISES

Review Questions

Review Questions test your understanding of the key concepts presented in this chapter.

1. The project is estimated to be completed within 10 weeks.
   (a) This is a valid estimate.
   (b) There is no way to estimate project completion time.
   (c) The validity of this estimate can be judged if the probability of completion in ten weeks is also known.
   (d) None of the above is true.

2. Effective software estimation involves
   (a) rigorous theoretical foundations
   (b) rules of thumb
   (c) both (a) and (b)
   (d) intuition

3. The importance of software estimation lies in
   (a) securing software project contracts
   (b) helping develop the software system
   (c) delivering quality software
   (d) all of the above

4. Good estimation involves
   (a) subjective judgements
   (b) estimates that are in the range of 25% of actual results 75% of the number of times
   (c) estimates with accuracy ±10%
   (d) b and c

5. Among the difficulties of making good estimates, the most important fact is that
   (a) changing Requirements are almost always there in real-life projects
   (b) accurate project information is always available
   (c) inaccuracies in the estimation process do not cause much problem
   (d) the organization's level of maturity does not matter

6. In the software development context, considering estimates and plans
   (a) both are the same
   (b) latter depends on the former
   (c) former depends on the latter
   (d) none of the above

7. Software estimates are based
   (a) exclusively on perception
   (b) exclusively on computation
   (c) on a combination of perception and computation
   (d) none of the above

8. Results from estimation from individual judgement versus estimation from group judgement highlight
   (a) the former is more accurate than the latter
   (b) the latter is more accurate than the former
   (c) there is no clear trend
   (d) both are equally inaccurate

9. Difficulties with the Delphi technique arise from

(a) political pressures and communicational styles of individuals
(b) experts' lack of judgement
(c) complicated calculations in combining experts' estimates
(d) none of the above

10. The edge of Wideband Delphi over the Delphi technique comes from
    (a) the role of the coordinator
    (b) frequent meetings
    (c) a number of experts
    (d) the element of anonymity in submitting and voting on estimates

11. Estimation by analogy is helpful when
    (a) past and current projects are executed by the same organization
    (b) past and current projects share similar assumptions
    (c) past and current projects have the same business domain
    (d) all of the above

12. Estimation by correlation uses
    (a) the idea of a proxy
    (b) the idea of analogy
    (c) both of these ideas
    (d) none of the above

13. Estimation by fuzzy logic
    (a) helps when you have exact understanding about the size being estimated
    (b) helps when you do not have exact understanding about the size but can classify into broad categories
    (c) helps when you are personally writing code
    (d) is the best way to estimate

14. In estimation by fuzzy logic, the difference between two contiguous categories—for example Very Small and Small—should vary by a factor

(a) 1 to 10

(b) 2 to 3

(c) 2 to 4

(d) as decided for a specific project

15. In estimation by story points, calibration

(a) is decided prior to the start of development

(b) stays the same across an organization's projects

(c) does not play an important part in the estimation process

(d) is refined across iterations of development

16. Estimation by story points

(a) is strongly dependent on historical data

(b) does not rely on historical data

(c) needs calibration based on historical data

(d) none of the above

17. In estimation by story points, stories are

(a) use cases

(b) Requirements

(c) features

(d) all of the above

18. Estimation by standard components

(a) is useful in the early stages of development

(b) is useful in the late stages of development

(c) is useful at any point in the development life cycle

(d) is not a very useful estimation method

19. COCOMO is

(a) single technique to estimate time

(b) a set of techniques to estimate effort

(c) a set of techniques to estimate cost

(d) a software development methodology

20. Function point calculation uses a set of 14 questions to determine

(a) complexity adjustment values

(b) the resource needs of the project

(c) how historical data can be used in estimation

(d) all of the above

21. In function point calculation, which of the following is not counted?

(a) User inputs

(b) User outputs

(c) User initiated bug reports

(d) Interface files

22. The ISBSG method discussed in this chapter helps estimate effort based on

(a) project size

(b) type of development environment

(c) maximum team size

(d) all of the above

23. The Basic Schedule Equation gives us

(a) Schedule In Months (SIM) in terms of the Staff Months (SM)

(b) Staff Months (SM) in terms of Schedule In Months (SIM)

(c) both (a) and (b)

(d) none of the above

24. The key parameters of the algorithmic models of software estimation are derived from

(a) expert judgement

(b) informal observation

(c) regression Analysis

(d) comparison

Reflective Questions

Reflective Questions require you to think more deeply about some of the ideas and come up with your own interpretations and answers.

1. We have mentioned weather forecasting as a very sophisticated estimation system that uses rigorous theoretical foundations and involved computation. Assuming we have all

the necessary resource at our disposal, could we have a software estimation process that is as sophisticated as the weather prediction model? Is there something unique to the nature of software that makes such a system possible or impossible?

2. The Case Study section underscores the differences between estimate, plan, and target. Take for example; we need to travel from place A to B, which are 100 k.m. apart. We are looking to travel by train, and a single locomotive can give us a speed of 40 k.m./hr. Each additional locomotive will increase the speed of the train by 20%. How much time will we need to reach B from A? What will you do if we must reach the destination in not more than 1.5 hours? Can you apply your solution for this situation to the Case Study's scenario? Would you ask Preeti to commit to deliver the 20 weeks of person-hour estimates in three weeks by increasing her team size? Justify your answers.

3. We mentioned calibration of historical data can facilitate the building of an estimation model. A model need not necessarily be complex; a simple relationship connecting entities of interest is also a model. For example, suppose we are trying to find out how much electrical current will flow through a given piece of wire, when certain voltage is applied at its ends. You have been given the following set of empirical observations $-(x,y)$ denotes x volts producing y amperes of current: (2,0.5), (3, 0.7), (5,1.2), (7,1.7), (10,2). Based on this data, can you construct a model to give the estimate of how much current will flow, given a voltage of 13 volts? Can you *model* the relationship between the voltage and current as an algebraic equation? Does the equation seem familiar?

From the point of view of calibration, what is the most important information contained in the equation? Can you construct a similar model for estimating software development effort, given the size of the software being developed? Are there any assumption in such a software estimation model?

4. What do you think are the three most important factors that can affect the accuracy of estimation by comparison? What can be done to neutralize the influence of these factors?

5. Estimation by comparison and estimation by correlation *seem* rather similar. What do you think is their key difference? Which one is likely to give more accurate estimates, and why?

6. Can estimation by story points be said to be an extension of estimation by fuzzy logic? Justify your answer.

7. Function points are an important metric for estimating size. However, function point based techniques are dependent on responses to the 14 questions. How relevant do you think are these 14 questions to the kinds of software systems now being developed widely? Do you recommend modifying/removing some questions and introducing others?

8. The ISBSG method for estimating size has eight different equations for different kinds of project. Suppose we want to derive a corresponding equation for open source projects. How will you go about doing it? What information do you need? Can you test your equation using information about open source projects available on the Web?

9. 'The Basic Schedule Equation has been feted as one of the most widely replicated

results in software engineering [Boehm, 1981]'—What do you think lies behind the popularity of the Basic Schedule Equation?

10. From what we have discussed in this chapter, it is evident that good software estimation needs many human qualities like judgement, experience, etc. Which of these attributes do you think will be most difficult to capture in an automated estimation tool?

Numerical Problems

Numerical Problems require you to do quick 'back-of-the-envelope' kind of calculations to arrive at answers to some simple but insightful problems.

1. You need to write a program to accept a string of characters from the user, and report the number of vowels, consonants, blank spaces, and special characters in the string. Estimate the Expected Case of how long a team of two developers will need to develop the program using the formula in Section 6.6.1.1. Can you also calculate the Magnitude of Relative Error?

2. With reference to the above problem, make the same estimate by fuzzy logic as explained in the Section 6.6.3.1. Which of the two estimates do you think will be more accurate? Does empirical evidence support your thinking?

3. In one of the Reflective Questions earlier, we asked you to think about how relevant estimation by function points is for open-source development. Go to http://sourceforge.net/projects/puggle/. Try and estimate the size of this open-source system using a function points based approach. Can you verify the accuracy of your estimate?

4. We have a project of 500 function points size; the development language is Java; the project enhances an existing application. Estimate the effort using the ISBSG method outlined in Section 6.8. Do you need any further information? If only 120 project focused hours per staff month can be assumed, will the estimate change?

5. Use the Basic Schedule Equation given in Section 6.9 to estimate the time for completing the project mentioned in the previous question. Did you see a reason to change the coefficient 3.0?

Programming Examples

Programming Examples require you to analyse or write a program or a program segment to understand a specific problem.

1. Write a program in Java that estimates effort using the ISBSG method given in Section 6.8 given user inputs. How will you determine which equation to use?

2. The most widely available software development artefact is source code. Can you write a program that automatically parses source code of a given program, and based on certain assumptions and given parameters, estimates the time that was taken to write the software program?

3. Integrated development environments (IDE) are now widely used for writing code for enterprise software systems. Estimate the time (for one developer) to write a program in Java to calculate factorials with and without using an IDE (such as the one available for free in www.eclispe.org). Write the program with and without using an IDE. How does the estimate compare with the actual time taken? Was an IDE helpful in this particular case? If not, what are the situations an IDE would really help?

REFERENCES

Albrecht, A. (1979), *Measuring Application Development Productivity*, Proc. Joint SHARE/GUIDE/IBM Application Development Symposium (October 1979), 83–92.

Boehm, B.W. (1981), *Software Engineering Economics*, Prentice Hall PTR.

Cohn, M. (2005), *Agile Estimating and Planning*, Prentice Hall PTR.

Conte, S., H. Dunsmore, and V. Shen (1986), *Software Engineering Metrics and Models*, Benjamin/Cummins.

Humphrey, W.S. (1995), *A Discipline for Software Engineering*, Addison-Wesley Professional.

ISBSG (2004), *Practical Project Estimation*, International Software Bench Grp LTD, 2nd ed.

Jones, T.C. (1993), *Assessment and Control of Software Risks*, Prentice Hall PTR.

Jones, C. (2007), *Estimating Software Costs*, McGraw-Hill Osborne Media, 2nd ed.

Laird, L.M. and M.C. Brennan (2006), *Software Measurement and Estimation: A Practical Approach*, Wiley-IEEE Computer Society Pr.

McConnell, S. (2006), *Software Estimation: Demystifying the Black Art*, Microsoft Press.

Pitterman, B. (2000), Telcordia technologies: The journey to high maturity, *IEEE Softw.*, 17(4): 89–96.

Pressman, R.S. (2000), *Software Engineering: A Practitioner's Approach*, McGraw Hill.

Putnam, L.H. and W. Myers (1991), *Measures For Excellence: Reliable Software On Time, Within Budget*, Prentice Hall PTR.

Putnam, L.H. and W. Myers (1997), *Industrial Strength Software: Effective Management Using Measurement*, IEEE Computer Society Press.

Yourdon, E. (2003), *Death March,* 2nd ed., Prentice Hall PTR.

Role of Metrics in Software Development

Learning Objectives

In this chapter, we discuss how metrics play an important role in software development. Specifically, we will cover:

- The need for measuring software
- Metrics as aids to management decisions
- Evolution of software metrics
- Art and craft of metrics making

7.1 MOTIVATION

Metrics are so deeply embedded in all conventional engineering disciplines that we hardly ever notice them. Whether we are trying to nail together pieces of wood to make a table, or designing a bridge, the first and the most frequent thing we do is to measure. Indeed, almost all of our activities, from buying groceries (5 kilograms of rice, 1 litre of milk), watching cricket (current vs required run rate), understanding state of the economy (rate of inflation, value of the Sensex), to taking care of our health (blood pressure, cholesterol level) are reflected upon by some metrics. Intuitively, a metric is an indicator, usually in terms of some number, of the degree or level of a particular entity of interest.

The role of metrics in software engineering is an interesting one. On one hand, using metrics across the software development life cycle can help us build better software in lesser time. On the other hand, the measurement of software presents some unique challenges. In this chapter, we will appreciate the scope and extent of such challenges, and see how they can be best addressed. The objective is to demonstrate how important metrics are in the successful development of large-scale software, and how, with a little forethought and discipline, we can build and apply our own metrics.

In the following sections, we will first highlight the need to measure in the software context, followed by discussions on how metrics go beyond merely measuring. The following section presents a chronological review of software metrics; we will try to discern the major trends in the evolution of software metrics over the last few decades. In the subsequent section, guidelines are given on how metrics can help streamline the development process. The art and craft of metrics making is discussed next and the chapter concludes with a case study.

7.2 NEED FOR MEASUREMENT

The need for objective, consistent, and precise measurements arise in every transaction. Humankind's earliest measurement needs may have come from its bartering days—when X quantity of commodity A was sought to be exchanged for Y quantity of commodity B. To ensure that X quantity of A was measured the same way no matter who measured it, systems of units and measurements evolved. We have all studied the MKS or SI (metre-kilogram-second), or FPS (foot-pound-second) systems of measurement in middle school. The fascinating part about these systems is that how only a few *fundamental* units help us build units of measurement for a plethora of simple and complex entities of interest.

The idea of measurement is so deeply entrenched in human experience that we use it without thinking, to understand situations and take decisions. Just knowing the current temperature gives us a *feel* of how hot or cold it is outside without having to go out and actually feel it. Measuring is easy when we are measuring physical 'things' like height or weight, or attributes like temperature that have physical manifestation (the length of a mercury column in a thermometer). The lack of 'physicality' of software is exactly what makes its measurement so difficult.

Earlier in the book, we have discussed Brooks' timeless insights on the very nature of software: Software is inherently complex, software must be made to conform to existing interfaces, software is constantly subject to change, and software is invisible and unvisualizable [Brooks 1995]. Out of these, the *invisibility* and *unvisualizability* aspects have the biggest bearing on the difficulty of measuring software. Software can be seen in many different ways, each equally apt for a particular perspective. The simplest way to see software is to see it as a set of instructions (software engineers often refer to it as 'code') that comprise a computer program which can be 'run' on a computer. Software from this view is easy to measure, just count the number of lines of code. As we shall discuss later in this chapter, lines of code is the oldest software metric, and perhaps the most widely used one. But it is certainly not the best way to measure software. Since we have just used the phrase 'software metric', let us see how it is formally defined.

The *IEEE Software Engineering Standards* (Standard 610.12-1990, pp. 47–48) defines metric as 'a quantitative measure of the degree to which a system, component, or process possesses a given attribute' [IEEE 1990]. According to Pressman, within the software engineering context, 'a measure provides a quantitative indication of the extent, amount, dimension, capacity, or size of some attribute of a product or process' and 'measurement is the act of determining a measure' [Pressman 2000].

These definitions, though helpful, often do not cover all that is important about software metrics. Measurement is indeed a central part of software metrics, but it is not the only part. This is in some ways due to the unique characteristics of software mentioned earlier, but also because metrics go beyond mere measuring. We elaborate on this theme in the next section.

7.3 METRICS GO BEYOND MERE MEASURING

For the remainder of this chapter, we will be using 'measure', 'measurement', and 'metric' interchangeably when talking about software. As we will use the term 'metric' most often, it is vital to understand what it will mean. Metrics give us some quantitative measure about the amount, level, or state of an attribute of interest; but they can do more than that. Metrics encapsulate much of the background thinking and intuition behind the measure. For example, a widely used indicator, *rate of inflation* is taken to indicate the state of the economy as it affects our daily purchases. Rate of inflation, as reported in the media, is a percentage score. But the process of arriving at that number is fraught with many subtleties, one among which is a careful selection of a 'basket' of commodities that is assumed to closely represent a common man's needs. The constituents of the basket may vary between different economies, just as one set of commodities may not be equally relevant to all socio-economic groups. So considerable thinking on economic, social, and even political issues underpins the 'number' representing the rate of inflation. To take another example closer to our hearts, *run rate*, i.e. runs scored per over bowled, is a hot metric in one day international cricket matches (and perhaps hotter, in the Twenty20 variant!). At any point in such shorter versions of the game, run rate gives a fair idea of how a particular team is batting. But for five-day test matches, run rate is not the most important metric. So the same run rate metric in different versions of the same game attracts different levels of importance, indicating the different thinking that goes into calculating and using the metric, in different contexts.

Software metrics—more often than not—resemble metrics like the rate of inflation or the run rate rather than absolute metrics such as the kilogram or the newton. This arises from the fact that metrics representing physical attribute are usually derived from laws of physics. For example, the measure of force in MKS (SI) system, newton, comes from Newton's second law of motion. But as discussed in earlier chapters, we do not have laws of software yet. So many of the metrics in software can be called *intuitive* metrics—constructs that seek to measure an attribute of interest in terms of our intuitive understanding. Intuition is inherent in all software metrics; this is what gives them their great power, but also attaches greater importance to their interpretation. We will now see how software metrics fare as 'management numbers'.

7.4 METRICS, MANAGEMENT, AND BEYOND

Before we start exploring the evolution of software metrics, we need to address a common misunderstanding regarding the role of metrics in software development. Software metrics are often confused with 'management numbers'. It is true that to be able to manage we need numbers—hard quantitative evidence on parameters we care about, like person-hours of effort, or profit after tax. Software metrics may sometimes contribute to such numbers, but that is *not* their primary utility. Indeed, seeing software metrics exclusively as management numbers carries the danger of not utilizing the metrics to their full potential, or using them for detrimental purposes.

The most important role of software metrics lies in guiding the software development process. As we shall see in later chapters, making and maintaining software is fraught with decision-making at every step. We need to apply our judgement at every juncture, often needing to rely primarily on experience, intuition, and nameless other 'gut-feelings'. Metrics help us make such choices with greater confidence and purpose. In many cases, the best way to understand a problem is to formulate a metric around it and use the metric value to weigh solution strategies against one another. So, it is essential to keep in mind that software metrics can serve as powerful development aids, in addition to being management numbers.

7.5 BRIEF REVIEW OF SOFTWARE METRICS

The study of software metrics has come a long way; yet it has a long way to go. Software engineering as a discipline is grappling with deepening complexity, more insightful metrics are being called upon to aid monitoring, feedback, and decision-making. In this section, we survey the study (and, to some extent, the practice) of software metrics [Datta 2007].

The major trends of development of software metrics are highlighted in Table 7.1 [Datta 2007]. We discuss the details in the following sections.

7.5.1 Early Perspectives

Any discussion of software metrics must begin with due deference to the first and probably still the most *visible* of all software measures, *lines of code* (LOC) or its inflated cousin, *kilo lines of code* (KLOC). LOC is so primal, no definite source can be cited as its origin. It is perhaps natural that counting lines of program instruction was the very first software metric. Lines of program instruction, after all, is the closest software gets to tangibility. LOC or KLOC gives software a 'size' in a very blunt sense of the term. Counting the number of bricks or stone units of the pyramids of Giza will give an idea of the monuments' size. But to anyone familiar with the structural brilliance of pyramids, the vacuity of such an idea will be apparent. Strangely, even somewhat sadly, to this day many measures

Table 7.1 Software metrics trends [Datta 2007]

Decade	Major theme
1970s	*Efforts at formulating 'laws' of software and complexity measures*
	Belady and Lehman scrutinize the behaviour of large systems and come up with their First, Second, and Third Laws of Program Evolution Dynamics [Belady and Lehman 1976, 1979] .
	McCabe introduces the *Cyclomatic Complexity* metric [McCabe 1976].
	Halstead's book, *Elements of Software Science*, brings in new vistas in the study of structure and behaviour of software systems [Halstead 1977].
1980s	*Building enterprise-wide metrics culture*
	Conte et al. present an extensive study of how metrics are used towards productivity, effort estimation, and defect detection [Conte et al. 1986] .
	Grady et al. report their endeavors in establishing a company-wide metrics program [Grady and Caswell 1987].
	DeMarco and Lister's book *Peopleware* argues strongly in favour of using metrics to enhance organizational productivity [DeMarco and Lister 1987].
1990s	*Object-oriented measures and quality concerns*
	Lorenz et al. present a set of metrics for the Design of object-oriented systems [Lorenz and Kidd 1994].
	Chidamber and Keremer propose the CK suite object-oriented metrics [Chidamber and Kemerer 1991, 1994].
	Whitemire's *Object-Oriented Design Measurement* builds a rigorous theoretical foundation for object-oriented measurements [Whitmire 1997].
2000s	*Measuring across the spectrum: product, people, process, project*
	Lanza introduces the *Evolution Matrix* to understand software evolution [Lanza 2001].
	COCOMO II, building on earlier COCOMO, is proposed as a model for estimating cost, effort schedule [CSE 2002].
	Solingen advocates measuring the ROI of SPI [van Solingen 2004].

Note: Only some of the important works are cited in Table 7.1. This is not an exhaustive list.

of software are sought to be normalized by somehow bringing in the LOC angle. LOC was a helpful metric when software systems in general were less complex, and there was shallower understanding of the dynamics of working software. In all but trivial systems of the present, there is almost nothing insightful LOC can measure. Misgivings about the LOC measure abound in current literature. As an example, Armour cautions against 'counting' LOC and highlights how it is high time now the so-called estimation of system size through LOC gave way to a more mature quest for measuring knowledge content [Armour 2004].

One of the reasons why software development is so less amenable to precise measurements is the absence of physical laws that underpin other sciences and engineering disciplines. As Humphrey explains so succinctly, 'Physicists and engineers make approximations to simplify their work. These approximations are based on known physical laws and verified engineering principles. The software engineer has no Kirchoff's law or Ohm's law and no grand concepts like Newtonian mechanics or the theory of relativity.' [Humphrey 2005]

Ironically, it was the very quest for laws of software that started initial explorations in software measurement. Belady and Lehman (1976, 1979) scrutinized the behaviour of large systems and came up with their First, Second, and Third Laws of Program Evolution Dynamics, respectively, as:

1. Law of continual change—A system that is used undergoes continual change until it is judged to be more cost-effective to freeze and recreate it.
2. Law of increasing entropy—The entropy of a system (its unstructuredness) increases with time, unless specific work is executed to maintain or reduce it.
3. Law of statistically smooth growth—Growth trend measures of global system attributes may appear to be stochastic locally in time and space, but, statistically, they are cyclically self-regulating, with well-defined long-range trends.

The beauty of these laws lies in the fact that even several decades after their conception, the ideas are still apt. The authors backed their assertions with adequate empirical data, and introduced sound techniques for understanding and documenting the behaviours of large systems.

In a paper evocatively titled *Metrics and Laws of Software Evolution—The Nineties View*, Lehman et al. review their earlier notions twenty years later. Using results from case studies, the authors conclude, 'The new analysis supports, or better does not contradict, the laws of software evolution, suggesting that the 1970s approach to metric analysis of software evolution is still relevant today'. [Lehman et al. 1997]

McCabe's *Cyclomatic Complexity* is one of the most widely referenced, (and strongly contended) quantitative formulations of software complexity [McCabe 1976]. This metric gives a measure of how difficult Testing for a particular module is likely to be; empirical studies have also established correlations between the McCabe metric and the number of errors in source code. The derivation of Cyclomatic Complexity is grounded in graph theory and it takes into consideration factors such as the number of independent paths through code. Based on practical project data, McCabe concluded that the value of 10 for Cyclomatic Complexity serves as a upper limit for modules. McCabe has given a more recent perspective on software complexity in a paper in *Crosstalk* [McCabe and Watson 1994].

Arguably, the very first book devoted entirely to metrics in software engineering is Tom Gilb's *Software Metrics* [Gilb 1977]. The author, seemingly aware of his pioneering position, comments in the preface, 'I have had few examples to build on and felt very alone during the preparation of the text'. As the first attempt at structuring a nascent discipline, the book does a very good job. It treats the subject with maturity, even touching upon areas such as 'motivational' metrics for human communication, and automating software measurement. The book ends with reflections on measuring such abstract notions as information, data, evolution, and stability. The author also provides copious code samples and examples to corroborate his points.

Halstead's book, *Elements of Software Science* [Halstead 1977] introduced significant new vistas in the study of structure and behaviour of software systems. The book highlights attributes such as program length (N), program volume (V), relations between operators and operands, and very interestingly, a quantification of 'Intelligence Content'. Taking n_1, n_2, N_1, and N_2 to respectively denote the number of distinct operators, number of distinct operands, total number of operator occurrences, total number of operand occurrences in a program, Halstead shows that

$$N = N_1 + N_2$$

and

$$V = N \log_2 (n_1 + n_2)$$

Program volume varies with programming language and indicates the volume of information in bits needed to describe a program. The work illustrates that theoretically a minimum volume must exist for a particular algorithm; volume ratio is defined as the most compact form of a program to the volume of the actual program. The rigor of this work's mathematical treatment is notable, and many of the ideas have remained surprisingly fresh, even after decades of scrutiny. However, consistent with the extant view of software being exclusively composed of computer programs, the author presents an overly algorithmic treatment. Thus some of the results have become dated in the light of more recent developments.

Cavano and McCall (1978) may be credited with the first organized effort towards a software quality metric. They identify quality dimensions as *Product*

Operations, *Product Revision*, and *Product Transition* and factors within these dimensions as correctness, reliability, efficiency, integrity, usability, maintainability, and testability. The major contribution of this work is the framework—though rudimentary—introduced for measuring software quality.

Albrecht (1979) proposed a function-oriented metric which has subsequently gained wide currency: *function point*. Function points are computed using the experimental relationship between the direct measures of the software's information domain and estimation of its complexity on a weighted scale. The information domain values are based on the following criteria: number of user inputs, number of user outputs, number of user inquiries, number of files, number of external interfaces. Once they are computed, function points are used in a manner similar to lines-of-code to normalize measures for software productivity, quality, and other attributes such as, errors per function point, defects per function point, etc. See Chapter 8 for the use of function points in estimation. *Feature points*—an extension of the function point idea, was suggested by Jones [Jones 1991]. This is a superset of the function point measure, and in a sense it expands the former's domain of applicability from business information system applications to general software engineering systems. In addition to the information domain values of function points, feature point identifies a new software characteristic—*algorithms*, which Jones defines as 'a bounded computational problem that is included within a specific computer program'. The main benefit of function and feature point based approaches is highlighted as their programming language independence. But detractors often point out that these techniques involve some 'hand-waving', i.e. there is notable influence of subjective judgement vis-à-vis objective Analysis. Exhibit 7.1 gives a piece of trivia on the relative sizes of operating systems.

7.5.2 A Maturing Discipline

Somewhat similar to the overall intent of the function point metric, the *bang* metric developed by DeMarco [DeMarco 1982] 'is an Implementation independent indication of system size'. Calculating the bang metric involves examining a set of *primitives* from the Analysis model—atomic elements of Analysis that cannot be broken down further. Following are some of the primitives that are counted: functional primitives, data elements, objects, relationship, states, Transitions, etc.

Exhibit 7.1   Interesting Info

In spite of its limitations, lines of code can give us a quick estimate of how large software systems are. According to estimates, Windows NT 3.1 (1993) has 6 million source lines of code, Windows XP (2001) has 40 million, and Linux kernel 2.6.0 has 5.2 million.

DeMarco asserts that most of the software can be differentiated into the types *function strong* or *data strong* depending on the ratio of the primitives, relationships, and functional primitives. Separate algorithms are given for calculating the bang metric for these two types of applications. After calculation of the bang metric, history of completed projects can be used to associate it with time and effort.

Conte et al. (1986) present an extensive study of the state of the art of software metrics in the mid-1980s. Expectedly, the introductory material covers arguments and counter-arguments regarding software as an engineering discipline vis-à-vis science. The only development methodology considered is the Waterfall model, and the authors base their metrics view on the *physical* attributes of code, such as size and volume. The book also introduces some models for productivity, effort estimation, and defect detection.

Grady and Caswell (1987) report their endeavors for establishing a company-wide metrics program at Hewlett-Packard in the 1980s. The book underscores many of the challenges large organizations face in producing industrial software, and how a consistent metrics culture can help deliver better solutions. This work remains memorable for the first exposition of the FURPS approach to classifying Requirements; this has since become a de-facto industry standard. Some extensions to this idea through a metrics-based technique can be found in the paper *Integrating the FURPS + Model with Use Cases – A Metrics Driven Approach* [Datta 2005].

DeMarco and Lister's book *Peopleware* [DeMarco and Lister 1987] is the fount of many a lasting wisdom of the software trade. The title of the first chapter has become a rubric, 'Somewhere Today, A Project Is Failing'. The book unravels the chemistry of diverse factors—technological, social, political, and inter-personal—that go into the making of successful software. Although not entirely devoted to software metrics, the authors have come up with useful ways to measure various dimensions of the development process. Often lacking in pedagogy, these measures are intuitive and easy to use none the less. The *Environmental Factor* or *E-Factor* is a good example. While discussing the effect of environmental factors on the quality of developer effort, the E-Factor is defined as a ratio of 'uninterrupted hours' to 'body-present hours'. Empirical data cited by the authors show large variation of E-Factor values from one site to another within the same organization, and higher values closely correspond to instances of higher personnel productivity. Similar insights make *Peopleware* a classic software engineering book.

7.5.3 Towards a Deeper Perception

Baker et al. (1990) calling themselves the 'Grubstake Group' in a jocular vein, present a serious view of the state of software measurements. The authors are convinced of the need to create an environment for software measures, which can only be done, '... if there exists a formal and rigorous foundation for software measurement. This foundation will not have to be understood by the users of the software measures, but it will have to be understood by those who define, validate, and provide tool support for the measures.' The paper applies notions of formal measurement theory to software metrics, stressing on the need for the identification and definition of:

- Attributes of software products and processes
- Formal models or abstractions which capture the attributes
- Important relationships and orderings which exist between the objects (being modeled) and which are determined by the attributes of the models
- Mappings from the models to number systems which preserve the order of the relationships

The authors also rue 'a general lack of validation of software measures' and highlight the role of sound validation schemes towards the reliability of a software measure. In summary, the paper establishes that software metrics should and can be developed within a measurement theory framework.

Card and Glass have defined three software Design complexity measures—*structural complexity*, *data complexity*, and *Design complexity* [Card and Glass 1990]. The structural and Design complexity measures use the *fan-out* idea which indicates the number of modules immediately subordinate to a module, i.e. which are directly invoked by the module. System complexity is defined as the sum of the structural and data complexities. The authors conjecture that as each of these complexities increase, overall architectural complexity of the system also increases, leading to heightened integration and Testing efforts.

Similar to Grady et al.'s report of initiating a metrics program at their organization (discussed earlier), Daskalantonakis has recounted the experience of implementing software measurement initiatives at Motorola [Daskalantonakis 1992]. Based on the practical issues faced during Implementation, the author concludes that metrics can expose areas where improvement is needed. Whether or not actual improvement comes about depends entirely on the actions taken on the results of analysing metrics data. This paper highlights the important learning that metrics are only the means to an end; the ultimate goal of improvement comes through measurement, Analysis, and feedback.

Extending the discussions of his earlier book on introducing metrics in a large organization, Grady points to the twin benefits of using metrics—expedient project management and process improvement [Grady 1992]. Grady first takes up the tactical application of software metrics in project management and follows it up by the strategic aspects in process improvement. The book gives a rare insight into the human issues of applying metrics in a chapter titled *Software Metrics Etiquette*. It has a number of enduring messages, most notably that metrics are not meant to measure individuals. Lack of understanding of this credo has led, and still leads, to the failure of many metrics initiatives, across organizations.

Layout appropriateness is a metric proposed by Sears [Sears 1993] for the Design of human-computer interfaces. The metric seeks to facilitate an optimal layout of graphical user interface (GUI) components, which is most suitable for the user to interact with the underlying software.

Davis et al. (1993) suggest a set of metrics reflecting on the following for gauging the quality of the Analysis model, based on corresponding Requirement specifications: *completeness, correctness, understandability, verifiability, internal and external consistency, achievability, concision, traceability, modifiability, precision,* and *reusability.* Many of these attributes are usually considered deeply qualitative. However, the authors establish quantitative metrics for each. As an example, specificity (i.e. lack of ambiguity) is defined as a ratio of the number of Requirements for which all reviewers had identical interpretation, to the total number of Requirements.

Summarizing his experiences with implementing metrics programs in a large organization, Grady puts forward a set of tenets in his article *Successfully Applying Software Metrics* [Grady 1994]. He highlights four main areas of focus which contribute substantially to the outcome of the overall metrics effort: project estimation and progress monitoring, evaluation of work products, process improvement through failure Analysis, and experimental validation of best practices. In conclusion, Grady gives the following three recommendations for project managers involved in a metrics initiative:

- Define your measures of success early in your project and track your progress towards them
- Use defect data trends to help you decide when to release a product
- Measure complexity to help you optimize Design decisions and create a more maintainable project

Paulish and Carleton report results of measuring software process improvement initiatives in Siemens software development organizations. The authors' recommendations include [Paulish and Carleton 1994]:

- Use of capability maturity model
- Conducting assessments to start software process improvement programs
- Selecting a few process improvement methods and implementing them diligently
- Paying equal or more attention to the Implementation of the method as to the method itself
- Recognizing the variation in the ease of introduction and Implementation across process improvement methods

Lorenz and Kidd present a set of metrics for the Design of object-oriented systems as well as projects that develop such systems [Lorenz and Kidd 1994]. Building from basic concepts such as inheritance and class size, the authors introduce metrics to better understand and control the development process. A selection of the metrics include *class size, number of operations overridden by a subclass, number of operations added by a sub class, specialization index,* etc. Some metrics are backed by empirical results from projects implemented in languages such as Smalltalk, C++, etc.

One of the most widely referenced sets of object-oriented metrics was put forward by Chidamber and Keremer in two related papers [Chidamber and Kemerer 1991, 1994]. The set has come to be called the *CK metrics suite* and consists of the six class-based Design metrics with explanatory names: *weighted methods per class, depth of inheritance tree, number of children, coupling between object classes, response for a class*, and *lack of cohesion in methods*. In the latter paper, the authors provide analytical evaluation of all the metrics and claim that, 'this set of six metrics is presented as the first empirically validated proposal for formal metrics for OOD'. The paper also mentions several applications of these metrics in the development of industrial software.

Weller tackles the practical yet contentious issue of using metrics to manage software projects [Weller 1994]. The author concludes that defect data can be used as a key element to improve project planning. However, he mentions the biggest bottleneck of any defect data based approach to be developers' reluctance for sharing such data with the management. This, and other human aspects of metrics based approaches, remain a lasting challenge for the software engineering discipline.

Fenton in his paper *Software Measurement: A Necessary Scientific Basis* [Fenton 1994] argues strongly in favour of adhering to fundamental measurement theory principles for software metrics. He also asserts, '...the search for general software complexity measures is doomed to failure' and backs his claim with detailed analysis. The paper reviews the tenets of measurement theory that are closely allied to software measurement, and suggests a 'unifying framework for software measurement'. Fenton also stresses on the need to validate software measures. The author mentions that in his observation, the most promising formulations of software metrics have been grounded in measurement theory.

Usually studies on software metrics tend to neglect post-delivery woes. Whatever happens in the realm of the loosely labelled 'maintenance' is seldom subjected to scrutiny. A notable exception is the IEEE suggested *software maturity index* (SMI) [IEEE 1994] that reflects on the level of stability of a software product as it is maintained and modified through continual post-production releases. Denoting the number of modules in the current release, the number of modules in the current release that have been changed, the number of modules in the current release that have been added, and the number of modules from the preceding release that were deleted in the current release, respectively, as M_T, F_c, F_a, F_d, the formula is given as SMI = [M_T − (F_a + F_c +F_d)] / M_T. As SMI approaches 1.0, the product begins to stabilize. Although maintenance issues can arise independent of the modules added or modified, such as lack of user awareness, failures in the operating environment etc., the SMI is a valuable formulation for quantifying post-delivery challenges.

Binder underscores the importance of metrics in object-oriented Testing [Binder 1994]. In fact, software testing, on account of its easily quantifiable inputs (effort in person-hours, number of units being tested, etc.) and outputs (number of defects, defects per unit of code, etc.), is one of the development activities most amenable to measurement.

Cohesion and coupling capture some of the significant characteristics of component interaction. In a way, they can be viewed as the *yin* and *yang* of software Design, contrary yet complementary forces that influence component structure and collaboration. Bieman and Ott have studied cohesion of software components in great detail [Bieman and Ott 1994]. They present a set of metrics, defined in terms of the notions of *data slice, data tokens, glue tokens, superglue tokens,* and *stickiness.* The authors develop metrics for *strong functional cohesion, weak functional cohesion,* and *adhesiveness* (the relative measure to which glue tokens bind data slices together). All of the cohesion measures have values between 0 and 1. Dhama (1995) proposes a metric for module coupling subsuming data and control flow coupling, global coupling, and environmental coupling. The module coupling indicator makes use of some proportionality constants whose values depend on experimental verification.

Basili et al. have adapted the Goal Question Metric approach to software development [Basili et al. 1994]. According to the authors, 'The approach is based upon the assumption that for an organization to measure in a purposeful way, it must first specify the goals for itself and its projects, then it must trace those goals to the data that are intended to define those goals operationally, and finally provide a framework for interpreting the data.' This measurement model has three levels: conceptual level (goal), operational level (question), and quantitative level (metric). The approach is ultimately 'a mechanism for defining and interpreting operational and measurable software.'

Churcher and Shepperd make an important point underlining the preoccupation with *class* as dominant entity of interest in object-oriented measurements [Churcher and Shepperd 1995]. 'Results of recent studies indicate that methods tend to be small, both in terms of number of statements and in logical complexity [Wilde et al. 1993], suggesting that connectivity structure of a system may be more important than the context of individual modules'. Lorenz and Kidd defines three simple metrics that analyse the characteristics for methods: *average operation size, operation complexity*, and *average number of parameters per operation* [Lorenz and Kidd 1994].

Berard examines the special place object-oriented (OO) metrics have in the study of software metrics. He identifies five points that set apart OO metrics [Berard 1995]:

- Localization
- Encapsulation
- Information hiding
- Inheritance
- Object abstraction techniques

In the introductory part of the article, the author asserts: Software engineering metrics are seldom useful in isolation, ' ... for a particular process, product, or person, three to five well-chosen metrics seems to be a practical upper limit, i.e. additional metrics (above five) do not usually provide a significant return on investment.'

As discussed in Chapter 5, Humphrey's *Personal Software Process* (PSP) [Humphrey 2005] and *Team Software Process* (TSP) [Humphrey 2006] have found wide currency in the industry as effective methodologies for enhancing the productivity of software development practitioners and teams. In a paper titled *Using a Defined and Measured Personal Software Process* [Humphrey 1996], Humphrey demonstrates how measurements can assist in the understanding and Implementation of individual skills and expertise. A cornerstone of Humphrey's techniques lies in continual monitoring of the development process, and metrics support this objective.

Garmus and Herron (1996) introduce functional techniques to measure software process. Their approach is based primarily on function point Analysis, which is customized towards process measurement.

Whitmire's *Object-Oriented Design Measurement* [Whitmire 1997] is a seminal work in the study of object-oriented metrics. Whitmire is rigorous in his treatment: putting measurement into context, building up the theoretical foundation, and capturing Design characteristics through his metrics. Whitmire proposes metrics to cover aspects of *size, complexity, coupling, sufficiency, completeness, cohesion, primitiveness, similarity*, and *volatility*. Within each area, motivations and origins, empirical views, formal properties, empirical relational structures, potential measures, etc. are discussed. The author presents an original perspective on many general concerns of software measurements. The most important contribution of Whitmire's book is the establishment of a sound mathematical framework for expressing and measuring software Design.

Harrison et al. have reviewed a set of object-oriented metrics [Harrison et al. 1998] referred to as the *MOOD Metrics Set* [Abreu 1995]. The set includes the metrics, *method inheritance factor, coupling factor, polymorphism factor,* etc. The reviewers examine the validity of these metrics in the light of certain criteria and conclude, 'as far as *information hiding, inheritance, coupling,* and *dynamic binding* are concerned, the six MOOD metrics can be shown to be valid measures ...'

In the keynote address titled *OO Software Process Improvement with Metrics* [Henderson–Sellers 1999], Henderson-Seller underlines vital links between product and process metrics. He also explores the interconnections of measurement and estimation, and outlines his vision for a software quality program. While summarizing his discussion, the author makes a very important point, '... instigating a metrics programme (sic) does not bring immediate "magical" answers to all software development. It cannot and should not be used to assess the performance of the developers themselves; nor can it create non-existent skills in developers ... A metrics programme provides knowledge and understanding; it does not provide quick fixes.'

Weigers, in an article titled *A Software Metrics Primer* [Wiegers 1999] explores some of the insights gleaned from software measurements. The author gives the following list of 'appropriate metrics' for three categories of software engineering practitioners:

- Individual developers—work effort distribution, estimated vs actual task duration and effort, code covered by unit Testing, number of defects found by unit Testing, code and Design complexity
- Project teams—product size, work effort distribution, Requirements status (number approved, implemented, and verified), percentage of test cases passed, estimated vs actual duration between major milestones, estimated vs actual staffing levels, number of defects found by integration and system Testing, number of defects found by inspections, defect status, Requirements stability, number of tasks planned and completed
- Development organization—released defect levels, product development cycle time, schedule and effort estimating accuracy, reuse effectiveness, planned and actual cost

Though far from an exhaustive list, this provides a valuable starting point for metrics orientation. Weigers also gives several 'tips for metric success: start small, explain why, share the data, define data items and procedures, understand trends'.

7.5.4 Metrics in the New Millennium

Demeyer et al. propose a set of heuristics for detecting refactorings by applying lightweight, object-oriented metrics to successive versions of a software system

[Demeyer et al. 2000]. The authors make the following assumptions regarding the implications of certain structural changes in the code:

- Method Size—A decrease in method size is a symptom for method split.
- Class Size—A change in class size is a symptom for a shift of functionality to sibling classes (i.e. incorporate object composition). Also, it is part of a symptom for the redistribution of instance variables and methods within the hierarchy (i.e. optimization of class hierarchy).
- Inheritance—A change in the class inheritance is a symptom for the optimization of a class hierarchy.

While these assumptions are not beyond contention—for example, a method may shrink in size due to the introduction of a smarter algorithm, not necessarily indicative of method split—the authors show important correlations between refactoring and Design drift and how metrics can aid in identifying and understanding them.

Pressman treats the discipline of software metrics deeply in his wide-ranging book *Software Engineering: A Practitioner's Approach* [Pressman 2000]—the standard text for many graduate courses. Pressman makes a distinction between the so-called *technical* metrics which seek to capture the progression and behaviour of the software product, vis-à-vis the metrics relevant for project management and process compliance. The book also devotes an entire chapter to metrics related to object-oriented systems.

Sotirovski underlines the inherent challenges of iterative software development, '... If the iterations are too small, iterating itself could consume more energy than designing the system. If too large, we might invest too much energy before finding out that the chosen direction is flawed. [Sotirovski 2001]' To tackle this quagmire, the author highlights the role of heuristics in iteration planning and monitoring. Successful metric efforts frequently lead to the encapsulation of their wisdom in heuristics. And in the absence of physical laws to fall back on, heuristics are often vital to software Design and Implementation.

Lanza takes an unconventional and interesting approach towards a metrics based understanding of software evolution [Lanza 2001]. The author proposes an *Evolution Matrix* which, '... displays the evolution of the classes of a software system. Each column of the matrix represents a version of the software, while each row represents the different versions of the same class.' Based on this construct, classes are categorized into groups with maverick names: Pulsar, Supernova, White Dwarf, Red Giant, Stagnant, Dayfly, and Persistent. Based on case study data, Lanza delineats phases in a system's evolution characterized by specific categories of classes. Though the paper points out several limitations of the approach, it gives a novel perspective on the mutation of software systems.

Understanding and mitigating the effects of change on enterprise software systems remain an important concern of software engineering. It is interesting to note how Kabaili et al. [Kabaili et al. 2001] have tried to interpret cohesion as a changeability indicator for object-oriented systems. The authors seek to establish a correlation between cohesion and changeability and have used empirical data from C++ projects to support their assertions. In conclusion, the authors comment that based on their studies, coupling vis-à-vis cohesion appears to be a better changeability indicator. This study presents novel insights on how Design characteristics may reveal more than they are initially intended to.

Mens et al. in their paper, *Future Trends in Software Evolution Metrics* [Mens and Demeyer 2001], underline the relevance of predictive Analysis and retrospective Analysis in studying software evolution. They mention the following areas as promising fields of future metrics research, in spite of the fact that some of them have already been closely examined:

- Coupling or cohesion metrics
- Scalability issues
- Empirical validation and realistic case-studies
- Long-term evolution
- Detecting and understanding different types of evolution
- Data gathering
- Measuring software quality
- Process issues
- Language independence

Ramil and Lehman study the relevance of applying measurements to long-term software evolution processes and their products [Ramil and Lehman 2001]. An example using empirical data from the *Feedback, Evolution, and Software Technology* (FEAST) program is presented. The example illustrates the use of a sequential statistical test on a suite of eight evolution activity metrics. The authors underline the need for precise definition of metrics, as small differences in definitions can lead to inordinately large divergence in the measured values.

Rifkin takes a perspective view of why software metrics are so difficult to put into practice, given the business needs enterprise software has to fulfill [Rifkin 2001]. Four different software development domains are reviewed and their attitudes to measurements compared: Wall Street brokerage house, civilian government agency, computer services contractor, and the non-profit world. The author advocates a measurement strategy suited to each type of organization, and concludes, 'We need to develop a whole new set of measures for all those customer–intimate and product–innovative organizations that have avoided measurement thus far'.

Fergus, in his book *How to Run Successful Projects III—The Silver Bullet,* [O'Connell 2001] in apparent allusion to Brooks' classic essay [Brooks 1995], discusses how measurement techniques can make great difference to the outcome of projects. His probability of success indicator (PSI) metric is especially insightful.

Software measurement initiatives in an organization usually focus on the concrete, such as lines of code, developer productivity etc. Buglione and Abran (2001) investigate how creativity and innovation at an organizational level can be measured. Based on the structure of commonly used software process improvement models such as CMMI and P–CMM, the authors discuss how both process and people aspects of creativity and innovation can be measured.

COCOMO and COCOMO II [CSE 2002] are primary among several models for estimating cost, effort, and schedule of software development activity. These are useful in the planning and execution of large software projects. They consist of three sub models: application composition, early Design, and post-architecture. The original COCOMO was first published by Boehm in 1981 [Boehm 1981] and this work still remains the best introductory reference to the model. The COCOMO model has been kept current by regular updates and refinements, as software engineering has undergone many paradigm shifts from 1981 till date (see Chapter 6 for more details on COCOMO).

Clark, in an article titled *Eight Secrets of Software Measurement* [Clark 2002], enumerates some tricks of making a software measurement scheme work. Some of the eight 'secrets', not unexpectedly, sound somewhat clichéd. But the author still makes some perceptive observations, such as '... measurement is not an end in itself; it's a vehicle for highlighting activities and products that you, your project team, and your organization value so you can reach your goals'.

Fenton et al. argue that the typical way of using software metrics is detrimental to effective risk management [Fenton et al. 2002]. They identify two specific roles of software measurement as quality control and effort estimation. They also point to most commonly used factors to assess software while it is being developed as complexity measures, process maturity, and test results. The problems with widely used regression models are discussed. The authors recommend a Bayesian network based defect prevention model and explain details of the *AID* (assess, improve, decide) tool built on it. The authors see the dawn of 'an exciting new era' in software measurement with wider applications of Bayesian networks.

Krutchen in his widely referenced book on the Rational Unified Process [Krutchen 2004] makes an important categorization of measures. He calls *measure* 'a concrete numeric attribute of an entity (e.g. a number, a percentage, and a ratio)' whereas *primitive measure* is 'an item of raw data that is used to calculate a measure'. The book only mentions measurement in the context of the project management discipline, which may be viewed as a constriction of the scope of metrics. Effective metrics, in addition to facilitating project management, can aid the planning and execution of developer and team activities.

In their book *Software by Numbers*, Denne and Cleland-Huang introduce the ideas of Incremental Funding Methodology (IFM), Minimum Marketable Feature (MMF), etc. to facilitate business decisions in enterprise software development [Denne and Cleland-Huang 2004]. This work makes a notable attempt at bridging the seemingly 'never the twain shall meet' chasm between those who build software and those who commission the building of software.

Eickelmann makes an interesting distinction between the measurements of maturity and process in the context of the CMM levels [Eickelmann 2004]. The author underlines that an organization's process maturity can be viewed from multiple perspectives.

Return on investment (ROI) and software process improvement (SPI) are two of the most audible buzzwords in the software engineering industry today, customarily called by their acronyms. Solingen addresses the cornerstone of ROI and SPI by establishing the practicality of measuring the former in terms of the latter [van Solingen 2004]. The author bases his discussion on the ROI numbers for several major software development organizations across the world.

Rico examines how the use of metrics by project managers and software engineers alike can lead to better return on investment on software process improvement [Rico 2004]. The book discusses investment Analysis, benefit Analysis, cost Analysis, net present value, etc. and integrates these ideas within the parameters of established methodologies such as Personal and Team Software Processes, Software Capability Maturity Model, ISO 9001. Although the author's focus is primarily on process improvement rather than the development process, there are interesting pointers to the positioning of metrics in the 'bigger picture' of the development enterprise.

Continuing on the ROI theme, Pitt's article *Measuring Java Reuse, Productivity, and ROI* [Pitt 2005] uses the *effective lines of code* (ESLOC) metric to measure the extent of reuse in Java code and the resultant return on investment achieved. The author reaches some expansive conclusions, but the choice of the ESLOC metric may not reflect all the significant nuances of a software system. Also, the author's remark that 'many elements are generated from an IT project, but arguably the most important element is the source code' is open to counter arguments. With increasing trends towards model-driven development, larger and larger portions of source code are being automatically generated; Analysis or Design artefacts (that finally drive code generation) can lay legitimate claims to being the so-called 'most important element'.

Bernstein's work [Bernstein and Yuhas 2005] embodies a modern outlook to software measurements: Metrics should not only reflect merely the countable aspects of a software product, such as lines of code, but must encompass the spectrum of people, product, process, and project that make up software engineering in totality.

The author presents interesting quantitative strategies on software development. However, some chapters present ready nuggets of wisdom, modulated as 'Magic Number' (An example: 'The goal for the architecture process is to reduce the number of function points by 40%', Page 142) which may not stand rigorous scrutiny.

Napier et al. in their book *Measuring What Matters: Simplified Tools for Aligning Teams and Their Stakeholders* [Napier and McDaniel 2006] discuss techniques for management to harness the potential of measurement seamlessly and painlessly. The book provides several interesting measurement templates and leverages the industry experience of the authors to significant effect.

7.6 ART AND CRAFT OF METRICS MAKING

An important objective of this chapter is to demonstrate the fact that *making your own metrics* is certainly an option in real-life software development projects. It is often the case that one does not find an existing metric that fits the problem at hand. In such cases, practitioners are strongly encouraged to make and use their own metrics. In the Case Study section of this chapter, we will actually see such a plan in action. But before we get there, let us highlight some tricks of the art and craft of metrics making.

- Whenever we want to think of a new metric, we must first think of the assumptions. No single metric will get us every insight we seek, so one must define a particular scope through reasonable assumptions. Very interestingly, basic laws of physics also have assumptions built into them. Newton's First Law of Motion talks about bodies continuing in their state of rest or uniform motion in a straight line, unless acted upon by an external, impressed force. The second part of the last sentence, (after 'unless') is the underlying assumption for the validity of the law; without it, the law will not hold. In our every day experience, bodies may continue in their state of rest, but invariably those in uniform motion either turn to non-uniform motion, or come to rest. So the assumption here is an *ideal* one, which is unlikely to be true anywhere other than under carefully controlled conditions. But even in spite of the ideality of this assumption, the First Law of Motion is a mighty useful law! The moral of the story: Do not be afraid to make assumptions about ideal situations when making metrics—it will only help clarify when and where the metrics are likely to work best.

- The formula behind a metric must have a simple, intuitive appeal. Complex formulas do not necessarily capture complex ideas, and may even obscure simple ones. Metrics which are simple and can be calculated easily have a far better chance of being widely understood, adopted, and used. To make simple metrics, we often need to break down the problem at hand into smaller, more manageable chunks. This is an old and very useful strategy for problem solving—break down a complex question into smaller and simpler ones, try and find simple answers to the simple questions, and combine the answers in a clever way to reach the final answer. This use is also very helpful in software engineering, in general.

- We will not be able to make a perfect metric in one go. So an iterative approach, expedient as it is for building large software systems, is also very important in metrics building. Using the outcome of applying one version of a metric, the metric itself can be refined further to better reflect what we are interested in. So, just as a software product gets built incrementally over iterations, getting refined over every cycle, successful software metrics also get formulated over cycles of trials and errors.

CASE STUDY

Earlier in this chapter we have sought to motivate the making of one's own metrics. Now we will demonstrate the idea through the formulation of the Morphing Index. The following discussion is inspired by the sections of the chapter *Analysis and Design—On Understanding, Deciding, and Measuring* of the book on metrics-driven enterprise software development [Datta 2007].

Earlier, we have introduced the 'context of the recurring case study in this book. We may recall Preeti, our protagonist. For the present, let us follow Preeti as she finds it necessary to build the Morphing Index metric.

Preeti joined the development team as a technical lead, two months after the project started. Officially, Requirement gathering and Analysis was over, Design was nearing the end, and Implementation was about to start. There was a general feeling that 'things would straighten out' once code would begin to be cut. During the first few meetings with the developers and the system analyst, Preeti found herself staring into reams of documentation. There were several versions of each document, often contradicting, and always more complicated than the previous one. It seemed even the basic directions of the Design were undergoing too many changes. Trying not to sound impolite or impolitic, Preeti asked her team why they thought things were changing so much. The analyst said customers could not make up their minds, adding 'as usual' with a deprecating smile; the developer said they went by the analyst's word. The project manager, whom Preeti briefed 'off-line', seemed to believe this was what iterative and incremental development was meant to be. Wasn't it all about letting customers change their minds? He generously added, he was of course no expert in these new technologies; (in his coding days real programmers wrote in assembly language) and he deferred to the tech lead's judgement. But how much was the Design changing, anyway?

Preeti had no answer to the how much. But without that how much, she knew she could never convince the manager that flags need to be raised to bring the customer and the development team on the same page. Let us see if we can help her with the 'how much' of change. We call the situation with a touch of lyricism, the shifting sands of Design.

Shifting Sands of Design

Preeti's plight should be familiar to all who have been in the trenches of software development. Software is perhaps the only industrial artefact that has to absorb change in its purpose, environment, circumstance of use, etc. and yet deliver the originally envisaged value at the same or increased level. To absorb this change, Design must also change during development.

Our objective is to derive a metric that will quantify the change in Design from one iteration to another. In the following sections, we develop the Morphing Index iteratively; a plain vanilla version first, seeing what it can give us, and then refining it further to make it more useful.

Making of the Metric

Morphing Index tries to capture the extent to which a particular Design is changing between pre-defined baselines. Before presenting the metric, it is important to settle the question of context.

Software Design tries to devise the optimal collaboration of system components so that user needs are best met. As we will discuss in Chapter 13, use cases play a major role in specifying user-system interaction. A scenario is a specific sequence of actions that illustrates an aspect of this interaction. Scenarios are to use cases as objects are to classes: A scenario is basically one instance of a use case [Booch et al. 2005]. Scenarios are starting points for Design artefacts such as sequence diagrams. The Morphing Index will reflect the change in Design for a particular scenario between two iterations.

Derivation—First Pass

To specify a metric, we often need to streamline the understanding of a familiar term. For this discussion, let us regard 'Design' as the interplay of components and their collaborations. This agrees with intuitive ideas of object-oriented Design (refer to Chapter 12); objects or class instances communicate amongst themselves via messages to get things done. Change in the number of classes and messages is a sure symptom of Design change. (For the remainder of this chapter, we will use 'class' and 'component' interchangeably.) Based on this observation, let us define the Morphing Index for the kth iteration, $RI(k)$ as follows:

Let

m = Total number of components
n = Total number of messages

Then

$$RI(k) = \frac{\sum_{i=1}^{m} C_i}{\sum_{j=1}^{n} M_j}$$

Evidently, C_i and M_j are the ith component and the jth message, respectively. Thus the *Morphing Ratio* for the k'th iteration is the ratio of the total number of components to the total number of messages.

With due deference to rigorists and carpers, ('So simple a ratio can hardly capture the subtleties of Design', 'Some classes are more equal than others, just counting classes does not help', 'Not all messages do useful things', ad nauseam!) we may reflect on what the Morphing Index does deliver. If the fraction has a value much greater than one; there are too many classes than the messages, indicating the so called island syndrome, classes trying to do things by themselves. This certainly clamors for better delegation of responsibilities. A very low value points to a gaggle of classes chatting back and forth through too many messages. As classes interact in new ways, variation in $RI(k)$ values gives an inkling of how the Design has changed relative to a previous iteration.

The Morphing Index in its present form is rather naive; but it tells us something, and something is better than nothing. Still we need to nose around for inadequacies. When would a flat headcount of the number of classes and messages not suffice? Or worse, give a misleading view of things? It is quite possible the number of classes and/or messages change without the intent of the underlying Design changing notably. In the initial stages this is rather common; going back and forth, trying out different combinations is inherent to the process of designing. And iterative and incremental development supports and encourages it. Let us see how we can make the Morphing Index more 'intelligent'.

As per the current definition, the Morphing Index would change each time a class/message was added or removed. All classes and messages do not do equally important things. Based on what they do, an element of differentiation should be introduced into the calculation of the Morphing Index. It is time for a second pass through the derivation.

Derivation—Second Pass

To differentiate between classes and messages, it is important to understand how and why they differ. Not all components fulfill equally important responsibilities; not all collaborations facilitate equally vital tasks.

In large-scale enterprise software systems, fulfilling business functions is of primary importance. Business logic may be varied, and not always very logical. It may include for example—deciding whether a customer is eligible for discount, based on the retailer's policies and the customer's purchase history or calculating the interest rate, depending on an account holder's account balance. Other activities used in implementing business logic are usually encapsulated in helper or utility classes. Based on this difference in their functions, we can segregate classes into the following categories:

- *Primary*—business objects
- *Secondary*—components encapsulating crosscutting functionality
- *Tertiary*—utility, helpers, etc.

What are the different kinds of things messages do? During Implementation messages translate to method calls, the services offered by one object when invoked by another object. A message can serve one of the following purposes:

- *Creational*—creating class instances; e.g. invoking a constructor in C++/Java.
- *Computational*—manipulating available data based on business logic; e.g. calculating the total price of a purchase.
- *Transmissional*—fetching or sending information from one component to another; e.g. database calls.

Let us now attach weights to the different types of classes and messages. For components, the weights are as follows:

- Primary—3
- Secondary –2
- Tertiary—1

Evidently, highest importance is ascribed to Primary components, least to Tertiary, and the Secondary ones in between. We assume each component has one clearly defined responsibility.

For messages, the weights are as follows:

- Creational—2
- Computational –3
- Transmissional—1

We take computation as being most important and transmission the least in the tasks messages carry out. Assigning weights is often a simple and effective way to determine the relative contributions of the factors contributing to a metric. There is nothing sacrosanct about these weights for the classes and messages. If in an application, the Tertiary components are far more complex than the Primary ones, or Transmissional messages serve a more meaningful purpose than Computational ones, feel free to flip the weights around. The key idea is that classes and messages vary in their importance in a particular scheme of Design, and weights bring out that difference.

We now redefine the Morphing Index for the kth iteration:

$$RI(k) = \frac{\sum_{i=1}^{m} w(C_i)}{\sum_{j=1}^{n} w(M_j)}$$

C_i and M_j are the ith component and the jth message respectively and $w(X_a)$ is weight assigned to the a'th component or message. Thus the Morphing Index for the k'th iteration is the ratio of the weighted sum of components to messages.

What does this refined Morphing Index buy us?

- There is now more 'granularity' in the measure. Variation in the number of components and messages, together with their relative importance affect the value of the metric.
- Morphing Index values over iterations can serve as a heuristic to gauge whether the system's functionality is evenly distributed across components or cluttered in a few.
- Changing values of the Morphing Index show the extent of Design change over iterations.

We will cut back to Preeti now, and examine some issues in applying the Morphing Index.

Back to Preeti

With the Morphing Index, Preeti hopes to convince her manager of the reigning chaos in the project, and its fallouts. After hours of sifting through Design diagrams to identify classes and messages between them, Preeti came up with her spreadsheet. It showed the Morphing Index values versus each of the use case scenarios for the four Design iterations already made. The graphs were expectedly jagged. Preeti presented the metrics data to her manager.

The project manager commended Preeti and came up with more questions. Preeti was asked to find out why the numbers spoke the way they did, and what could be the underlying causes and probable cures. It was time some more of the metrics wisdom dawned on Preeti:

- One single metric, by itself, is hardly ever helpful—The Morphing Index gives some insight, but raises more questions than it answers. If the Design has gone back and forth so much, without converging, has there been some problem with the Analysis? Were the Requirements elicited, understood, specified, and documented correctly? Did the users have a clear idea of the system they want? Would plunging headlong into coding help? For Preeti, the need of the hour was a set of metrics covering the development process, each revealing a facet, and together giving the sense of what is and what needs be changed.

- Metrics seldom, if ever, work as afterthoughts—The jagged graphs on Preeti's spreadsheet could not recoup the time already lost on the project. More disturbingly, when presented to the team, they were taken to insinuate incompetence, and raised rancour. There are major political and inter-personal implications for any metrics initiative, and these must be addressed upfront.

- To be most effective, metrics need to be a part of the development process. The software development life cycle is deeply interconnected; every artefact stands to be affected by, and affects, upstream and downstream artefacts. Only when metrics align with this continuum can they hope to make a difference.

We now take a look at a metric which examines an idea very similar to the 'morphing' we have dealt with so far.

An Allied Metric—Whitmire's Volatility Index

Whitmire says, 'Volatility is the likelihood or probability that a change will occur. It is an important characteristic of applications, domains, and Design components and must be watched carefully' [Whitmire 1997]. He then derives a measure of volatility. As mentioned earlier in this chapter, Whitmire combines measurement theory and the theory of objects to build up a rigorous foundation for object-oriented metrics.

Seeing the Morphing Index and Whitmire's measure in each another's light gives a key insight: There are always different ways of formulating metrics of similar scope and focus. Each has its pros and cons; and the choice is influenced by experience and intuition.

SUMMARY AND TAKE-AWAYS

In this chapter, we introduced the fascinating world of software metrics. Here are some of the key ideas:

- There are no laws of software (at least none has been discovered yet); thus software metrics need to be based more on intuition and ingenuity than physical phenomenon.

- Although metrics can serve as useful management aids, software metrics have a very important role to play in guiding the software development life cycle.
- Development of software metrics is an ongoing process with new metrics being continuously tried out.
- For a particular situation of interest, any established metric can be used. If none is found to be appropriate, a new metric can be devised.
- Creating one's own metric helps understand a problem and evaluate candidate solutions.

WHERE TO LOOK FOR MORE

- Software Engineering Institute—http://www.sei.cmu.edu/: An important repository for software engineering related information, including metrics.
- Software Metrics Sites—http://measurement.fetcke.de/: A useful guide to software metrics related information.

EXERCISES

Review Questions

Review Questions test your understanding of the key concepts presented in this chapter.

1. Software metrics help
   (a) as management aids
   (b) guide the software development process
   (c) both (a) and (b)
   (d) none of the above

2. Which of the following metrics seek to measure the complexity of a software program?
   (a) Program Volume
   (b) Lines of code
   (c) McCabe's Cyclomatic measure
   (d) Morphing Index

3. Function points are computed using
   (a) experimental relationships between the direct measures of the software's information domain and estimation of its complexity on a weighted scale
   (b) lines of code

   (c) number of components
   (d) all of the above

4. Weighted methods per class, depth of inheritance tree, number of children, etc. are examples of
   (a) management numbers
   (b) object-oriented metrics
   (c) graph parameters
   (d) Design metrics

5. Software maturity index
   (a) indicates the level of maturity of an organization
   (b) helps in estimation
   (c) reflects on the level of stability of a software product as it is maintained and modified through continual post-production releases
   (d) none of the above

6. Data slices, data tokens, glue tokens, superglue tokens, and stickiness are metrics to describe
   (a) yin and yang

(b) cohesion

(c) coupling

(d) both (b) and (c)

7. PSP and TSP by W.S. Humphrey are

(a) software metrics

(b) programming paradigms

(c) both (a) and (b)

(d) methodologies for enhancing productivity of practitioners and teams

8. COCOMO and COCOMO II are

(a) code Construction models

(b) models for estimating cost, effort, and schedule of software development activity

(c) two versions of an award winning software program

(d) metrics for measuring programmer productivity

9. Lines of code does not always capture interesting aspects of a software system because

(a) interesting software systems have too many lines of code

(b) according to Brooks, software is 'invisible' and 'unvisualizable'

(c) interesting software systems may not have a single line of code

(d) all of the above

10. A group of related software metrics rather than a single metric is more likely to capture a phenomenon of interest.

(a) true

(b) false

(c) sometimes true, sometimes false

(d) cannot be said without more detail

Reflective Questions

Reflective Questions require you to think more deeply about some of the ideas and come up with your own interpretations and answers.

1. Point out three deficiencies of the Morphing Index metric. How would you modify the metric to make it more useful?

2. You are the technical lead of a team of software engineers working on a project. You are building a software system using the iterative and incremental methodology. How would you measure how much the incremental release from each iteration is contributing towards completion of the project? Can you devise a metric towards this end? What are the assumptions you may need to make?

REFERENCES

Abreu, F.B (1995), 'The MOOD Metrics Set', Proc. ECOOP'95 Workshop on Metrics, 1995.

Albrecht, A. (1979), 'Measuring Application Development Productivity', Proc. Joint SHARE/GUIDE/IBM Application Development Symposium (October 1979), pp. 83–92.

Armour, P.G. (2004), 'Beware of counting loc.', *Commun, ACM*, 47(3): 21–24.

Baker A.L., J.M. Bieman, N. Fenton, D.A. Gustafson, A. Melton and R. Whitty (1990), 'A philosophy for software measurement', *J. Syst. Softw.*, 12(3): 277–281.

Basili, V.R., G. Caldiera, and H.D. Rombach (1994), *The Goal Question Metric Approach:*

Encyclopedia of Software Engineering, Wiley and Sons, Inc.

Belady, L.A. and M.M. Lehman (1976), 'A zmodel of large program development', IBM, *IBM Systems Journal*, Volume 15, Number 3, p. 225.

Belady, L.A. and M.M. Lehman (1979), 'The characteristics of large systems', In Research Directions in Software Technology, pp. 106–138, MIT Press.

Berard, E.V. (1995), 'Metrics for object-oriented software engineering', http://www.ipipan.gda.pl/~marek/objects/TOA/moose.html, last accessed on May 19, 2010.

Bernstein, L. and C.M. Yuhas (2005), *Trustworthy Systems through Quantitative Software Engineering*, Wiley-Interscience.

Bieman, J.M. and L.M. Ott (1994), 'Measuring functional cohesion', *IEEE Trans. Softw. Eng.*, 20(8): 644–657.

Binder, R.V. (1994), 'Object-oriented software testing', *Commun. ACM*, 37(9): 28–29.

Boehm, B.W. (1981), *Software Engineering Economics*, Prentice Hall.

Booch, G., J. Rumbaugh, and I. Jacobson (2005), *The Unified Modeling Language User Guide,* 2nd ed., Addison-Wesley.

Brooks, F.P. (1995), *The Mythical Man-Month: Essays on Software Engineering,* 20th Anniversary Edition, Addison-Wesley.

Buglione, L. and A. Abran (2001), 'Creativity and innovation in spi: An exploratory paper on their measurement?', In *IWSM'01: International Workshop on Software Measurement*, pp. 85–92, Montreal, Quebec, Canada.

Card, D.N. and R.L. Glass (1990), *Measuring Software Design Quality*, Prentice-Hall.

Cavano, J.P. and J.A. McCall (1978), A framework for the measurement of software quality, *SIGSOFT Softw. Eng. Notes*, 3(5): 133–139.

Chidamber, S.R. and C.F. Kemerer (1991), 'Towards a metrics suite for object oriented Design', In *OOPSLA '91: Conference proceedings on Object-oriented programming systems, languages, and applications*, pp. 197–211, New York, NY, USA. ACM Press.

Chidamber, S.R. and C.F. Kemerer (1994), 'A metrics suite for object oriented Design', *IEEE Trans. Softw. Eng.*, 20(6): 476–493.

Churcher, N.I. and M.J. Shepperd (1995), 'Towards a conceptual framework for object oriented software metrics', *SIGSOFT Softw. Eng. Notes*, 20(2): 69–75.

Clark, B. (2002), 'Manager: Eight secrets of software measurement', *IEEE Softw.*, 19(5): 12–14.

Conte, S.D., H.E. Dunsmore, and V.Y. Shen (1986), *Software Engineering Metrics and Models*, The Benjamin/Cummins Publishing Company, Inc.

CSE (2002), Cocomo, http://sunset.usc.edu/research/COCOMOII/. last accessed on May 19, 2010.

Daskalantonakis, M.K. (1992), 'A practical view of software measurement and Implementation experiences within Motorola', *IEEE Trans. Softw. Eng.*, 18(11): 998–1010.

Datta, S. (2005), 'Integrating the furps+ model with use cases — a metrics driven approach', In *Supplementary Proceedings of the 16th IEEE International Symposium on Software Reliability Engineering (ISSRE2005), Chicago, IL, November 7–11, 2005*, pp. 4–51—4–52.

Datta, S. (2006), 'Crosscutting score: an indicator metric for aspect orientation', In *ACM-SE 44:*

Proceedings of the 44th annual southeast regional conference, pp. 204–208, New York, NY, USA. ACM Press.

Datta, S. (2007), *Metrics-Driven Enterprise Software Development: Effectively Meeting Evolving Business Needs*. J. Ross Publishing.

Davis, A., S. Overmyer, K. Jordan, J. Caruso, F. Dandashi, A. Dinh, G. Kincaid, G. Ledeboer, P. Reynolds, P. Sitaram, A. Ta, and M. Theofanos (1993), 'Identifying and Measuring Quality in a Software Requirements Specification', in *Proceedings of the 1st International Software Metrics Symposium*.

DeMarco, T. (1982), *Controlling Software Projects*, Yourdon Press.

DeMarco, T. and T. Lister (1987), *Peopleware: Productive Projects and Teams*, Dorset House Pub. Co.

Demeyer, S., S. Ducasse, and O. Nierstrasz (2000), Finding refactorings via change metrics, *SIGPLAN Not.*, 35(10): 166–177.

Denne, M. and J. Cleland-Huang (2004), *Software by Numbers: Low-risk, High-return Development*, Prentice Hall PTR.

Dhama, H. (1995), 'Quantitative models of cohesion and coupling in software', In *Selected papers of the sixth annual Oregon workshop on Software metrics*, pp. 65–74, New York, NY, USA, Elsevier Science Inc.

Eickelmann, N. (2004), 'Measuring maturity goes beyond process', *IEEE Softw.*, 21(4): 12–13.

Fenton, N. (1994), 'Software measurement: A necessary scientific basis', *IEEE Trans. Softw. Eng.*, 20(3): 199–206.

Fenton, N., P. Krause, and M. Neil (2002), 'Software measurement: Uncertainty and causal modeling', *IEEE Softw.*, 19(4): 116–122.

Garmus, D. and D. Herron (1996), *Managing the Software Process: A Practical Guide to Functional Measure*, Prentice Hall.

Gilb, T. (1977), *Software Metrics*, Winthrop Publishers, Inc.

Grady, R.B. (1992), *Practical Software Metrics for Project Management and Process Improvement*, Prentice Hall.

Grady, R.B. (1994), 'Successfully applying software metrics', *Computer*, 27(9): 18–25.

Grady, R.B. and D.L. Caswell (1987), *Software metrics: establishing a company-wide program*, Prentice Hall.

Halstead, M.H. (1977), *Elements of Software Science*, Elsevier North-Holland, Inc.

Harrison, R., S.J. Counsell, and R.V. Nithi (1998), 'An evaluation of the mood set of object-oriented software metrics', *IEEE Trans. Softw. Eng.*, 24(6): 491–496.

Henderson-Sellers, B. (1999), 'Oo software process improvement with metrics', In *METRICS '99: Proceedings of the 6th International Symposium on Software Metrics*, p. 2, Washington, DC, USA. IEEE Computer Society.

Humphrey, W.S. (1996), 'Using a defined and measured personal software process', *IEEE Softw.*, 13(3): 77–88.

Humphrey, W.S. (2005), *PSP: A Self-Improvement Process for Software Engineers*, Addison-Wesley.

Humphrey, W.S. (2006), *TSP: Leading a Development Team*, Addison-Wesley.

IEEE (1990), 'IEEE software engineering standards', standard 610.12-1990, pp. 47–48. standards.ieee.org/software/.

IEEE (1994), *Software Engineering Standards,* 1994 edition. IEEE.

Jones, C. (1991), *Applied Software Measurements*, McGraw-Hill.

Kabaili, H., R.K. Keller, and F. Lustman (2001), 'Cohesion as changeability indicator in object-oriented systems', In *CSMR '01: Proceedings of the Fifth European Conference on Software Maintenance and Reengineering*, p. 39, Washington, DC, USA, IEEE Computer Society.

Krutchen, P. (2004), *The Rational Unified Process: An Introduction,* 3rd ed., Addison-Wesley.

Lanza, M. (2001), 'The evolution matrix: Recovering software evolution using software visualization techniques', In *IWPSE '01: Proceedings of the 4th International Workshop on Principles of Software Evolution*, pp. 37–42, New York, NY, USA, ACM Press.

Lehman, M., J. Ramil, P. Wernick, and D. Perry (1997), 'Metrics and laws of software evolution: The nineties view', http://citeseer.ist.psu.edu/lehman97metrics. html. last accessed on May 19, 2010.

Lorenz, M. and J. Kidd (1994), *Object-oriented Software Metrics: A Practical Guide*, PTR Prentice Hall.

McCabe, T. (1976), 'A software complexity measure', In *IEEE Trans. Software Engineering, vol. SE-2, December 1976*, pp. 308–320.

McCabe, T. and A. Watson (1994), 'Software complexity', In *Crosstalk, Vol. 7, No. 12, December 1994*, pp. 5–9.

Mens, T. and S. Demeyer (2001), 'Future trends in software evolution metrics', In *IWPSE '01: Proceedings of the 4th International Workshop on Principles of Software Evolution*, pp. 83–86, New York, NY, USA, ACM Press.

Napier, R. and R. McDaniel (2006), *Measuring What Matters: Simplified Tools for Aligning Teams and their Stakeholders*, Davies-Black Pub.

O'Connell, F. (2001), *How to Run Successful Projects III: The Silver Bullet*, Addison-Wesley.

Paulish, D.J. and A.D. Carleton (1994), 'Case studies of software-process-improvement measurement', *Computer*, 27(9): 50–57.

Pitt, W.D. (2005), 'Measuring java reuse, productivity, and roi', *Dr.Dobb's Journal*, July 2005.

Pressman, R.S. (2000), *Software Engineering: A Practitioner's Approach*, McGraw-Hill.

Ramil, J.F. and M.M. Lehman (2001), 'Defining and applying metrics in the context of continuing software evolution', In *METRICS '01: Proceedings of the 7th International Symposium on Software Metrics*, p. 199, Washington, DC, USA, IEEE Computer Society.

Rico, D.F. (2004), *ROI of Software Process Improvement: Metrics for Project Managers and Software Engineers*, J. Ross Pub.

Rifkin, S. (2001), What makes measuring software so hard?, *IEEE Softw.*, 18(3): 41–45.

Sears, A. (1993), 'Layout appropriateness: A metric for evaluating user interface widget layout', *IEEE Trans. Softw. Eng.*, 19(7): 707–719.

Sotirovski, D. (2001), 'Heuristics for iterative software development', *IEEE Software*, 18(3): 66–73.

van Solingen, R. (2004), 'Measuring the roi of software process improvement', *IEEE Softw.*, 21(4): 32–34.

Weller, E.F. (1994), 'Using metrics to manage software projects', *Computer*, 27(9): 27–33.

Whitmire, S.A. (1997), *Object-Oriented Design Measurement*, Wiley Computer Pub.

Wiegers,K.E.(1999),'A software metrics primer', http://www.processimpact.com/articles/ metrics_primer.html. last accessed on May 19, 2010.

Wilde, N., P. Matthews, and R. Huitt (1993), Maintaining object-oriented software, *IEEE Softw.*, 10(1): 75–80.

Software Project Management

Learning Objectives

In this chapter, we outline project management as applied to software projects. The very nature of software engineering makes it imperative that software engineers—even those entering the profession at the beginning of their careers—have an understanding of the basic elements of project management. We will specifically address the following areas:

- Place of management in the enterprise of software development
- The four Ps of software development: people, project, product, and process
- Typical project life cycle
- Process groups and knowledge areas of project management
- Principles of software project management
- The IEEE Software Project Management Plan (SPMP)
- Dynamics of software teams
- Important project management activities
- Managing versus leading

8.1 MOTIVATION

What do we need to build good software? We are not talking about *great* software—systems of breathtaking size or complexity; by *good* we merely mean software that reasonably satisfies user needs, is easy to maintain and enhance, and is delivered within time and budget. To build such systems—not once by chance, but consistently—we need efficient programmers, sound processes, a reliable infrastructure; what else? It is difficult to pinpoint what else, but we know we are missing something. It is not merely enough to bring together smart programmers, consummate processes, fast computing resources, and expect good software out of the mix. We need *some way* to ensure commitments about schedule; cost and quality are made and kept; when risks arise (as they always will) they are addressed with due importance; people are recruited and motivated; and multifarious other issues that inevitably seem to crop up when a group of people get together to do something are attended to. This 'some way' is difficult to catch in a single, simple definition; this is the realm of management.

There is management in almost everything we do in our lives. Even when an individual works all by himself or herself, there is need to manage time and resources, so that the best results are obtained at the least expense. This becomes particularly important when we are commissioned to do something—someone else is paying to get something done through the application of our knowledge or skills. The need for planning, monitoring, supervising, and motivating arises: Management starts off by addressing these concerns, and good management goes far beyond just these. In commercial software production, some customer hires the development team to build a software system. Since the customer organization is paying to get the software built, it is natural it will want the best returns on its money. It is the responsibility of management to ensure these ends are met.

This is a book on software engineering, and our primary goal is to discuss how superior software can be designed, built, deployed, maintained, and enhanced. A software engineer joining the industry fresh from college will be more involved in the Construction of software than the management of software projects. However, engineering and management share a close (and at times difficult!) relationship. Many engineers go on to take formal courses in management, and almost all engineers need to practise management informally, in their everyday work. Management is an interesting, complex, and exciting field of study, research, and practice; we certainly do not have the scope here to either plumb its depths or plough its width. This chapter thus gives us a preliminary flavour of management as applied to software projects.

In the next section, we underscore a unique characteristic of management.

8.2 THAT ELUSIVE SOMETHING

Management is something which one hardly notices, if it is done well. Just as good writing gets across ideas without the reader tripping over words, good management gets things done without imparting a feeling of being manipulated or supervised. On the other hand, symptoms of bad management are easy to recognize: bossing around, blame games, fixing people; most of us have seen (and felt) it some time or other. Good management is an *elusive something*, difficult to pinpoint; a novel combination of skills, insights, attitudes, and qualities.

In Jerome K. Jerome's *Three Men in a Boat: (To Say Nothing of the Dog)* [Jerome 2006] we find a pithy and amusing reflection on an aspect of management.

The author (who narrates the story in the first person, identifying himself as 'J') and his two friends, George and Harris are packing for a holiday on a boat on the river Thames. J said he would pack, and George and Harris readily acquiesced. Then the former went on to recline on an easy chair and the latter put up his legs on a table. These gestures surprised the narrator. In his words,

'What I had meant, of course, was, that I should boss the job, and that Harris and George should potter about under my directions, I pushing them aside every now and then … .

He goes on to add, 'There is nothing that irritates me more than seeing other people sitting about doing nothing when I'm working' [Jerome 2006].

Those who manage projects often end up being as chagrined as J. Getting other people to do things at your bidding is tricky business. No single strategy works for all people and all situations. Superintending and supervision are very much about intentions and visions and less about merely acting 'super'. At the heart of good management lies an understanding of human traits that make humans what they are—intelligent, resourceful, and hardworking. Bad management can most readily produce very contrary symptoms in the same people.

The 'elusive something' that is management has been approached as a science, as an art, or even a craft. Some say management principles can be best learnt in business schools, while others swear by picking up management skills on the job, whereas still others are convinced being able to manage successfully is a talent one needs to be born with; either you have it, or you don't.

With this general background, we will now see where management fits into the four Ps of software development.

8.3 FOUR PS OF SOFTWARE DEVELOPMENT: PEOPLE, PROJECT, PRODUCT, AND PROCESS

The 'management spectrum' consists of the four Ps of people, project, product, and process, necessarily in that order [Pressman 2000]. *People* are the actual human beings—developers, users, customers, other stakeholders—for whom the bells of a software system toll. A *project* is the 'organizational element' which gives the context for managing software development. The released software product is the end result of a project. *Product* encompasses artefacts such as source code, executables, and documentation generated in the course of a project. *Process* is the definition of activities needed to transform user Requirements into a software product [Jacobson et al. 1999] (see Chapter 14 for related discussion).

Project management can be thought to be the key ingredient that goes into this mix—people, project, product, and process—to transform it into something that can satisfy the needs the software system has to cater to. In the following sections, as we discuss the management functions in a software project in more detail, it will help to have these four Ps in perspective. Whatever software project management seeks to achieve is within the spectrum of people, project, product, and process. There is no straightforward relationship between the constituents of this spectrum, often one is used to leverage others, but the objective is always to facilitate the delivery of effective software systems.

Project management functions are closely linked to the project life cycle, and some activities are more important than others at specific junctures in a project's life. Thus, before going into the project management processes and knowledge areas, it will help to highlight the project life cycle.

8.4 PROJECT LIFE CYCLE

Every project starts off with an *idea*, that gets clarified in a *charter*, and then in a *scope statement*. Project management then needs to produce a *plan* which when agreed upon becomes a *baseline*; subsequently *progress* is tracked, the product being produced is *accepted* and *approved* by the customers, and it is finally *handed over*. All this while, time has been running and it is interesting to note how resources are expended. Figure 8.1 is the curve for resource expenditure versus time for a typical project. Resource need for a project starts off low, then increases steadily, and finally tapers off. It has been found that for large projects the *Rayleigh distribution* approximates the variation of resource consumption R_c with time t:
$R_c = \dfrac{t}{k^2} e^{-t^2/2k^2}$, where $0 \leq t \leq \infty$. The parameter k is a constant and denotes the time at which the resource consumption is at its peak, and $e \approx 2.718$, the base of natural logarithms [Schach 2005].

Along with the variation of resource needs over time, another aspect is of interest to the management of software projects. As depicted in Figure 8.2, cost of change as well as stakeholder influence varies in opposite ways with the passage of time in a project [PMI 2004]. Cost of change is the time, money, or other resource that needs to be expended to make a change, be that due to a newly discovered bug, or a change in Requirement, or some other issue that was not seen upfront. The trend of its variation with time is rather intuitive, as the earlier you find what is wrong with the program, the lesser pain it would cause all around to fix it. So the cost of change is less in the early part of the project life cycle and increases towards the end. Stakeholder influence is something more subtle. Simply put, stakeholders are people whose material interests stand to be affected, for better or for worse, by an

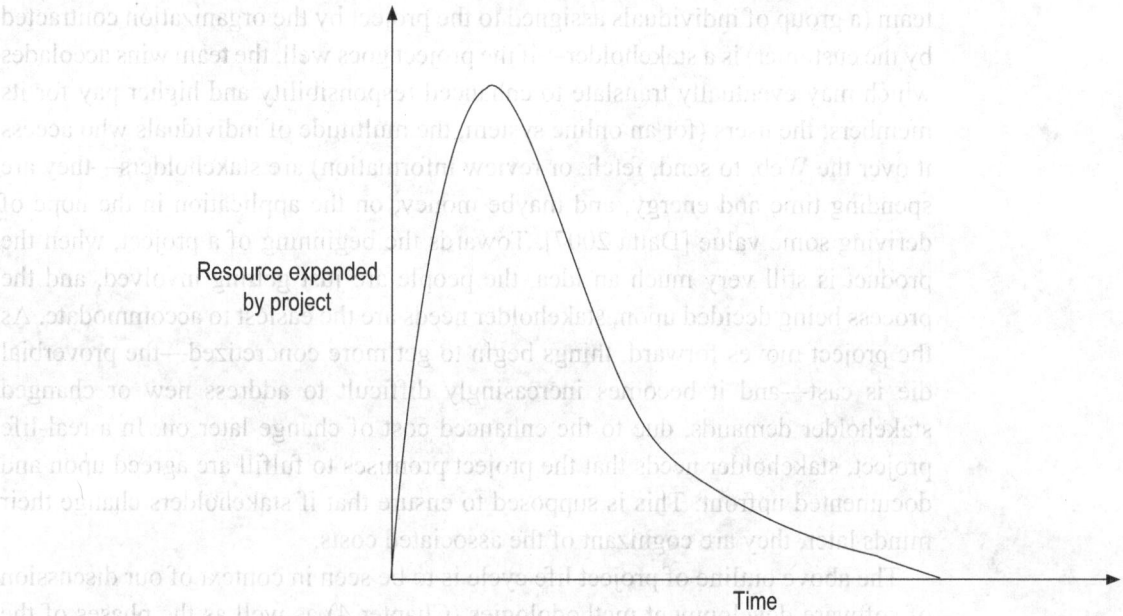

Resource expended
by project

Time

Figure 8.1   Resource expenditure versus time in a typical project

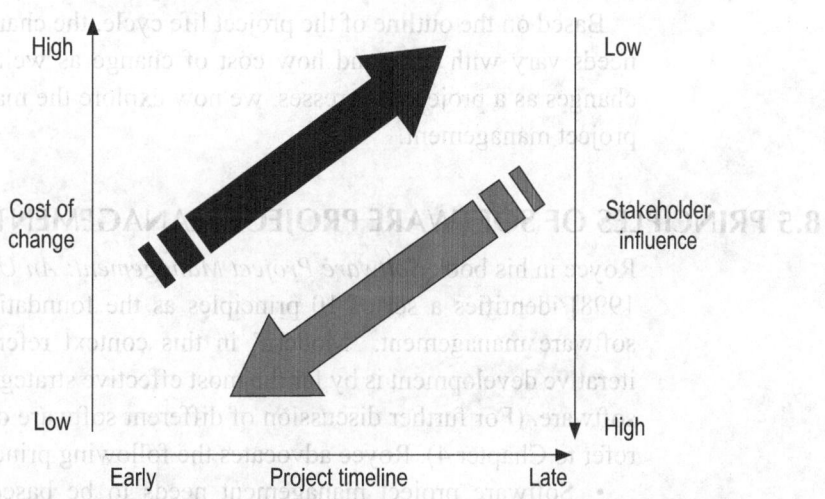

High   Low

Cost of
change

Stakeholder
influence

Low   High

Early   Project timeline   Late

Figure 8.2   Variation of the cost of change and stakeholder influence with time

undertaking they associate with. For a software project, stakeholders consist of a large variety of individuals and groups—the customer (represented by a group from the customer organization) is a stakeholder—that are investing money into the project with the hope that it will help bring back more money; the development team (a group of individuals assigned to the project by the organization contracted by the customer) is a stakeholder—if the project goes well, the team wins accolades which may eventually translate to enhanced responsibility and higher pay for its members; the users (for an online system, the multitude of individuals who access it over the Web, to send, fetch, or review information) are stakeholders—they are spending time and energy, and maybe money, on the application in the hope of deriving some value [Datta 2007]. Towards the beginning of a project, when the product is still very much an idea, the people are just getting involved, and the process being decided upon, stakeholder needs are the easiest to accommodate. As the project moves forward, things begin to get more concretized—the proverbial die is cast—and it becomes increasingly difficult to address new or changed stakeholder demands, due to the enhanced cost of change later on. In a real-life project, stakeholder needs that the project promises to fulfill are agreed upon and documented upfront. This is supposed to ensure that if stakeholders change their minds later, they are cognizant of the associated costs.

The above outline of project life cycle is to be seen in context of our discussion of software development methodologies (Chapter 4) as well as the phases of the software development life cycle (Chapter 14). To avoid repetition, we have not mentioned features of the software development life cycle here.

Based on the outline of the project life cycle, the characteristic of how resource needs vary with time and how cost of change as well as stakeholder influence changes as a project progresses, we now explore the major principles of software project management.

8.5 PRINCIPLES OF SOFTWARE PROJECT MANAGEMENT

Royce in his book *Software Project Management: An Unified Framework* [Royce 1998] identifies a set of 10 principles as the foundational elements of *modern* software management. 'Modern' in this context refers to the recognition that iterative development is by far the most effective strategy for building commercial software. (For further discussion of different software development philosophies, refer to Chapter 4). Royce advocates the following principles [Royce 1998]:

- Software project management needs to be based on an architecture-first approach: With reference to our discussion in Chapter 11, we recognize the central role architecture plays in the conception, Construction, and entire lifetime of a software product. It comes as no surprise that sound project management encourages, indeed mandates, architecture be the first and foremost concern in the development of a software product.

- Iterative development has to be adopted so that risks are met and resolved early. We underscore the effectiveness of iterative development in Chapter 4. Successful project management in software has to be tuned to the iterative paradigm of software development.

- Component-based development needs to be encouraged: Component-based development is antithetical to what Royce calls a 'line-of-code mentality' [Royce 1998]. Management is far more effective when the unit of software is taken as a component, with interfaces and specific tasks, rather than lines of code.

- A change management environment has to be established: As we describe in Chapter 20, in the real-world of software development, change is the pre-eminent fact of life. Software has to be designed, developed, deployed, and maintained within a continuum of change. For a software project to succeed, a streamlined approach to managing change must be in place.

- Tools supporting round trip engineering need to be used: Round trip engineering ensures all the artefacts of a project from source code to documentation are automatically kept synchronized. This task is extremely laborious, to the point of being untenable, if attempted manually.

- Design has to be documented in formal model-based notation: As we discuss in Chapter 13, a consistent and effective modelling notation is vital for Design. A formal model-based notation allows better opportunities for sound Design.

- The quality of the product as well as progress of the project has to be objectively assessed: In our discussion of software metrics (Chapter 7), we underscored that *some metric is always better than no metrics*, and how simple, intuitive measures go a long way towards more objective observations. Management depends heavily on numbers; good managers use them objectively and with discretion, and the opposite can be said about the not so good ones.

- Intermediate artefacts need to be assessed through a demonstration-based approach: All that is released to the user needs to function in an operational context, rather than just paper and pretty diagrams.

- Intermediate releases should be planned with evolving levels of detail: As the software system is released in increments, each succeeding release needs to address the latest details that have been added to the development team's understanding of the system being built.

- A configurable process that is economically scalable needs to be established: As far as processes go, there is no one size that fits all. A process succeeds when it is tuned to the other Ps—people, project, and product; and fails, often spectacularly, otherwise. In this context 'economy' is used in the sense of *economies of scale*, which implies that processes need to allow themselves to be tailored to suit a particular project's need.

The above list of software management credos is by no means exhaustive, nor are all of them equally effective in every circumstance. Note how frequently we refer to other chapters in our discussion of the bullet points above. This is no accident. Management is not an esoteric pursuit; it comes out of a combination of many of the best practices of the discipline, in this case the enduring ideas in software methodologies, architecture, Design, etc.

From the broad perspective of general principles, we focus on the specific concerns of project management in the next section.

8.6 PROJECT MANAGEMENT: PROCESSES GROUPS AND KNOWLEDGE AREAS

The Project Management Body of Knowledge (PMBOK) 'is the sum of knowledge within the profession of project management' [PMI 2004]. The Project Management Institute (PMI) takes the initiative in helping evolve, compile, and keep up-to-date this body of knowledge through voluntary participation of contributors who study and practice project management. PMI also runs certification programs such as the Project Management Professional (PMP), Certified Associate in Project Management (CAPM), which are important career development steps for project managers. PMI's *A Guide to the Project Management Body of Knowledge, Third Edition*, PMBOK Guide in short, is taken to be a definitive reference on project management.

The PMBOK Guide underscores the multi-dimensional nature of project management and describes a set of *process groups* and *knowledge areas* that cover its essential elements. The point made repeatedly in the PMBOK Guide is that process groups and knowledge areas have several overlaps and reflect the diverse nature of the project management discipline, which draws from a wide range of expertise and experience.

The project management process groups are identified as the following [PMI 2004]:

- *Initiating Process Group:* Aims at defining and authorizing the project or a project phase.
- *Planning Process Group*: Aims at defining and refining objectives, plans, and course of action towards fulfilling the scope and target of the project.
- *Executing Process Group*: Aims at integrating people and other resources to ensure that the project management plan is carried out properly.

- *Monitoring and Controlling Process Group*: Aims at regularly measuring and monitoring progress to detect variances from the project management plan such that remedial action may be taken as necessary.
- *Closing Process Group*: Aims at ensuring project deliverable is formally accepted and bringing the endeavour of the project or project phase to an organized closure.

According to the PMBOK Guide, the following are the project management process areas [PMI 2004]:

- *Project Integration Management*: This includes the process and activities that are required to 'identify, define, combine, unify, and coordinate' [PMI 2004] the various processes and activities with the process groups.
- *Project Scope Management*: This includes the processes that are needed to ensure all the work that the project needs to meet its goals successfully are addressed.
- *Project Time Management*: This includes the processes needed to complete the project on time.
- *Project Cost Management*: This includes the processes involved in 'planning, estimating, budgeting, and controlling costs' [PMI 2004] for the project to be completed within budget.
- *Project Quality Management*: This includes all the activities that determine quality policies, objectives, and responsibilities towards satisfying the needs the project seeks to address.
- *Project Human Resource Management*: This includes processes that organize and manage the project team. We devote a later section of this chapter to software teams.
- *Project Communications Management*: This includes the processes needed to ensure 'timely and appropriate generation, collection, distribution, storage, and retrieval' [PMI 2004] of project information.
- *Project Risk Management*: This includes processes to foresee, detect, and address risks associated with a project.
- *Project Procurement Management*: This includes the processes to acquire goods or services from outside the project team that are essential for the completion of the project.

The PMBOK Guide also defines three major project documents, each with a specific purpose [PMI 2004]. These may be thought of as the why, what, and how, respectively, of any given project:

- *Project Charter*: Gives a formal authorization for the project, often with justification as to why the project was commissioned in the first place.

- *Project Scope Statement*: Specifies what the project needs to accomplish and what deliverables need to be produced.
- *Project Management Plan*: Describes how the project's work will be performed.

The project management plan is a detailed blueprint of how a project will go from Inception to completion and is very important for understanding, at any point in the project's life, which way it is heading. In the next section, we look at a specific template for such a plan.

8.7 SOFTWARE PROJECT MANAGEMENT PLAN

The *IEEE Standard for Software Project Management Plans* (IEEE Std 1058-1998) is a reliable starting point for software project management. It was drawn up in consultation with experts from industry as well as the academia, and seeks to distill the experience accumulated over many successful and failed projects. The IEEE project management plan is meant to be used for projects with diverse life cycle models. The plan is more of a framework than a prescription; it has to be customized for needs specific to organizations, domains, or technologies. Although the plan is not restrictive, it is particularly well-suited to the Unified Software Development Process (refer to Chapters 5, 13, and 14). Requirements control and risk management are given central importance in it. The IEEE project management plan also facilitates process improvement in terms of the key process areas of the Capability Maturity Model (refer to Chapter 17). However, as with all plans and frameworks of this type, it is imperative that the IEEE project management plan be applied with discretion. For example, some aspects of the plan may not be relevant for small projects, and it should not end up merely increasing overheads without adding benefits.

The IEEE project management plan provides placeholders for specifying a project's purpose, scope, objectives, assumptions, constraints, deliverables, and budget summary. It defines the important words and phrases related to a project; for example, *software project* is defined as 'the set of work activities, both technical and managerial, required to satisfy the terms and conditions of a project agreement. A software project should have specific starting and ending dates, well-defined objectives and constraints, established responsibilities, and a budget and schedule' (IEEE Std 1058-1998). The plan specifies a project's organization in terms of its external interfaces, internal structure, roles, and responsibilities. The three major areas of *managerial process plans, technical process plans,* and *supporting process plans* are outlined in detail, as well as *additional plans* are provided for, which includes plan components specific to a project, such as security plans, installation plans, etc. Figure 8.3 gives a summary of the IEEE project management plan framework.

1. Overview	1.1 Project summary	1.1.1 Purpose, scope, and objectives
		1.1.2 Assumptions and constraints
		1.1.3 Project deliverables
		1.1.4 Schedule and budget summary
	1.2 Evolution of the project management plan	
2. Reference materials		
3. Definitions and acronyms		
4. Project organization	*4.1 External interfaces*	
	4.2 Internal structure	
	4.3 Roles and responsibilities	
5. Managerial process plans	*5.1 Start-up plan*	*5.1.1 Estimation plan*
		5.1.2 Staffing plan
		5.1.3 Resource acquisition plan
		5.1.4 Project staff training plan
	5.2 Work plan	*5.2.1 Work activities*
		5.2.2 Schedule allocation
		5.2.3 Resource allocation
		5.2.4 Budget allocation
	5.3 Control plan	*5.3.1 Requirements control plan*
		5.3.2 Schedule control plan
		5.3.3 Budget control plan
		5.3.4 Quality control plan
		5.3.5 Reporting plan
		5.3.6 Metrics collection plan
	5.4 Risk management plan	
	5.5 Project close-out plan	
6. Technical process plans	*6.1 Process model*	
	6.2 Methods, tools, and techniques	
	6.3 Infrastructure plan	
	6.4 Product acceptance plan	
7. Supporting process plan	*7.1 Configuration management plan*	
	7.2 Testing plan	
	7.3 Documentation plan	
	7.4 Quality assurance plan	
	7.5 Reviews and audits plan	
	7.6 Problem resolution plan	
	7.7 Subcontractor management plan	
	7.8 Process improvement plan	
8. Additional plans		

Figure 8.3   Summary of the IEEE software project management plan (inspired by Schach 2005)

8.8 TEAM DYNAMICS

All the discussion so far in this chapter has had an implicit and very important assumption. We assumed software development to be inherently a group activity. Thus a vital component of software project management comes down to managing teams. True, great software has been developed by individual effort. And many a time malfunctioning teams lead to a project succeeding or failing solely on account of individual heroics (or martyrdom!). But most often, development of large and complex software systems need successful team effort.

Humphrey defines a team in terms of the following four underlying features [Humphrey 2006]:

- A team consists of *at least two people*.
- The team members work towards a *common goal*.
- Each member of the team is assigned a *specific role*.
- Reaching the common goal needs some form of *dependency* among the team members.

Humphrey mentions *self-directed teams* as the acme of collaborative group work, and points out such teams to be characterized by a sense of membership and belonging, a commitment towards the common goal, ownership of process and plan, the skill to make a plan and the discipline to adhere to it, and finally and perhaps most importantly, a dedication to excellence [Humphrey 2006]. These are all lofty ideals. Although not all teams will be able to fulfill all of them at all times, they are worth striving for. Humphrey has defined the *Team Software Process* (TSP), which lays down a roadmap for successful software development through team effort. Refer to Chapter 5 for a deeper discussion of TSP.

With software development having become an increasingly global enterprise, software teams too have acquired a global character. It is not unusual now for a team to be expected to function smoothly even when its members are strewn across several continents and many time zones. Team dynamics have also changed drastically, it is often not feasible now for all members of a team to huddle into a room and stay put until a problem is solved. Instead, one needs to rely on e-mails, conference calls, and instant messages for communication. In Chapter 21, we discuss the implications of global software development.

So far in this chapter, we have focused on the nuts and bolts of management functions. Let us briefly review some important project management activities next. Then we will place management in the overall context of leadership.

8.9 IMPORTANT PROJECT MANAGEMENT ACTIVITIES

In this section, we briefly describe some of the important project management activities. This is by no means an exhaustive discussion. As underscored earlier, project management skills and techniques involve many layers of detail and subtlety. The following only addresses few of the responsibilities a project manager is expected to fulfill.

To set the context of these activities, we must recognize the key concerns of a project manager. It is critical to understand the dependencies between the different tasks that constitute the project, to plan for timely completion of these tasks, to determine the value each task adds to the overall project, and to diagnose and monitor errors in the completion of these tasks. The following sections outline the activities oriented towards each of these concerns [Pressman 2005].

8.9.1 Defining a Task Network

Assuming that a "task" defines a unit of work in a project, a task network represents how tasks are dependent among one another. It is a graph structure, where nodes (or vertices) denote tasks, and links (or edges) denote dependencies. Figure 8.4 shows the task network for a hypothetical tea-making project. Adding tea and sugar have the dependency of the water having been boiled before, and stirring has the dual dependency of tea and sugar having been added before.

8.9.2 Scheduling

The principal components of a project ecosystem are (a) tasks—what needs to be completed, (b) time—by when the tasks need to be completed, and (c) resources—who will complete the tasks. For a project to succeed, the interaction of these components must be optimized. This is the objective of scheduling. Effective scheduling seeks to minimize delays (due to one task waiting for the completion of other task(s) it depends on) and maximize parallelization (independent tasks

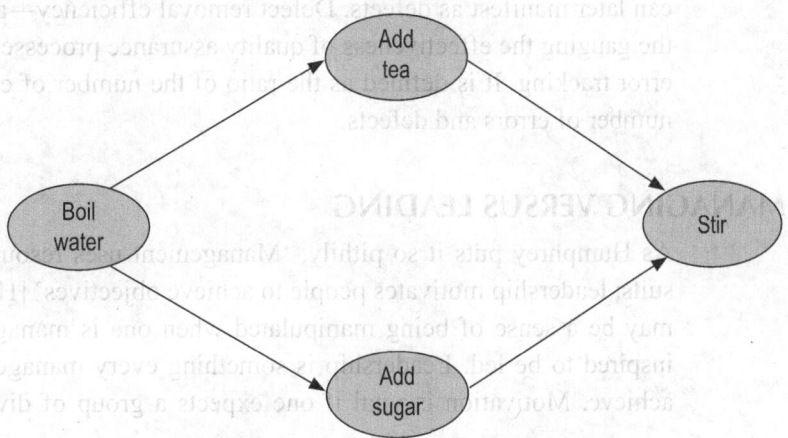

Figure 8.4   Task network for a tea-making project

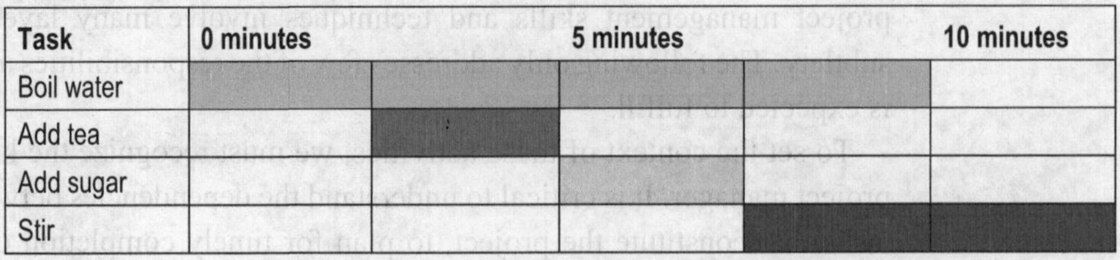

Task	0 minutes	5 minutes	10 minutes
Boil water			
Add tea			
Add sugar			
Stir			

Figure 8.5   Timeline chart for a tea-making project

proceeding simultaneously). The most popular project scheduling methods are Program Evaluation and Review Technique (PERT) and Critical Path Method (CPM). A critical path is the sequence of events that define the duration of the project. Time line charts (or their variation, Gantt charts) are useful artefacts for scheduling. Figure 8.5 shows the time line chart for our tea-making project, whose task network is shown in Fig. 8.4.

8.9.3 Earned Value Analysis

Earned value Analysis is a technique for measuring a software project's progress by providing a common value scale across the project's tasks. The quantitative insights available from earned value Analysis helps determine what percentage of the project has been completed. Earned value Analysis is essential for initiating feedback and taking corrective action.

8.9.4 Error Tracking

Error tracking is the process of identifying and resolving errors in the project's execution path. Errors may be unearthed through reviews of project artefacts or Testing of the end product. Errors that are not identified during error tracking can later manifest as defects. Defect removal efficiency—an important metric for the gauging the effectiveness of quality assurance processes—is closely related to error tracking. It is defined as the ratio of the number of errors to the sum of the number of errors and defects.

8.10 MANAGING VERSUS LEADING

As Humphrey puts it so pithily, 'Management uses resources to accomplish results; leadership motivates people to achieve objectives' [Humphrey 2006]. There may be a sense of being manipulated when one is managed, but one must feel inspired to be led. Leadership is something every manager aspires for, and few achieve. Motivation is vital if one expects a group of diverse individuals to be aligned to a common cause. But unfortunately, there are no sure shot recipes to impart motivation to a group of varied individuals. Leadership is often associated with personality traits unique to an individual—charisma, flair, and that unknown something which makes people willing to follow a leader. Some people are said to be born leaders. Management at its most sublime may meld into leadership. But even when that does not happen, effective management of software projects can be brought about by systematic use of principles discussed in this chapter.

SUMMARY AND TAKE-AWAYS

In this chapter, we have outlined some basic principles of project management, as applied to software development. Management is a complex and challenging field of study and practice; the discussion in this chapter only gives an overview of some of its areas. Here are the highlights:

- The four Ps of software development—people, project, product, and process—are foundational to the study and practice of management.
- Resource needs vary across the project life cycle; the characteristic of change is approximated by the Rayleigh distribution for large projects.
- An architecture-centric and iterative approach is central to the principles of modern software management.
- The Project Management Body of Knowledge specifies a set of process groups and knowledge areas to capture the essential elements of project management.
- The IEEE Software Project Management Plan (SPMP) is a framework that needs to be customized for specific projects.
- Teams, especially globally distributed teams, and their dynamics play a major role in successful software development of the present.
- Defining a task network, scheduling, earned value Analysis, and error tracking are some of the important project management activities.
- Management is about utilizing resources optimally towards specific ends, leadership is about motivating people to accomplish results.

WHERE TO LOOK FOR MORE

- www.pmi.org—The Project Management Institute website—offers many valuable resources for the project management discipline.

======== **EXERCISES** ========

Review Questions

Review Questions test your understanding of the key concepts presented in this chapter.

1. A software engineer joining the industry fresh from college will most likely be involved in
   (a) Construction of software to a larger extent than management of software projects
   (b) management of software projects to a larger extent than Construction of software
   (c) both to equal extents
   (d) neither

2. Software project management
   (a) is fundamentally different from managing other engineering projects
   (b) sometimes similar, sometimes dissimilar to managing other engineering projects
   (c) follows the same general principles as the management of other enterprises
   (d) none of the above

3. The management spectrum of software projects consists of
   (a) producers, providers, process, project
   (b) producers, consumers, process, project
   (c) people, project, product, process
   (d) producers, providers, people, project, product, process

4. Resource consumption versus time in a large project can be approximated by the
   (a) normal distribution
   (b) Binomial distribution
   (c) Poisson's distribution
   (d) Raleigh distribution

5. The cost of change in a project
   (a) decreases as the project progresses
   (b) increases as the project progresses
   (c) remains the same as the project progresses
   (d) is not related to project progress

6. Stakeholder influence on a project should ideally
   (a) decrease as the project progresses
   (b) increase as the project progresses
   (c) increase or decrease depending on the project
   (d) remain the same

7. The principles of software management
   (a) are always true
   (b) represent the best practices from different areas of software development
   (c) represent an exhaustive list
   (d) all of the above

8. The Project Management Body of Knowledge is
   (a) a process for successful project management
   (b) a textbook on project management
   (c) a course on project management
   (d) a collection of project management best practices

9. The PMBOK Guide describes the essentials of project management in terms of
   (a) process areas and knowledge groups
   (b) process areas and group knowledge
   (c) process groups and knowledge areas
   (d) process knowledge

10. The elements of project management as described in the Project Management Body of Knowledge
    (a) are distinct and separate
    (b) are distinct but overlapping
    (c) are similar and overlapping
    (d) depend on the project

11. As defined in the PMBOK Guide, Project Charter is a
    (a) document
    (b) a process group
    (c) a knowledge area
    (d) none of the above
12. The IEEE Software Project Management Plan (SPMP) is
    (a) specific management plan for a particular project
    (b) a framework that has to be tailored according to a project's needs
    (c) a new way of managing software projects
    (d) none of the above
13. Management process plans, technical process plans, and supporting process plans are
    (a) different types of processes
    (b) different management methodologies
    (c) elements of the SPMP
    (d) can be any of the above depending on the project
14. According to the definition of a team given by Humphrey, a team consists of
    (a) one or more individuals
    (b) at least three individuals
    (c) at most five individuals
    (d) two or more individuals
15. According to Humphrey, the most important characteristic of a self-directed team is
    (a) commitment to a common goal
    (b) sense of belonging
    (c) dedication to excellence
    (d) guarantee to finish the project on time
16. An effect of the increasingly global nature of software development on software teams has been
    (a) decrease in the size of teams
    (b) increase in the size of teams
    (c) distribution of team members across geographic locations
    (d) none of the above
17. Management is about consummate use of resources towards an end whereas leadership is about motivating people. This statement is
    (a) true
    (b) false
    (c) sometimes true, sometimes false
    (d) this statement can only be judged in a specific context

Reflective Questions

Reflective Questions require you to think more deeply about some of the ideas and come up with your own interpretations and answers.

1. 'The management spectrum consists of ... necessarily in that order'. Why do you think the ordering is important?
2. What in your opinion best describes software project management: science, art, or craft?
3. Why do you think an architecture-first approach is so essential for good software management?
4. What will be the most important difference in the management approach of a project that follows the Waterfall model versus the one that follows the iterative and incremental model?
5. Consider the following roles: The leader of a software development project, the chief executive officer of an organization, the prime minister of a country. Do all these roles need management skills and leadership talent in equal measures? Justify your answer.
6. Do you feel management of distributed software teams is easier than that of teams localized in one place? Justify your answer.

Numerical Problems

Numerical Problems require you to do quick 'back-of-the-envelope' kind of calculations to arrive at answers to some simple but insightful problems.

1. If for a particular project, the resource consumption versus time curve is approximated more closely by the normal distribution instead of the usual Rayleigh distribution, will the peak demand arise earlier or later than a typical project? State any assumption you may need to make.

REFERENCES

Datta, S. (2007), *Metrics-Driven Enterprise Software Development: Effectively Meeting Evolving Business Needs*, J. Ross Publishing.

Humphrey, W. S. (2006), *TSP: Leading a Development Team*, Addison-Wesley.

Jacobson, I., G. Booch, and J. Rumbaugh, (1999), *The Unified Software Development Process*, Addison-Wesley.

Jerome, J. K. (2006), *Three Men in a Boat: (To Say Nothing of the Dog*, Dover Publications, Dover ed edition.

PMI (2004), *A Guide to the Project Management Body of Knowledge*, Project Management Institute, 3rd edition.

Pressman, R. S. (2005), *Software Engineering: A Practitioner's Approach*, 6th ed., McGraw-Hill.

Royce, W. (1998), *Software Project Management: A Unified Framework*, Addison-Wesley Professional, 1st edition.

Schach, S. (2005), *Object-oriented and Classical Software Development,* Sixth Edition, McGraw-Hill International Edition.

Human Aspects of Software Development

> ## Learning Objectives
>
> In this chapter we will discuss the influences of human aspects on software development. Specifically, we address the following:
>
> - Viewing software development as a means to serving needs of *real* users, i.e. human beings with likes, dislikes, preferences, prejudices, talents, and limitations
> - Why users of software systems always seem to change their minds and what can be done about it
> - Need for software engineers to help users know their needs
> - Why language and communication are so important in software development
> - Importance of effective human computer interaction
> - Building software systems that are usable

9.1 MOTIVATION

The first eight chapters of this book show that software engineering is different from programming a computer. (If you are still unconvinced, I recommend going over the first eight chapters again!) What do you think is the principal difference between computer programming and software engineering? This is one of those questions which can have many answers, and many shades to each answer. At the level of perception (vis-à-vis execution), programming is concerned more with algorithmic soundness, performance, syntactic elegance, and the like, with the objective of solving a computational problem. But software engineering's primary concern is to Design, develop, deploy, and maintain (and in many cases enhance) a software system that brings some value to *end-users*. As indicated in italics, the key phrase is 'end-users'. In almost all cases, the end-user of a commercial software system is a human being. A software system serves the needs of humans by assisting, educating, entertaining, or guiding them. In this chapter, we explore some of the *human aspects*—factors arising from the fact that usual 'consumers' of the software development enterprise are human beings. With increasing penetration of software systems into our lives, understanding and accounting for the human aspects of software development become increasingly important. But unfortunately, these issues are seldom dealt with in software engineering textbooks, being left to be learnt 'on the job'. Learning on the job is a very helpful way of learning, and an essential one too for engineers to stay up-to-date with the progress of their profession. But learning on the job becomes much easy if one has some idea of what antennae to keep open before the job begins. This is a chapter about tuning such antenna on how human aspects influence software development.

At the outset, it would help to clarify that in this chapter 'human aspects' do not necessarily refer to *humane* aspects. We are not considering ways and means to make software development less onerous, or more forgiving. (As books like Yourdon's *Death March* [Yourdon 2003] establish, there is often significant scope for making software projects more humane.) Software systems handle critical tasks, often involving 'life and death' issues in their most literal meaning. Thus, the stakes for professional software development remain commensurately high. Our discussions of the human aspects in this chapter seek to illustrate how such factors affect the way software is built and how their better appreciation can lead to the building of better software.

It can be argued—and certainly not without merit—that every engineering activity ultimately serves some human end-user. Bridges are not built for androids, or houses for robots; at least not yet. But there is an inherent difference between a bridge or house and a software system, especially in terms of how end-users view and exercise their respective utility. We highlight this point in the next two sections. In the subsequent section, we underscore how essential it is for software engineers to help prospective users of the software system know what they need. (I concede this sounds a bit strange—after all aren't we supposed to know our needs in the first place? But wait till we get into how this may not be as trivial as it sounds in the software context). Closely related to the reality of helping users unravel what they want, is the idea of co-evolution—a simultaneous maturing of the problem to be solved as well as the solutions being devised for it, which we will briefly review. We then move into how language and communication play vital roles in addressing many of the human aspects of software development. Human-computer interaction (HCI) is an area made very interesting by the intersection of many diverse disciplines; we will outline how HCI often influences the usability of software systems. And this will help us reflect on one of the holy grails of software development—creating software that users will want to use.

We start our journey into the human aspects of software development by focussing on software for real users.

9.2 SOFTWARE FOR REAL USERS

Talking about something 'real' these days, we are tempted to consider its 'virtual' counterpart. So, in this section, are we looking to distinguish software's use between real and virtual users? Actually, not. 'Real' in this context means users who expect to derive some value from using the software system; who are free *not* to use the system if it does not give them any value; and who are equally free to put forward changed expectations from the system, or just move to a new system. So, real users are those who will use a software system only if they like it. Why are we making such a big deal of the so-called 'real' users? What other kinds of users can there be?

When we learn software engineering in the classroom, the users we build our systems for are certainly not real users. How do you tell a real user from one who is not? The best and quickest test is to check whether the users are coming back with changed Requirements while or after they use the system. Only real users, whose stakes are linked to how well a software system works (or does not work), will expect more from a system, or want the system to do something differently.

One of the most significant human aspects of software development is to identify the real users of a system. *Software for real users* should be a motto every software engineer practices his or her profession by, but it seldom is. In this chapter, we seek to establish this credo; through considering factors that impart that reality to users. In Chapter 20, we go deeper into the nuances of building software for the real-world and for real users.

As we just mentioned, a distinguishing characteristic of real users is that they change their minds about what a software system should do for them. The extent and frequency of such change, very often is significant. Are users of a software system, especially capricious? If so, why?

9.3 CAPRICIOUS USERS

When a building is commissioned—say, for an office, hospital, school, home—the users have a clear idea of what the finished product needs to be, how they will use it, even how they should not use it. This subconscious understanding comes from the fact that humans have been using buildings for centuries, if not millennia. This same understanding exists to varying but significant extents for users of other engineering artefacts such as a bridge or a car. So the *cognitive gap* [Datta 2007] between what users want engineered, and what can be engineered for them, is manageably narrow. But it has been only a few decades that people who cannot build software themselves have been using software in a big way. So, not knowing what they want is a common characteristic of the common software consumer. As we shall see in the next section, helping users know their needs is one of the primary tasks of a software engineer. But for the moment, it is important to appreciate that this cognitive gap is one of the pre-eminent reasons for the usual capriciousness shown by users of a software system. What a software system can and cannot do is very often vaguely understood or totally misunderstood by those who commission or use the system, and this leads to repetitive cycles—the proverbial infinite loop—of changes going back and forth between a system's development team and its users.

This issue is compounded by the fact that software systems often replace an old way of doing a particular task, with promises of making it easier, smoother, and faster. But the old ways of doing things are often wired in the users' brains, which is then projected into the expectations from the software system. This is something which can be referred to as 'legacy thinking' [Datta 2007]. What is this legacy thinking, which contributes greatly at times to the capriciousness of users? Let us illustrate by an example. This is a real-life 'story', heard first-hand by the author [Datta 2007].

A software engineer was once working on a project to develop a Web-based system which gave credit card holders online access to their transaction histories. Before the Web, cardholders accessed their data differently; they had a piece of client software installed on their personal computers which periodically downloaded the information from the credit card company's servers. Once the data was on their local machine, the client software gave ways to search and sort the data. Gradually major problems came up with this client-server mechanism—there had to be many versions of the client for different platforms, often the cardholders had issues installing and configuring the client, and so on. All this was why the Web-based system was commissioned. Much hype went out to the cardholders announcing the new system, glossing over how easy it was now becoming to access one's data— anywhere, anytime; just get onto the Web, and bingo! When released the new system helped build did exactly what it was meant to. But within the first month, there were many complaints from the users. Sorting the data took time, searching was worse on the web pages, reports took a long time to display if one tried pulling the last twenty months of data; the list was long and ugly. This was classic legacy thinking in play. The users were so used to the legacy system (accessing the data from their desktop clients), their thinking was tuned to its quirks. Sifting data on one's local machines is always faster than reaching for it over a network, and when one tries to pull details of several thousand transactions, the difference shows. It took quite some coaxing, and finally the threat of discontinuing support to the desktop clients to shepherd the cardholders to the online system. So legacy thinking is all about seeing a new (software) system in the light of an older one. No software engineering theory, to the best of my knowledge, teaches you to tackle legacy thinking. And a practicing software engineer often regrets that he or she has not been taught what to do with legacy thinking.

In the Case Study section, our protagonist Preeti tackles a situation where the cognitive gap of the software system's user group causes the development team a lot of pain. Figure 9.1 highlights the ideas of cognitive gap and legacy thinking in the context of software development: thinking in terms of an abacus while using a calculator, or a building when using a software system.

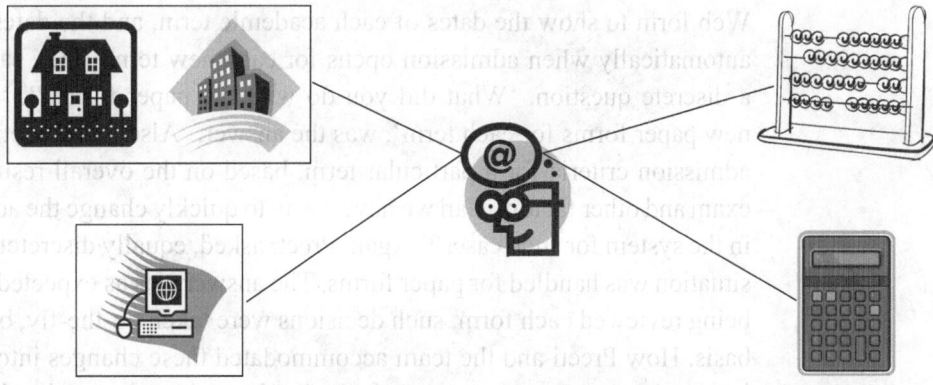

Figure 9.1 Cognitive gap and legacy thinking

CASE STUDY

This time, Preeti leads a software development team which is working on a college admission system. The system is transitioning from paper to online. Earlier students had to fill up paper forms, hand them over to the admissions desk, and wait for the admission decision to be communicated to them over postal—'snail'—mail. In this cyber age, all this seemed rather outdated, so the college commissioned Preeti's organization to build a Web-based system for them which would let students apply online—with details such as their exam grades, course preferences, etc.—and get a quick decision on whether they can be admitted or not.

Initial meetings with the customer were very smooth. 'Just take the paper form and make it online', Preeti was told, with the addendum, 'We want the system to function just as it did for paper forms, but faster and easier for all concerned'.

Preeti's team got to work, building a three-tier system, the Web-based front-end mimicking the paper form, the middle-tier containing the logic to decide whether a student is eligible for admission, and the back-end databases to hold and verify student information. The first release of the system rolled out; and the fun started.

In systems like these, there are various levels of users. In this case, the students who will use the system to apply for admission are at one level; whereas the ones who decide on the admission criteria—college authorities who hired Preeti's organization—are at another level. Often the latter group is referred to as 'customers' group; they control the purse strings, whereas the former is called 'user' or 'end-users' group. The system Preeti and her group built was used widely; students, i.e. the end-users, found parts of it useful, some complained about the slow response, for some the system just 'hung' when they were so anxious to know whether they had been admitted. All this is usual feedback that comes when a new software is launched. But the college authorities had a different take. 'We want the Web form to show the dates of each academic term, and the dates should change automatically when admission opens for each new term', they said. Preeti put in a discrete question, 'What did you do with the paper forms?'. 'We just printed new paper forms for each term', was the answer. 'Also, we sometimes change the admission criteria for a particular term, based on the overall result of that year's exam and other factors. Can we have a way to quickly change the admission criteria in the system for such cases?' Again Preeti asked, equally discretely, how a similar situation was handled for paper forms. The answer was as expected; when a human being reviewed each form, such decisions were taken on-the-fly, on a case-to-case basis. How Preeti and the team accommodated these changes into the system is a long and tortuous story, very typical of real- world software development, and we will not go into the details here.

But whatever we have heard of the story is very illuminating. It points to many of the topics we have discussed so far, like cognitive gap and legacy thinking, in a real-life situation. On the face of it, Preeti's project was very simple. Automate the functionality of students applying for college admission. It seemed that translating the paper forms into a web interface, with some back-end processing, would suffice. But what makes such a project so challenging are the human factors. The paper form way of doing things was so ingrained in the students' as well as the college authorities' psyche that they saw the new Web-based system in the same light. Automatically updating the Web form with each term's start and end dates calls for significant behind-the-scenes processing, something far more involved than ordering new forms to be printed for each term. Again, relaxing or tightening of the admission criteria when done via human processing is easy and intuitive. But when the same facility has to be built into a software system, it needs quite some Design forethought. So, software development often need to take into account factors that are truly human, like looking at a new system through the glass of older experience.

Things would have been significantly easier if the users of this system knew their needs fully before Preeti and her team built it. This brings us to our next section, which talks about helping users know their needs.

9.4 HELPING USERS KNOW THEIR NEEDS

I once asked a physician friend of mine, what he saw as the biggest challenge of a doctor's profession. To provoke him a bit, I added, 'With such smart diagnostic tests and advanced medicine, isn't all you do is to prescribe some tests and then some broad spectrum antibiotics?' My friend's answer was very insightful, and it helped me see a surprising link between the practices of medicine and software engineering. In essence, this is what he said: The most difficult part of treating a patient is often to place his or her symptoms in the proper context, to be able to *sense* what actually is wrong from the myriad aches and pains the patient may be talking about. So the doctor's job is not just: Given a particular ailment, treat it. But much more like, first find what is wrong with the patient and then help the patient get healed. This sounds very reasonable in the medical context. We go to a doctor not knowing what is wrong, recite our complaints; and hope to come back cured. But this is not the way we are taught how software projects should be run.

In software development, we assume Requirements come all ready and reflect the actual needs of those who will use the system. The software engineer just has to take them as given, and Design and build a system based on them. This assumption is seldom if ever valid; and that is why we need to help users know their needs. But there is no formula for going around helping users know their needs. And that is why we invest a whole section about it in this chapter.

The act of gathering Requirements is sometimes referred to as *Requirements Elicitation*. 'Elicit' is defined as a verb—'evoke or draw out (a response or reaction)' (http://www.askoxford.com/concise_oed/elicit?view=uk). So the phrase 'Requirements Elicitation' accepts that there is an element of evoking or drawing out of Requirements (from users), vis-à-vis Requirements being handed to the development team cut and dried. How do we draw out Requirements? Or this question should perhaps be rephrased as—how do we draw out those Requirements that the users care most about?

Gilb has popularized what he calls the *Juicy Bits First* principle [Gilb and Gilb 2006]: If you deliver the functionality that users care about most (the juicy bits), you can expect users to be less carping about other pieces of delivery that come in later. It sometimes helps to try and sense what the juiciest bit is for the users of a particular system, i.e. what functionality is the most important piece of utility users seek to derive from the system? There may not be one single such bit, but it is important to help users identify all such bits. Prototyping—building a quick and dirty system that shows the functioning of some parts of the final system, often superficially—is sometimes a very useful way of getting users to know what they want. (For a more detailed discussion of prototyping, see Chapter 8). Some kind of visual aid, even one as simple as boxes and arrows, which shows the overall flow of the application, forces developers and users alike to home in on points which the latter may feel are at odds with their cognitive gap or legacy thinking.

Requirements must be written down, very frequently in natural language, but users seldom read and tell developers upfront that they do not agree with what is written. Only when the system is built will users come back and say this is not what they wanted. So getting users to evoke their true expectations from the system (not merely what gets written down as Requirements) takes a lot of insight into human psychology, and building up of trust and openness between users and developers. This is a skill that is whetted by experience, but until that happens, one needs to learn via trial and error. Like a good doctor, a good software engineer needs to uncover the true pain points of the user, and do something to help. All good software systems give their users relief from some pain. And this promise of pain relief is what drives customers to invest in commercial software development, and lets software engineers earn a living!

Developers and users exploring Requirements jointly is not just a goodwill exercise, it takes us to the very interesting idea of co-evolution, which we look at in the next section.

9.5 CO-EVOLUTION: INTERACTION OF THE PROBLEM AND SOLUTION DOMAINS

On the face of it, users changing their minds just as they see a completed software system seem something of a nuisance. But, like many other nuisances, this is also very much a fact of life. Towards addressing this situation, Brooks underlines the relevance of the co-evolution model, 'The effective problem space evolves, as the solution space evolves by being explored' [Brooks 2000]. Very interestingly, the co-evolution paradigm was suggested in the context of computer-aided Design for civil engineering; Maher et al. have underlined how the 'Problem Space' and 'Design Space' interact in an evolutionary process over the time spectrum [Maher and Poon 1996]. Figure 9.2 illustrates the idea of co-evolution in simple terms; the state of the problem P_{t+1} is influenced by the problem state P_t and solution state S_t as a recurring pattern.

Maher et al.'s work goes far beyond just suggesting the co-evolution model, it also implements the model using genetic algorithms. Such detail is beyond the scope of our discussion here, but the aptness of the co-evolution model is appreciable. A software system is seldom if ever built on a pre-defined, cast-in-stone problem space; as the system is built, the solution becomes manifest and the problem also morphs into newer dimensions. In essence, the solution drives the problem just as the problem drives the solution. It is very helpful—at times, essential—for software engineers to recognize co-evolution when they are eliciting Requirements. Instead of fretting over why users do not come with a clearly defined problem, ready to be solved, it is important to allow users to evolve the problem to a large extent in parallel with maturing of the solution. Among the key benefits of the iterative and incremental model is that it allows for the inevitable changes in the problem domain, and helps tune the solution domain with the latest realities of the problem, thus giving a built-in scope for co-evolution. (For a detailed discussion of the iterative and incremental development model, refer to Chapter 4.) Evidently, the scope and significance of co-evolution is a deeply human attribute of software

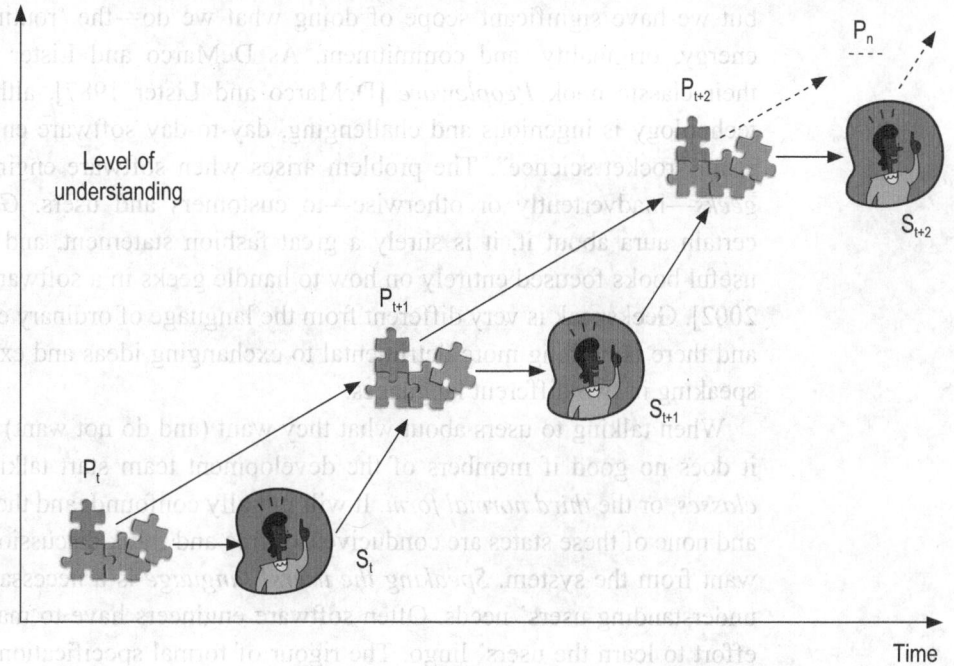

Figure 9.2   Idea of co-evolution [Maher and Poon 1996]

(P_t and S_t represent states of the problem and solution at time t)

development. The situation we outlined in the Case Study section is very much a symptom of co-evolution. The trick is to view co-evolution not as a detriment, but more as an opportunity towards a better maturing of what users really want, and to leverage that towards building a system that users will ultimately use. Co-evolution has been discussed briefly in a related context in Chapter 3.

To be able to leverage co-evolution to its maximum, software engineers need to communicate effectively with users, and the language in which they do play a major role in effective communication. We take this up in the next section.

9.6 LANGUAGE AND COMMUNICATION

Software as a medium of problem solving—supporting large-scale business processes, and making fundamental differences in the ways we see, live, and think—certainly feels very new, very powerful, and very magical at times. As software engineers, it is a great privilege and responsibility for us to be associated with this medium. But, nothing puts off customers and users more than a software engineer's *affectation* for doing something great and sublime in their everyday work. Most of us do not do great and sublime things in our everyday work, but we have significant scope of doing what we do—the 'routine stuff'—with energy, originality, and commitment. As DeMarco and Lister pointed out in their classic book *Peopleware* [DeMarco and Lister 1987], although software technology is ingenious and challenging, day-to-day software engineering work is no "rocket science". The problem arises when software engineers appear as *geeks*—inadvertently or otherwise—to customers and users. *Geekiness* has a certain aura about it, it is surely a great fashion statement, and there are even useful books focused entirely on how to handle geeks in a software project [Glen 2002]. Geekspeak is very different from the language of ordinary communication, and there is nothing more detrimental to exchanging ideas and expectations than speaking in two different languages.

When talking to users about what they want (and do not want) from a system, it does no good if members of the development team start talking about *inner classes*, or the *third normal form*. It will initially confound, and then irritate users, and none of these states are conducive to a free and frank discussion of what users want from the system. *Speaking the users' language* is a necessary condition of understanding users' needs. Often software engineers have to make a conscious effort to learn the users' lingo. The rigour of formal specification languages and modelling languages (such as UML) is fine for documenting and communicating ideas between developers. But end-users of a software system can hardly make much sense out of these. Communication between developers and users needs to happen in natural language—English is now by far the universal choice—and in a jargon users are most familiar and comfortable with. However, when development is finally over, users will need to communicate with the computer, to see how well their needs are met. This brings us to human-computer interaction.

9.7 HUMAN-COMPUTER INTERACTION

According to the Association for Computing Machinery's curricula for Human-computer interaction (HCI), 'Human-computer interaction is a discipline concerned with the Design, evaluation, and Implementation of interactive computing systems for human use and with the study of major phenomena surrounding them' (http://www.sigchi.org/cdg/cdg2.html). Simply put, this is the study of how human beings interact with computer systems. As all software ultimately runs on a computer (or some other computing device), human-computer interaction influences the experience users have from using the software system. HCI represents the confluence of several distinct fields such as computer science, communication theory, linguistics, social sciences, cognitive psychology, and industrial design.

In the most usual scenario, human beings communicate with the computer via mouse, keyboard, and screen. Newer modes of communication, via touch screens, voice, or scratch pads are becoming common these days. Human interaction with computers works on an action-reaction or request-response model, and software systems have to use this paradigm to deliver the functionality users are looking for. For Web-based systems, the graphical user interface or GUI plays an important role in the user experience and it needs to be carefully designed for maximum ease of use. Web pages have come a long way from being static placeholders of information to more dynamic interactive mediums. Some basic ideas of web page Design often help a lot in ensuring user satisfaction. Software engineers can hardly be expected to be experts in all the nuances of human-computer interaction. But recognizing the importance of these studies in enhancing the usability of a system helps a lot in creating more usable software systems. This brings us to the ultimate goal of discussing human aspects of software development—to build usable software systems.

9.8 TOWARDS USABLE SOFTWARE SYSTEMS

We may paraphrase the well-known adage 'You can bring a horse to water, but you cannot make it drink', in the software development context as, 'You can build a software system for users, but you cannot make them use it'. The message from the original adage can be taken as: A horse will drink from a particular source of water only if it feels its thirst will be quenched. Similarly, *putative* users of a software system become *real* users not because a software system is built and given to them (bringing the horse to the water) but because using it will meet their needs (their thirst will be quenched).

The very essence of our discussion about the human aspects of software development relates to the quest for usability in software systems. Usability is recognized as an important criteria for software systems—it is the 'U' in the FURPS+ model [Datta 2005], but often lip service is all that it gets paid. What does the elusive *usability* represent for a software system?

Jurisat et al. call usability 'quality in use', and go on to explain that usability 'reflects how easy the software is to learn and use, how productively users will be able to work, and how much support users will need' [Juristo et al. 2001]. They add a very telling point, that usability is not merely about the user interface but it 'relates closely to the software's overall structure and to the concept on which the system is based' [Juristo et al. 2001]. As we have discussed so far in this chapter, the human aspects of software development have a significant influence on how the software's overall structure is developed from the underlying concepts. Usability has been divided into the following attributes: learnability, efficiency, user retention over time, error rate, and satisfaction [Ferre et al. 2001]. Ferre et al. also outline a *usability process,* which in tandem with the software development process seeks to ensure the usability of software systems. Figure 9.3 illustrates the attributes of usability.

Use cases (discussed in detail in Chapters 14 and 15) take the spirit of usability to drive Analysis and Design. By encouraging developers to think in terms of user actions and system responses, and closely reviewing these action-reaction sequences, use cases offer an *outside-in* view into the software system. As we shall see in later chapters, use cases are powerful and widely used mechanisms for designing user-oriented software systems.

Usability is both a deep and wide idea. As we explore in Exhibit 9.1. Issues around usability, or lack thereof have cropped up in situations of much historical significance.

With reference to Exhibit 9.1, what was wrong with Daulatabad? Evidently, it failed a vital usability test; no city can survive if it has inadequate water supply. Very interestingly, a similar move to a new 'synthetic' capital has happened in recent times, with the government of Myanmar (erstwhile Burma) deciding to move from the well-established capital city of Yangon (erstwhile Rangoon) to the new capital at the remote location of Pyinmana (http://news.bbc.co.uk/2/hi/asia-pacific/4416960.stm). Only time will tell, whether the new capital will stand the usability test!

In software development too, often sizeable time and resources are invested in building a new system, and having users migrate to it, only to find serious usability issues later, and to have users move back to the old way of doing things.

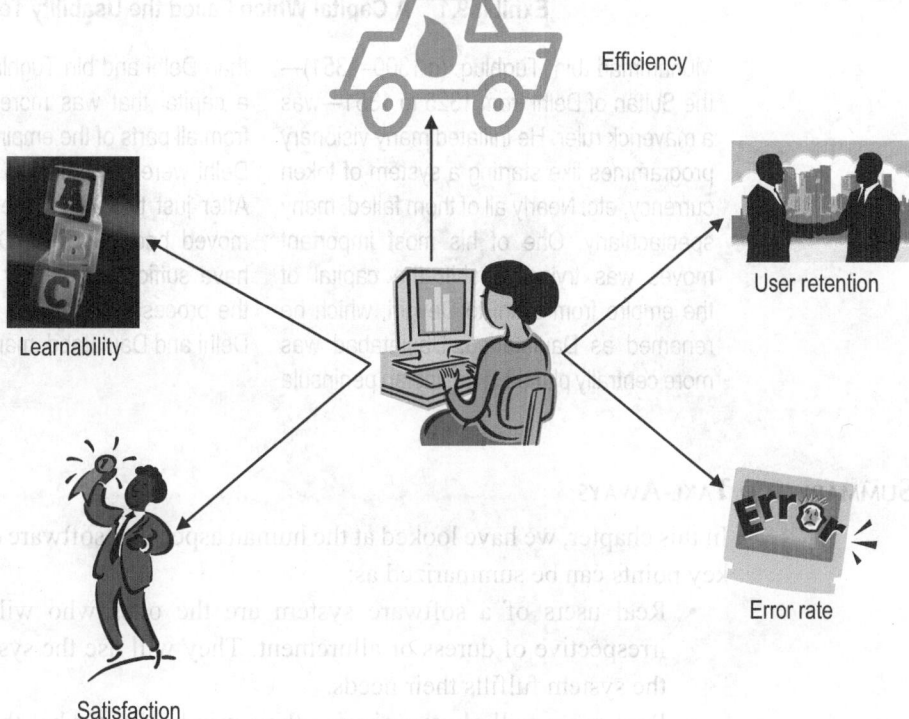

Figure 9.3   Attributes of usability

We bring the discussion of this chapter to a closure by considering the elusive 'human factor' in the next section.

9.9 THE HUMAN FACTOR

In his article, *Growth of human factors in application development*, Cockburn puts forward a tenet that dominance of human factors in software development have remained constant over the last few decades, but the recognition of the human factor has changed—'Social processes, Design techniques, and technology have come to support the weaknesses of people, as well as draw upon their strengths' (http://alistair.cockburn.us/index.php/Growth_of_human_factors_in_application_development).

In several chapters of this book in addition to the current one, we have specially highlighted the importance of human factors in software development. As software is built more and more by distributed teams, the influence of these factors will become further amplified. No matter how advanced we are in technologies, human aspects will continue to affect software development; whether for better or for worse depends on how well they are understood and leveraged.

> **Exhibit 9.1   A Capital Which Failed the Usability Test**
>
> Muhammad bin Tughluq (c.1300–1351)—the Sultan of Delhi from 1325 to 1351—was a maverick ruler. He initiated many visionary programmes like starting a system of token currency, etc. Nearly all of them failed, many spectacularly. One of his most important moves was trying to shift the capital of the empire from Delhi to Devgiri, which he renamed as Daulatabad. Daulatabad was more centrally placed in the Indian peninsula than Delhi and bin Tughlaq wanted to have a capital that was more easily accessible from all parts of the empire. The residents of Delhi were compelled to go to Daulatabad. After just two years, the capital had to be moved back to Delhi. Daulatabad did not have sufficient water for its inhabitants. In the process of this back and forth between Delhi and Daulatabd, many perished.

SUMMARY AND TAKE-AWAYS

In this chapter, we have looked at the human aspects of software development. The key points can be summarized as:

- Real users of a software system are the ones who will use the system irrespective of duress or allurement. They will use the system only because the system fulfills their needs.
- Real users will always change their minds about what they want from the system when they actually get to use the system. The ideas of cognitive gap and legacy thinking help us understand why real users are so capricious.
- Software engineers cannot assume they would have pre-defined, frozen Requirements to work with. An important software development activity lies in helping users know what they need.
- The co-evolution model represents how the problem and solution domains evolve hand-in-hand, one influencing the other. This is an important pointer to the influence of the human aspects of software development.
- Speaking in the users' language is an essential condition for effective communication between developers and users.
- Studies in human-computer interaction have opened up new perspectives in understanding the human aspects of software development.
- The study of the human factors of software development aim at a better understanding of the elusive attribute of usability.
- Human factors have remained a central concern of software development over the past few decades, and changing technologies have not undermined their importance.

WHERE TO LOOK FOR MORE

- ACM SIGCHI Curricula for Human-Computer Interaction—http://www.sigchi.org/cdg/cdg2.html
- Software Usability Research Laboratory—http://www.surl.org/

EXERCISES

Review Questions

Review Questions test your understanding of the key concepts presented in this chapter.

1. The output of a piece of software is always used by a human being. This is
   (a) a true statement
   (b) a false statement
   (c) a statement whose veracity can only be judged if the output is known
   (d) a statement whose veracity can only be judged if more information is known about the human being who is using the output

2. Human aspects of software development involve
   (a) ways and means to make software development less demanding
   (b) understanding some of the factors that influence the building of software systems to be used by human beings
   (c) (a) and (b)
   (d) none of the above

3. 'Real' users of a software system are those
   (a) who are forced to use the software system
   (b) who build the software system themselves
   (c) who will use the software system only if it gives them some value
   (d) who do not use the system in virtual reality

4. The cognitive gap for the users of a software system arises from
   (a) the relative novelty and unfamiliarity of the software medium
   (b) lack of software engineering education in users of a software system
   (c) general reduction in the cognitive skills of human beings
   (d) (a) and (c)

5. Which of the following can be thought to be a manifestation of legacy thinking
   (a) expecting the solution from a new technology to have the same features as the solution from an old technology
   (b) the disparity between old and new dress fashions
   (c) the symptoms of generation gap
   (d) none of the above

6. Eliciting Requirements from users is closest to which of the following
   (a) a teacher questioning a pupil
   (b) an employer interviewing a potential employee
   (c) a doctor discussing symptoms with a patient
   (d) a salesperson advertising a product

7. The 'juicy bits first' principle advocates
   (a) delivering parts of the system first that are easiest to build
   (b) delivering parts of the system first that are technically most challenging

(c) delivering parts of the system first that users want most

(d) delivering parts of the system first that developers feel should best serve the users' interests

8. Co-evolution is the
   (a) conjoint evolution of the problem and solution domains with one influencing the other
   (b) conjoint evolution of user and developer skills
   (c) (a) or (b) depending on the specific project
   (d) none of the above

9. Which of the following are most often used for communication between developers and users
   (a) formal specification languages
   (b) a natural language like English
   (c) Design diagrams
   (d) all of the above

10. One of the uses of a better understanding of human–computer interaction principles for software engineers is
    (a) designing better user interfaces
    (b) designing better test cases
    (c) designing better constraints
    (d) writing better Requirements

11. Usability relates to
    (a) a software system's overall structure
    (b) concepts on which the system is based
    (c) how fast a system responds
    (d) (a) and (b)

12. The following software development artefact helps drive Analysis and Design based on usability concerns
    (a) CRC cards
    (b) Sequence diagrams
    (c) Class diagrams
    (d) Use cases

13. The importance of human factors in software development over the past decades have
    (a) decreased with the advent of new technology
    (b) remained constant
    (c) lost its relevance
    (d) depends on the system being developed

14. Other than software, the idea of usability has ready relevance in
    (a) town planning
    (b) cars
    (c) electronic devices
    (d) all of the above

Reflective Questions

Reflective Questions require you to think more deeply about some of the ideas and come up with your own interpretations and answers.

1. We have made a distinction between human aspects and humane aspects as applied to software development. Could you give two examples of humane aspects related to software development?

2. Do you think cognitive gap and legacy thinking are ideas which may be relevant beyond contexts described in this chapter? If yes, identify two such situations where they are relevant. If no, justify your answer.

3. In the software development scenario outlined in the Case Study section, was co-evolution at play? If it was, was it helping or hindering the development team?

4. You are asked to Design a web page for users to check the status of flights. How do you think the page should look? Sketch a picture of the web page with locations of buttons, menus, etc.

5. Usability has been called 'quality in use'. Which aspects of quality do you feel are relevant to software usability?

6. Among the several human aspects of software development discussed in this chapter, which do you think are most influential in
   (a) ensuring ease of end-user use
   (b) ensuring ease of development

Programming Examples

Programming Examples require you to analyse or write a program or a program segment to understand a specific problem.

1. We have focused on usability from the end-users' point of view. As a computer programmer, do you find a particular language more usable than any other? What are the criteria upon which you will judge a particular language for its usability? Write a simple program to calculate the factorial of a number in Java and C++ and evaluate each language's usability. Will the criteria change if you are writing a larger program?

REFERENCES

Brooks, F. P. (2000), The Design of Design, http://www.siggraph.org/s2000/conference/turing/index.html, last accessed on May 19, 2010.

Datta, S. (2005), Integrating the FURPS+ model with use cases — a metrics driven approach, in *ISSRE 2005: Supplementary Proceedings of the 16th IEEE International Symposium on Software Reliability Engineering*, pages 4.51—4.52, IEEE Computer Society.

Datta, S. (2007), *Metrics-Driven Enterprise Software Development: Effectively Meeting Evolving Business Needs*, J. Ross Publishing.

DeMarco, T. and T. Lister (1987), *Peopleware: Productive Projects and Teams*, Dorset House Pub. Co.

Ferre, X., N. Juristo, H. Windl, and L. Constantine (2001), Usability basics for software developers, *IEEE Softw.*, 18(1):22–29.

Gilb, T. and K. Gilb (2006), Gilb community, http://www.gilb.com/community/tiki-page.php?pageName=Methods, last accessed on May 19, 2010.

Glen, P. (2002), *Leading Geeks: How to Manage and Lead the People Who Deliver Technology*, Jossey-Bass.

Juristo, N., H. Windl, and L. Constantine (2001), Guest editors' introduction: Introducing usability, *IEEE Software*, 18(1):20–21.

Maher, M. and J. Poon, (1996), Modelling Design exploration as co-evolution.

Yourdon, E. (2003), *Death March, Second Ed.*, Prentice Hall PTR.

Role of Automation in Software Development

Learning Objectives

In this chapter, we discuss the role of automation in software development. Specifically, the following areas will be addressed:

- The need to automate some of the software engineering activities
- CASE (computer-aided software engineering)
- The use of automation in software engineering: a brief overview
- The why, what, and how of automating software development
- The utility and challenges of automating Testing, Implementation, Design, Analysis, and Requirements specification

10.1 MOTIVATION

The popular comic-series character Tintin once arrives in America and is awed by the highly mechanized society. He sees a meat processing plant where a cow enters a machine through a conveyer belt and sausages pop out of a tap at the other end [Herge 2003a]. This is end-to-end automation at its acme. How we hope software engineering had a similar automatic machine, where we put in our Requirements, and outcomes the finished software—smooth and perfect! But software engineering of today is still very far from such sublime levels of automation. At every step of software development, we seem to need active human intervention to monitor, guide, and make sense of matters.

This state of affairs is rather different from other engineering industries. From making electronic transistor 'chips' to cars, automation is used in a big way in industrial production. It is customary now for large and formidable shop-floors to function like clockwork without humans running around. The reasons why software production has not reached these levels of automation are many, varied, and closely linked to the innate differences between software and other engineering branches (as discussed at length in earlier chapters). The aim of this chapter is not to lament software's lack of automation maturity. Instead, we will explore issues such as how automation helps software engineering and which aspects of software development are most amenable to automation.

Before we go deeper into the *role* of automation in software development, it will be helpful to put into perspective the *place* of automation. The aim of automation in software making is more towards complementing human agency than replacing it. As we shall see later in this chapter, effective software automation techniques provide insights and information to help humans make better choices or do things in more effective ways. We may say that good automation takes care of some of the drudgery of software development, thus freeing the human mind for innovation and creativity.

The idea of automation is hardly new in software engineering. Ever since software came to be built on large scale, software engineers have dreamt of software being built by software. Computer-aided software engineering, or CASE, has been a buzzword in the industry for quite some time. In the next section, we outline the broad contours of CASE and then move on to how specific automation techniques often serve very helpful purposes. We will then explore the progression of automation from the early 1970s. Next we consider the key dimensions of automation—the why, the what, and the how. This will lead us to an important consideration; which activities of software development give us the best returns from automation. Finally to underscore the point that automation need not necessarily be something elaborate or complex, we present an example.

10.2 COMPUTER-AIDED SOFTWARE ENGINEERING (CASE)

Computer-aided software engineering (CASE) refers to the software used to support software development activities such as Requirements engineering, Design, program-development, and Testing. Sommerville [2004] mentions that CASE tools can include Design editors, data dictionaries, compilers, debuggers, system building tools etc. This highlights the fact that ideally CASE should blend into the computing infrastructure that supports the writing, compiling, and running of computer programs, as well as the planning and execution of software development.

The scope of CASE depends on the sense the term is used in particular contexts. CASE can be a single tool supporting a specific activity or can be a complex 'environment' subsuming tools, processes, people, and hardware. CASE tools actually sit atop a structure comprising the underlying architecture, hardware platform, operating system, portability services, and integration framework.

CASE tools can be classified according to a number of different criteria: A *functional perspective* classifies according to the functions the tools fulfill, a *process perspective* classifies as per the process activities the tools facilitate, and an *integration perspective* classifies according to how the tools are combined and organized into integrated units, towards supporting specific processes or activities. For example, the taxonomy of the functional classification would include planning tools (estimation, scheduling), editing tools (text, diagrams, models), Testing tools (automated testers, oracles, data generators), language processing tools (compilers, interpreters), debugging tools etc. Evidently, there is significant overlap between these different classes: Something like an integrated development environment or IDE can spread across much of editing, language processing, debugging, and Testing and a number of other functional areas.

Another widely used classification schema for CASE is the distinction between *tools*—that support individual process tasks like compilation, or running a test case; *workbenches*—that support workflow activities like specification, Design, Implementation etc. (we discuss how easy or difficult each of these activities are in a later section); or *environments*—that support end-to-end processes or sub processes, and may comprise several integrated workbenches. This classification brings to light the fact that CASE need not be something large, complex, and involved; a simple (and essential) task such as compilation falls within CASE's purview as does something so ambitious (and difficult) as automating the entire development process. Figure 10.1 shows these different views of CASE.

CASE is something of a second-order utility—it looks to automate the automation brought about by software systems. CASE remains very effective in large and complex development scenarios, and every software engineering task today has an element of CASE embedded in it. In a way, CASE has become so ubiquitous, it is no longer seen as a special element in software development. No development organization today would seriously consider running an advertisement which says that they build better software since CASE is used in a big way in their projects!

But CASE can also compound the problem of the proverbial *overkill* in software development situations. On many occasions, learning and applying very involved CASE tools can consume valuable time and resources, which could have been better applied to building the software system in the first place. Another difficulty with CASE is that it often raises unrealistic expectations. Good software can only be constructed if software engineers are skilled, committed, and perceptive. CASE can help such software engineers do their tasks faster, and with fewer errors. But CASE cannot help bad software engineers create good software. More disturbingly, it may

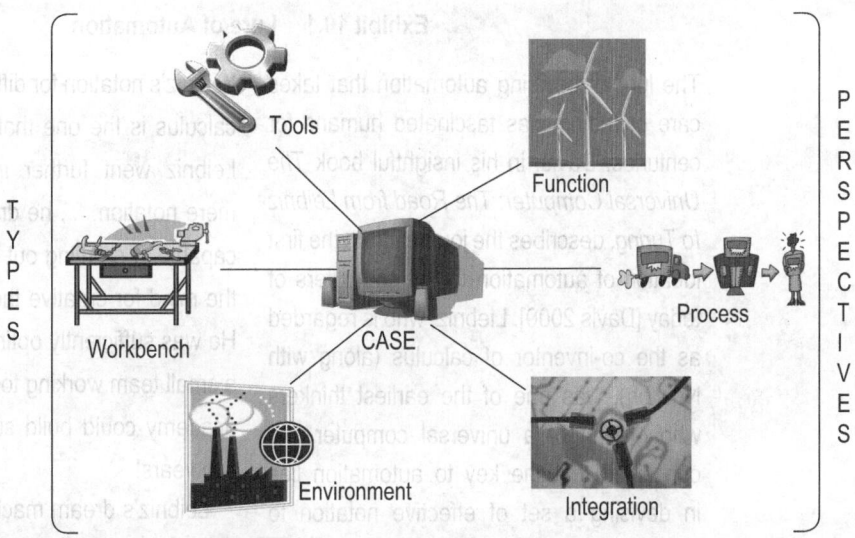

Figure 10.1   CASE can be viewed from different points of view, not always overlapping

even happen that bad software engineers will do a far worse job if given a free run of sophisticated CASE tools. Brooks' enduring 'no silver bullet' prophecy holds true for CASE (just as it does for many other technological advances in software development). Whereas it can be very effective when used with discretion and purpose, it is foolish to expect orders of magnitude improvements merely through the use of CASE.

Not all software projects have the scope in terms of time, money, or even skills to use heavyweight CASE tools. But there is scope—indeed a necessity at times— to harness the central idea behind CASE in many tasks of software development, irrespective of the size or budget of the project. This central idea is *the use of automation in complementing human effort and judgement in the development of software*. In the following sections we consider the history, extent, and limitations of automation in software.

10.3 THE ODYSSEY OF AUTOMATION

In this section we briefly review efforts so far at automating different activities of software development.

Freeman's paper, *Automating Software Design*, is one of the earliest expositions of the ideas and issues relating to Design automation [Freeman 1973]. The author starts from the basics by defining terms such as *program, software, Design*, and *creation* and explores two paradigms of semi-automated and automated

Exhibit 10.1   Lure of Automation

The lure of devising automation that takes care of tedium has fascinated humans for centuries. Davis, in his insightful book *The Universal Computer: The Road from Leibniz to Turing*, describes the journey from the first ideation of automation to the computers of today [Davis 2000]. Liebniz, who is regarded as the co-inventor of calculus (along with Newton), was one of the earliest thinkers who dreamt of a universal computer. He conceived that the key to automation lies in devising a set of effective notation to express ideas and their relationships, so that manipulating symbols mechanically would result in the carrying out of complex calculations without the need for conscious and deliberate thinking at each step.

Leibniz's notation for differential and integral calculus is the one that we use today. But Leibniz went further in his dreams than mere notation, '... he dreamed of machines capable of carrying out calculations, freeing the mind for creative thought' [Davis 2000]. He was sufficiently optimistic to believe that a small team working together at a scientific academy could build such a machine in a few years!

Leibniz's dream machine was not built in those few years, but he accomplished something far more lasting, laying the foundation for the automation of problem solving, and exploring many aspects of human cognition.

software creation. Important distinctions between Design automation and program automation are made, and the needs for knowledge representation and structuring of problem-solving skills highlighted. Although the paper hardly delves deeply into the automation issues that have subsequently assumed importance, it is a significant work in terms of an overview of automation ideas of the 1970s.

Karimi et al. [1988] report their experiences with the Implementation of an automated software Design assistant tool. The authors accept that the software Design process is difficult to generalize, as it depends to a large extent on personal judgement and individual styles and preferences. However, in their opinion, striking similarities exist in spite of the unique nature of each Design effort. Several 'manual' Design methodologies are first reviewed, followed by the description of the process of development of a computer-aided tool that offers 'intelligent assistance' in the 'determination of program modulus in the Design of software'. The paper describes the mechanism used to derive a set of quantifiable measures towards a 'scientific' basis for automated Design assistance. The results cited in the paper were reached through the use of a process structuring workbench called *Computer-Aided Process Organization* (CAPO) which seeks to derive '... a nonprocedural specification of modules, given the logical model of a system'. The empirical data from the use of CAPO leads to the specification of program modules with more cohesion, less coupling, and what the authors claim follows consequently, more maintainable systems. Although the advent of the object-oriented paradigm has resulted in a significantly different 'view' of software system organization, this paper presents a thorough and detailed discussion of building and applying an automated Design tool.

As software systems are extended, enhanced, and modified to accommodate changing Requirements, the original *intent* of Design is often subverted, leading to serious problems in their maintenance. Ciupke, in the paper *Automatic Detection of Design Problems in Object-Oriented Reengineering* [Ciupke 1999], presents a tool-based technique for analysing legacy code to detect Design problems. In this scheme problems are specified as queries on the Design model; the author illustrates the formalization of Design rules using the Prolog language. The catalogue of queries are based on established Design heuristics and from the author's own experience. Although this is an important contribution towards automating highly tedious tasks of software development such as scrutinizing legacy code for hidden Design flaws, useful formulation of the queries can pose serious challenges, given the diversity of software Design idioms.

O'Keeffe et al. [2003] state that 'All but the most trivial programming decisions can be considered Design decisions, and all such decisions are made with a view to maximizing certain properties in our designs'. They present an approach towards automatically improving Java Design through *simulated annealing* and results from using the *Dearthoir* prototype tool to validate the simulated annealing concept. The tool takes Java code as input, builds and manipulates parse tree, and outputs altered Java code. The paper points to exciting opportunities of code improvement without human intervention; however the tool as described can only effectuate a small number of refactoring schemes, which are limited in scope and usefulness.

Daniel Jackson's group at the Massachusetts Institute of Technology (MIT) are working on the *Alloy Analyser* tool for analysing models written in *Alloy*, a simple structural modelling language based on first order logic [Jackson 2006a], [Jackson 2006b]. The tool employs 'automated reasoning techniques that treat a software Design problem as a giant puzzle to be solved'. The initial results from using Alloy as reported are promising. But it seems an essential prerequisite for applying Alloy is to create a very detailed model of the Design, precisely clarifying what the author calls 'moving parts and specific behaviours, both desired and undesired, of the system and its components'. Developers' bandwidth for this kind of effort early in the software development life cycle cannot always be taken for granted. Also more often than not, changing Requirements preclude the crystallization of the Design to the extent that *all its* finer features can be specified upfront.

Automating software development activities have been the focus of attention since the mid 1970s. That no widely applicable technique has yet emerged speaks volumes regarding the challenging nature of the task. We next discuss the why, how, and what of automation.

10.4 AUTOMATION: WHY, HOW, AND WHAT

As we saw in an earlier section, CASE is a strongly organized, formal approach towards using computers to help build software systems. 'Using computers to help build software systems' sounds a rather naive statement now; but when CASE first came around—roughly two decades ago—it seemed a rather sleek way of doing things. That is how fast fashions move in the world of software!

Fully recognizing its many benefits, it must still be said that in several situations CASE is something of an overkill, or a distraction from the primary objective due to its overhead. Say, you are faced with a particular problem in Design, Implementation, or Testing which seems would be easier to crack using automation. You may look at established CASE tools, techniques, and workbenches. What if the problem is so unique, or peculiar, or original, that you find nothing in off-the-shelf CASE components that fits the bill? This is where automation, outside the strict purview of CASE, but within the ambit of using computers to help software development, becomes very helpful.

In this section, we will examine some key questions of automation: *Why* we seek to automate, *how* does one go about automation, and most significantly, *what* can be automated? Let us take up the *why* first.

Since the beginning of civilization, human beings have invented machines to do tasks for them that are too onerous, too dangerous, or just plain too boring. It is now common for industrial robots to lift heavy loads, work in conditions that are too hot or too cold or otherwise inhospitable, or to do repetitive work, especially when it needs high precision. Thankfully, software engineering does not call for lifting of heavy loads or working in extreme temperatures, or many of the penances other engineerings. But this also means software engineers need to be really creative about how automation can help them. One important place for the use of automation in software development is for activities that once defined, do not need human inputs for regular operation and may also be too tedious to be done on a large scale. Testing is one such activity. We discuss in Chapter 17 how involved the act of exercising even a reasonable number of test cases can be. Automation thus greatly helps in such situations.

Interestingly, often it is just not about overcoming tedium through automation. There are situations where only if you are able to crunch through a very large data set or very many combinations of situations can certain insights be gleaned. This occurs very frequently in scientific computing. In commercial software development also, there are cases when being able to explore all options gives a deeper understanding of the problem at hand. Automation comes in very handy in such cases. For example, as we will see later in the chapter, automation can help decide how best to delegate responsibilities to software components.

From *why* let us move now to *how*. Evidently, automating software development involves writing the programs that perform the automation in the first place. This is definitely an added overhead whose returns come only with widespread use of these automation aids; which is not immediately. Such returns are also difficult to quantify. Automation in software development needs to consider the *economies of scale*. Automation becomes really useful with large-scale development and when the problem-solution pattern is well-established. It is sometimes much easier to think through a problem if the scale is not sufficiently large, rather than trying to build an all-purpose automation framework to solve it and many of its yet unborn variants. However, although automation helps large-scale problems, the scale of automation does not always need to be large to be helpful. Simple, intuitive ideas can just as well be put to automation. Each software engineer may develop his or her own automation toolkit or use established ones to aid specific tasks. This brings us to the question, which software engineering tasks are most amenable to automation and which the least, that is, *what* can be automated.

The point we have made at several places in this book is that software engineering does *not* have the certitude of the physical laws to fall back upon. To some extent software development remains intuitive at best and ad hoc at worst. How easy it is to automate a particular software engineering activity draws from this characteristic. As we will see in Chapter 17, deciding what to test for in a software system remains a significant challenge. But once we have somehow made that decision, Testing boils down to performing a pre-defined set of activities in sequence, noting the results, and comparing the actual and expected results. This sounds like something which is easy to automate, and it is. Implementation, on the other hand, is slightly more involved. It is not entirely about writing syntactically correct code, but also making sure the code makes *sense;* that is, does what it is meant to do, nothing more and nothing less. No matter how much we dream to have code writing software some day, automating Implementation is trickier than automating Testing. Next comes the issue of automating Design.

I studied electrical engineering as an undergraduate. It was a fascinating subject, but designing a transformer as a third year project was heavy work. I will not bore you with details of the hardship or what came of it. But the way I went about such *Design* was starting off with some equations, tweaking parameters here and there but still making sure the equations held, and then hoping for the best. The truly disturbing aspect of software Design is that there are not even any equations to start off with. To the tyro software designer, it all seems to be hand-waving and gut-feeling. With experience, one is able to discern clearer patterns and trends. But even at their clearest, these patterns and trends are hardly tangible, and harder still to automate. Automating software Design—as we mentioned earlier in this chapter—remains one of the holy grails of software engineering. Although no one can claim we are there yet, we are making progress. Going further upstream, we come next to Analysis and Requirement elicitation. How easy is it to automate them? Certainly it is more difficult than Testing, Implementation, and even Design.

In the next few subsections we discuss the automation of software engineering life cycle activities, from the easy to the hard. Figure 10.2 shows how the level of difficulty of automation varies across the workflows and phases of software development. As the project life cycle progresses, the software system turns more concrete from abstract and the challenges of automation ease up to a large extent. It should be noted however, that the curves of Figure 10.2 are impressionistic only, and they are not derived from empirical data.

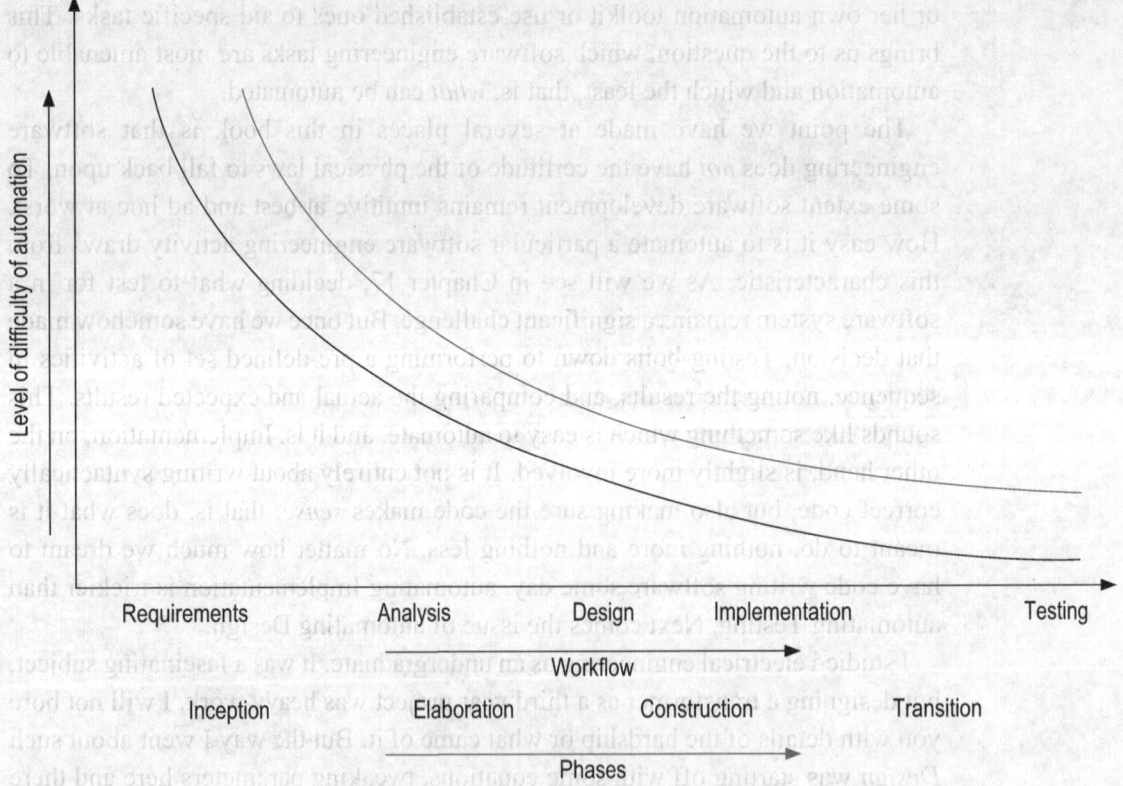

Figure 10.2 The variation of the difficulty of automation, with the progression of the software development life cycle

10.4.1 Test Automation

As we will discuss in Chapter 17, Testing can occur at many levels. At the most basic, unit Testing—which should ideally be part of a developer's personal discipline—can be automated by writing the test cases before coding, and using something like JUnit [1] to run them whenever new code is written or old code modified. Performance Testing and stress Testing are two other types of Testing which automation helps in a big way. It needs to be underscored that automated Testing can mainly help with repetitive Testing, or Testing a large number of scenarios, or running a diverse set of test cases, provided some human agency has formulated the test and the criteria of its success *a priori*. Automated Testing cannot fully answer the central question of *what needs to be tested*. That is something the human mind can figure out best.

10.4.2 Implementation Automation

Computers are very good at following rules, just the rules, and nothing but the rules. This comes in handy when checking for syntax errors in coding. Industrial code is now seldom written with general purpose word processors; instead integrated development environments, or IDEs, play a major role in making the act of coding faster and less error prone. Key features of an IDE include automated syntax checking, powerful context menus for Application Programming Interface (API) help, smart debuggers etc. Note how many of these IDE features were mentioned in our earlier discussion of CASE. This further supports the observation that CASE is now less of an esoteric add-on and more deeply integrated in the common framework of software development. The coding-compiling-running cycle can be drastically reduced in time and effort by the effective use of IDEs. The Eclipse platform [2] and its plug-ins provide an intuitive, easy to use, yet powerful IDE for development in major programming languages. They even provide support for more involved tasks like refactoring, dead-code elimination etc.

Automation can also take the act of coding to another level. Model-driven-development aims at integrating models more closely with code by facilitating smooth—ideally transparent—Transitions from modelling to coding. From heavyweight commercial tools to open source ones (such as StarUML [3]), there are many which promise to create the code structure, given the model. The code generated by such tools, can serve as a helpful starting point for programmers. Although we are still some distance from software tools writing complete code entirely on their own on an industrial scale, the level of automation we have now can help Implementation to a large extent.

---

1   www.junit.org
2   www.eclipse.org
3   www.staruml.com

10.4.3 Design Automation

With Design, we move from the reactive to the more reflective tasks of software development. Design means different things in different contexts, and software Design is very different from the Design of other engineering products. As we said earlier, in software Design there are no equations to get started or fall back upon; everything seems to be in the air. But the software that comes out of software Design is concrete, and it makes tangible impact on our lives. One way of addressing the challenge of software Design automation is to (somehow) *concretize* the essence of a Design problem, identify its objective, and apply pre-defined techniques towards fulfilling the objective. Design has by far the largest scope for originality amongst all software development activities. Design automation thus presents some unique challenges.

10.4.4 Automation of Specification and Analysis

But even more difficult than automating Design is the automation of Requirements Elicitation, Specification and Analysis. Here we are get deeper into cognition, psychology, and other inter-personal, social, and even political issues. The key to understanding and analysing Requirements is communication with stakeholders, which involves varieties of individuals expressing themselves in varieties of ways in varieties of situations. Discerning common patterns, which are basic to the foundations of automation are extremely difficult in these cases.

However, once Requirements have been elicited, their specification through a clear, unambiguous, and expressive idiom is a task that must be done for software systems to fulfill user needs effectively. And this is where automated frameworks like the *Alloy* tool mentioned earlier in this chapter can play an important role. The 'Z' specification language, as explained in Chapter 13, can be used for some aspects of automated Analysis. Truly effective and widespread use of automation in specification and Analysis will be helped by inter-disciplinary advances in computer and cognitive sciences.

10.4.5 Spectrum of Automation

In this section we have considered the place of automation in the different activities of software development. As we shall see in more detail in Chapters 14 and 15, the software development space is visualized as a two-dimensional matrix of *workflows* and *phases*. Workflows correspond to what we usually view as *activities*—Requirements, Analysis, Design, Implementation, Testing; whereas phases correspond to what are seen as *stages* in the life cycle—Inception,

Table 10.1   Key aspects of the role of automation across the workflows of software development

Workflow	Intent	Challenges and benefits of automation	Level of automation difficulty
Requirements	To understand user needs.	A robust and multi-purpose framework for automatically understanding user needs is difficult, if not impossible, to build with current technologies.	High
Analysis	To establish specifications that can be translated into the Design for a software system.	Automated specification needs to use highly structured specification languages such as 'Z'. Difficult to be applied on large-scale industrial systems.	High
Design	To devise software components and their interactions that best deliver the functionality expressed in the specifications within given constraints.	Design involves many decisions that arise from experience, intuition, and other gut-feelings. Additionally, software Design cannot start from laws and equations. Although end-to-end automation of software Design is still some distance away, automation can help complement a designer's judgement in many cases.	Medium
Implementation	To develop computer programs that mirror the structure and functions of the components designed.	Integrated development environments and Model-Driven Development tools help the act of coding by generating skeletal code, helping in compiling, debugging, and Application Programming Interface (API) support for specific languages. Yet, the Implementation of involved algorithms requires human inputs to a large extent.	Medium
Testing	To ensure the computer programs perform their tasks according to specifications.	Automation can be widely used in different types of Testing; however formulation of test cases need considerable human judgement.	Low

Elaboration, Construction, and Transition. In Table 10.1, we highlight the key aspects of the role of automation across the phases and in Table 10.2 across the workflows.

Table 10.2 Key aspects of the role of automation across the phases of software development

Phase	Intent	Challenges and benefits of automation	Level of automation difficulty
Inception	To decide the overall vision and scope of the system.	System has not yet been defined, so there are not enough concrete tasks that can be automated.	High
Elaboration	To extend vision and scope into specific goals of what needs to be delivered.	System is being understood and defined; automation can help complement human judgement.	Medium
Construction	To build and test the system according to the goals agreed upon by stakeholders.	Widespread use of automation is possible given that the basic processes and activities are clearly understood and specified.	Low
Transition	To tune the system for best performance and release it to users.	Widespread use of automation is possible given that the basic processes and activities are clearly understood and specified.	Low

10.5 AUTOMATING ONE ASPECT OF DESIGN: AN EXAMPLE*

As reflected earlier in this chapter, automating Design remains one of the most challenging tasks of software development. But challenging does not mean it will not be attempted. In this section, we outline one such attempt as illustrated in [Datta 2008]. The following discussion should be taken in the light of some of the ideas discussed elsewhere in this book, most notably in Chapters 12, 13, and 16. This section uses some basic concepts of set theory and linear programming, and is aimed at advanced readers.

Every software component exists to perform specific tasks, which may be called its *responsibilities*. The canons of good software Design recommend that each component be entrusted with one primary responsibility. Components may end up being given more than one task, but it is important to try and ensure they have one primary responsibility. Whether components have one or more responsibilities, they cannot perform their tasks entirely by themselves, without interaction with other components. This is especially true for the so-called *business objects*—components containing the business logic of an application. The extent to which a component

---

* Material in Section 10.5 has been reproduced with kind permission of Springer Science and Business Media from the materials of the following paper:
COMP-REF: A Technique to Guide the Delegation of Responsibilities to Components in Software Systems by Subhajit Datta and Robert van Engelen in Lecture Notes in Computer Science published by Springer Berlin/Heidelberg, ISSN 0302-9743 (print), 1611-3349 (online), Volume 4961/2008 in the book *Fundamental Approaches to Software Engineering*, DOI 10.1007/978-3-540-78743-3-25, Copyright 2008, ISBN 978-3-540-78742-6, Pages 332–346.

has to interact with other components to fulfill its core functionality is an important consideration. If a component's responsibilities are strongly focused on a particular line of functionality, its interactions with other components can be expected to be less disparate. Let us take *aptitude* to denote the quality of a component that reflects how coherent its responsibilities are. Intuitively, the *Aptitude Index* measures the extent to which a component (one among a set fulfilling a system's Requirements) is coherent in terms of the various tasks it is expected to perform.

The essence of software Design lies in the collaboration of components to collectively deliver a system's functionality within given constraints. While it is important to consider the responsibility of individual components, it is also imperative that inter-component interaction be clearly understood. Software components need to work together in a spirit of harmony if they have to fulfill Requirements through the best utilization of resources. Let us take *concordance* to denote such cooperation amongst components. How do we recognize such cooperation? It is manifested in the ways components share the different tasks associated with fulfilling a Requirement. Some of the symptoms of less than desirable cooperation are replication of functionality—different components doing the same task for different contexts, components not honouring their interfaces (with other components) in the tasks they perform, one component trying to do everything by itself etc. The idea of concordance is an antithesis to such undesirable characteristics—it is the quality which delegates the functionality of a system across its set of components in a way such that it is evenly distributed, and each task goes to the component most well-positioned to carry it out. Intuitively, the metric *Concordance Index* measures the extent to which a component is concordant in relation to its peer components in the system.

Now let us define the metrics formally.

Considering a set of Requirements $Req = \{R_1, ..., R_x\}$ and a set of components $Comp = \{C_1, ..., C_y\}$ fulfilling it, we define the metrics in the following sub-sections.

10.5.1 Aptitude Index

The *Aptitude Index* seeks to measure how coherent a component is in terms of its responsibilities.

To each component C_m of *Comp*, we attach the following *properties* [Datta 2006]. A *property* is a set of zero, one or more components.

- *Core* - $\alpha(m)$
- *Non-core* - $\beta(m)$
- *Adjunct* - $\gamma(m)$

$\alpha(m)$ represents the set of component(s) required to fulfill the primary responsibility of the component C_m. As already noted, sound Design principles suggest the component itself should be in charge of its main function. Thus, most often $\alpha(m) = \{C_m\}$.

$\beta(m)$ represents the set of component(s) required to fulfill the secondary responsibilities of the component C_m. Such tasks may include utilities for accessing a database, date or currency calculations, logging, exception handling, etc.

$\gamma(m)$ represents the component(s) that guide any conditional behaviour of the component C_m. For example, for a component which calculates interest rates for bank customers with the proviso that rates may vary according to a customer *type* ('gold', 'silver' etc.), an *Adjunct* would be the set of components that help determine a customer's type.

The Aptitude Index AI(m) for a component C_m is a relative measure of how much C_m depends on the interaction with other components for delivering its core functionality. It is the ratio of the number of components in $\alpha(m)$ to the sum of the number of components in $\alpha(m)$, $\beta(m)$, and $\gamma(m)$

$$AI(m) = \frac{|\alpha(m)|}{|\alpha(m)| + |\beta(m)| + |\gamma(m)|}$$

10.5.2 Requirement Set

The Requirement Set RS(m) for a component C_m is the set of Requirements that need C_m for their fulfillment.

$$Rs(m) = \{R_p, R_q, ...\}$$

where C_m participates in the fulfillment of p, q etc.
Evidently, for all C_m, $RS(m) \subseteq Req$.

10.5.3 Concordance Index

The Concordance Index CI(m) for a component C_m is a relative measure of the level of concordance between the Requirements being fulfilled by C_m and those being fulfilled by other components of the same system.

For a set of components $Comp = \{C_1, C_2,..., C_{y-1}, C_y\}$ let
$W = RS(1) \cup RS(2) \cup ... \cup RS(y-1) \cup RS(y)$
For a component C_m ($1 \le m \le y$), let us define
$X(m) = (RS(1) \cap RS(m)) \cup ... \cup ((RS(m-1) \cap RS(m)) \cup$
$((RS(m) \cap (RS(m+1)) \cup ... \cup ((RS(m) \cap (RS(y)))$

Thus $X(m)$ denotes the set of Requirements that are not only being fulfilled by C_m but also by some other component(s).

Expressed as a ratio, the *Concordance Index* $CI(m)$ for component m is:

$$CI(m) = \frac{|X(m)|}{|W|}$$

Based on these metrics, COMP-REF is a technique to guide Design decisions towards allocating responsibilities to a system's components. For successful collaboration, software components are expected to carry out their tasks in a spirit of cooperation such that each component has clearly defined and specialized responsibilities, which it can deliver with reasonably limited amount of support from other components. *Aptitude Index* measures how self-sufficient a component is in carrying out its responsibilities and *Concordance Index* is a measure of the degree of its cooperation with other components in the fulfillment of the system's Requirements. Evidently, it is desired that cooperation across components be as high as possible, within the constraint that each Requirement be fulfilled by a limited number of components. This observation is used to formulate an objective function and a set of linear constraints whose solution gives a measure of how much each component is contributing to maximizing the concordance across the entire set of components. If a component is found to have low contribution (low value of the a_n variable corresponding to the component in the LP solution as explained below), *and* it is not significantly self-sufficient in carrying out its primary responsibility (low *Aptitude Index* value) the component is a candidate for being de-scoped and its tasks (which it was hardly executing on its own) distributed to other components. This results in a more compact set of components fulfilling the given Requirements.

The goal of the COMP-REF technique is identified as *maximizing* the *Concordance Index* across all components, for a given set of Requirements, in a particular iteration of development, within the constraints of *not* increasing the number of components currently participating in the fulfillment of each Requirement.

A new variable a_n ($a_n \in [0,1]$) is introduced corresponding to each component C_n, $1 \leq n \leq N$, where N = the total number of components in the system. The values of a_n are arrived at from the LP solution. Intuitively, a_n for a component C_n can be taken to indicate the extent to which C_n contributes to maximizing the *Concordance Index* across all components. As we shall see later, the a_n values will help us decide which components to merge.

The LP formulation can be represented as:

$$\text{Maximize} \sum_{n=1}^{y} CI(n)a_n$$

Subject to: $\forall R_m \in \mathrm{Re}q, \sum_{n=1}^{y} a_n \leq p_m / N, a_n$ such that $C_n \in CS(m).p_m = |CS(m)|.$

(As defined in [Datta 2006], the *Component Set* $CS(m)$ for a Requirement R_m is the set of components required to fulfill R_m.)

So, for a system with x Requirements and y components, the objective function will have y terms and there will be x linear constraints.

The COMP-REF technique is then summarized as: Given a set of Requirements $Req=\{R_1,\ldots,R_x\}$ and a set of components $Comp=\{C_1,\ldots,C_y\}$ fulfilling it in iteration z of development,

- STEP 0: Review *Req* and *Comp* for new or modified Requirements and/or components compared to previous iteration.
- STEP 1: Calculate the *Aptitude Index* for each component.
- STEP 2: Calculate the *Requirement Set* for each component.
- STEP 3: Calculate the *Concordance Index* for each component.
- STEP 4: Formulate the objective function and the set of linear constraints.
- STEP 5: Solve the LP formulation for the values of a_n
- STEP 6: For each component C_n, check:
  - Condition 6.1: a_n has a low value compared to that of other components? (If yes, implies C_n is not contributing significantly to maximizing concordance across components.)
  - Condition 6.2: $AI(n)$ has a low value compared to that of other components? (If yes, implies C_n has to rely heavily on other components for delivering its core functionality.)
- STEP 7: **If** *both* conditions 6.1 and 6.2 hold TRUE, GOTO STEP 8, **else** GOTO STEP 10
- STEP 8: For C_n, check:
  - Condition 8.1: Upon merging C_n with other components, in the resulting set $\tilde{C}omp$ of q components (say), $CI(q) \neq 0$ for all q? (If yes, implies resulting set of q components has more than one component).
- STEP 9: **If** condition 8.1 is TRUE, C_n is a candidate for being merged; after merging components C_n GOTO STEP 0, starting with *Req* and $\tilde{C}omp$, **else** GOTO STEP 10.
- STEP 10: Wait for the next iteration.

Figure 10.3 gives a flow chart representation of the COMP-REF technique.

In the following Case Study section we will illustrate the use of the COMP-REF technique through an example.

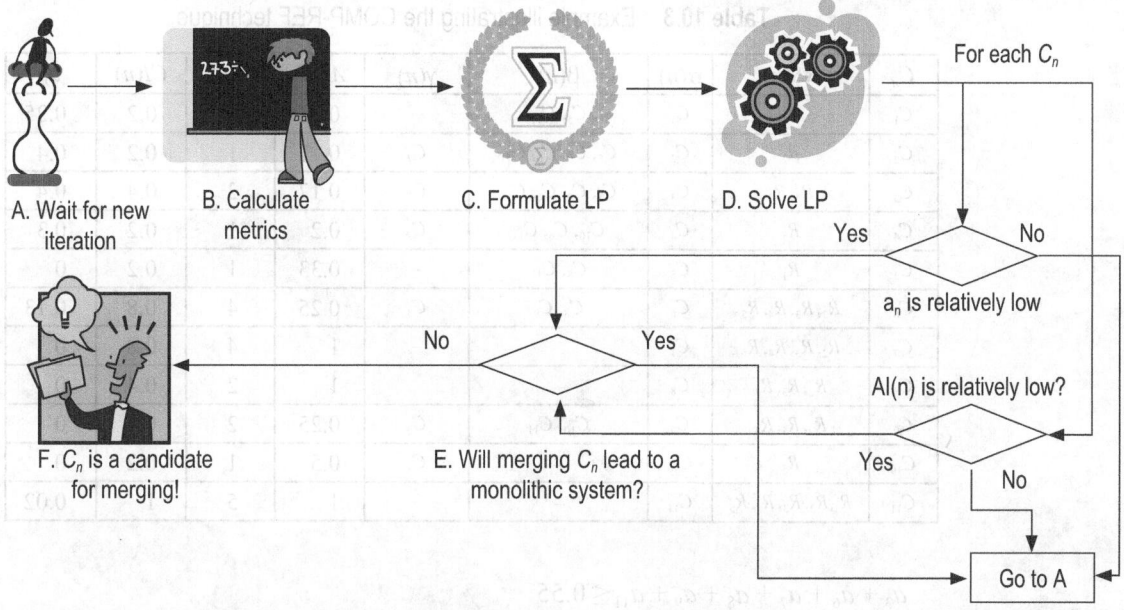

Figure 10.3   Flow chart representation of the COMP-REF technique

CASE STUDY

Let us see an application of the COMP-REF technique, also illustrated in [Datta 2008]. For a hypothetical software system, there are five Requirements, $R_1, \ldots, R_5$ and 11 components, $C_1, \ldots, C_{11}$. The metrics calculation is shown in Table 10.3 as well as the LP solutions.

The $RS(n)$ column of Table 10.3 shows the *Requirement Set* for each component. Evidently, $W = \{R_1, R_2, R_3, R_4, R_5\}$ and $|W| = 5$. The $AI(n)$ and $CI(n)$ columns of Table 10.3 give the *Aptitude Index* and the *Concordance Index* values respectively for each component.

From the Design artefacts, it was noted that R_1 needs components C_1, C_5, C_{11}, ($p_1 = 3$), 2 needs $C_2, C_6, C_7, C_8, C_9, C_{11}$ ($p_2 = 6$), R_3 needs $C_3, C_6, C_7, C_8, C_9, C_{11}$ ($p_3 = 6$), R_4 needs $C_3, C_6, C_7, C_8, C_9, C_{11}$ ($p_4 = 6$), and R_5 needs $C_4, C_6, C_7, C_{10}, C_{11}$ ($p_5 = 5$) for their respective fulfillment. Evidently, in this case $N = 11$.

Based on the above, the objective function and the set of linear constraints was formulated as:

Maximize

$$0.2a_1 + 0.2a_2 + 0.4a_3 + 0.2a_4 + 0.2a_5 + 0.8a_6 + 0.8a_7 + 0.4a_8 + 0.4a_9 + 0.2a_{10} + a_{11}$$

Subject to

$$a_1 + a_5 + a_{11} \leq 0.27$$
$$a_2 + a_6 + a_7 + a_8 + a_9 + a_{11} \leq 0.55$$

Table 10.3 Example illustrating the COMP-REF technique

C_n	$RS(n)$	$\alpha(n)$	$\beta(n)$	$\gamma(n)$	$AI(n)$	$|X(n)|$	$CI(n)$	a_n
C_1	R_1	C_1	C_5, C_{11}	–	0.33	1	0.2	0.25
C_2	R_2	C_2	C_8, C_9, C_6, C_{11}	C_7	0.17	1	0.2	0.4
C_3	R_3, R_4	C_3	C_8, C_9, C_6, C_{11}	C_7	0.17	2	0.4	0.4
C_4	R_5	C_4	C_{10}, C_6, C_{11}	C_7	0.2	1	0.2	0.3
C_5	R_1	C_5	C_8, C_{11}	–	0.33	1	0.2	0
C_6	R_2, R_3, R_4, R_5	C_6	C_8, C_9	C_3	0.25	4	0.8	0.13
C_7	R_2, R_3, R_4, R_5	C_7	–	–	1	4	0.8	0
C_8	R_2, R_3, R_4	C_8	–	–	1	2	0.4	0
C_9	R_2, R_3, R_4	C_9	C_{10}, C_{11}	C_3	0.25	2	0.4	0
C_{10}	R_5	C_{10}	–	C_7	0.5	1	0.2	0
C_{11}	R_1, R_2, R_3, R_4, R_5	C_{11}	–	–	1	5	1	0.02

$$a_3 + a_6 + a_7 + a_8 + a_9 + a_{11} \leq 0.55$$
$$a_3 + a_6 + a_7 + a_8 + a_9 + a_{11} \leq 0.55$$
$$a_4 + a_6 + a_7 + a_{10} + a_{11} \leq 0.45$$

Using the automated solver, GIPALS (General Interior-Point Linear Algorithm Solver)[4], the above LP formulation was solved (values in the a_n column of Table 10.3).

Let us examine how the COMP-REF technique can guide Design decisions. Based on the a_n values in Table 10.3, evidently components $C_5, C_7, C_8, C_9, C_{10}$ have the least contribution to maximizing the objective function. So the tasks performed by these components may be delegated to other components. However, as mandated by COMP-REF, another factor needs be taken into account before deciding on the candidates for merging. How self-sufficient are the components that are sought to be merged? We next turn to $AI(n)$ values for the components in Table 10.3. We notice, $AI(5) = 0.33$, $AI(7) = 1$, $AI(8) = 1$, $AI(9) = 0.25$, and $AI(10) = 0.5$. Thus C_7, C_8 and C_{10} have the highest *Aptitude Index* values. These are components delivering functionalities of general utility, user input validation, and database access logic respectively—facilities used across the application. Thus it is expedient to keep them localized. But C_5 and C_9, as their relatively low values of $AI(n)$ suggest, need to interact significantly with other components to carry out their task. And given their negligible contribution to maximizing concordance; a helpful Design choice would be to merge them with other components. A smaller set of high concordance components is preferred over a larger set of low concordance ones, as the former

---

4 http://www.optimalon.com/

has lesser inter-component interaction, thereby leading to better resilience, to the modification of particular components due to Requirement changes.

Thus one cycle of application of the COMP-REF technique suggests the reduction of the number of components from eleven to nine in fulfilling the set of Requirements for the example system.

Having seen how the COMP-REF technique is formulated as well as how it can be applied, we are now in a position to consider how it helps in automation of an aspect of software Design. Evidently, COMP-REF has a number of steps: calculating the metrics, formulating and solving the LP, and then interpreting the results to guide some Design decisions. The information needed to calculate the metrics can be automatically extracted from Design artefacts and code, the LP can be solved using automated solvers. The interpretation does require an element of *judgement* whose automation is not that trivial, and it is precisely where the human mind can complement automation. But the most important point about the COMP-REF technique is that it seeks to capture some qualitative aspects of software Design into quantitative measures, and based on the measures, tries to guide the Design process.

SUMMARY AND TAKE-AWAYS

In this chapter we have considered the role of automation in software development. Automation is staple in the production process of other engineering products but software's unique features influence the extent and effectiveness of automation in building software systems. The salient points of this chapter are given below.

- Computer-aided-software-engineering (CASE) involves the use of software to help the development of software. CASE tools can be classified in many ways, and many of the classification categories are overlapping. Although CASE can be used for an elaborate, end-to-end automation of software development, it can also help in the automation of simple, stand- alone tasks. The essence of CASE lies in using automation to help software engineers build better software systems.
- The use of automation in software engineering has been of interest since the 1970s. Several ideas, techniques, and tools have been put forward, which in spite of not being within the strict purview of CASE, are none-the-less helpful in specific situations of software development.
- Not all activities of software development are equally amenable to automation. In increasing order of difficulty, they are: Testing, Implementation, Design, Analysis, and Requirements specification.
- Although software Design is relatively less easy to automate, the COMP-REF technique gives an example how designers can leverage automation to

complement their judgement. The formulation of the COMP-REF technique, as well as a detailed example illustrating its use are given to establish the point that reasonable forethought and discipline can lead to useful steps in automation.

- In this chapter, our main interest has been trying to find out how automation can help software development. We have not considered the related but much deeper question of what ultimately computers can (or cannot) do for us [Harrel 2000]. Use of automation in software development can significantly help software engineers do their tasks better; but it is limited by what computers can do at one level, and the unique characteristics of software at another.

WHERE TO LOOK FOR MORE

- http://www.sei.cmu.edu/legacy/case/case_whatis.html – Although somewhat dated, this page contains valuable discussion of the background of CASE.
- http://www.eclipse.org – The Eclipse platform serves as a very useful test bed for developing automated tools and techniques to help software development.

EXERCISES

Review Questions

Review Questions test your understanding of the key concepts presented in this chapter.

1. Which of the following seems the most probable reason for studying the role of automation in software development?
   (a) to understand how we can have running software being produced automatically from specifications
   (b) to understand the limitations of automated software development
   (c) to understand how automation can complement human judgement and effort in developing superior software
   (d) all of the above

2. Compared to another engineering activity like the industrial production of cars, the current level of automation in software engineering
   (a) is higher

   (b) is lower
   (c) varies on a case to case basis
   (d) cannot be compared

3. In the context of software development, which of the following tasks is/are most likely to be helped by automation?
   (a) running a very large number of test cases
   (b) exploring many Design alternatives
   (c) generating skeletal code
   (d) all of the above

4. Which of the following software engineering activities is easiest to automate?
   (a) gathering Requirements
   (b) Testing
   (c) writing code
   (d) Design

5. In which of the following aspects of Testing, automation is least helpful?
   (a) running test cases

(b) deciding what to test

(c) both (a) and (b)

(d) none of the above

6. Integrated development environments (IDEs) may help in
   (a) syntax checking
   (b) semantics checking
   (c) unit Testing
   (d) integration Testing

7. Which of the following statements is true about the interest in automating software development activities?
   (a) it is a recent phenomenon
   (b) there is no longer any interest in it in the industry
   (c) there has been interest since the early 1970s
   (d) None of the above

8. In the context of this chapter, what is *Alloy*?
   (a) a mixture of two metals
   (b) a type of metal used for making airplanes
   (c) a simple structural modelling language based on first order logic
   (d) none of the above

9. The COMP-REF technique can be used to automate
   (a) Testing
   (b) Implementation
   (c) Analysis
   (d) Design

10. The COMP-REF technique recommends
    (a) splitting of components
    (b) merging of components
    (c) refactoring of components
    (d) all of the above

11. CASE is a
    (a) specific software development scenario
    (b) type of use case
    (c) type of test case

(d) software product used to support the development of software

12. CASE classifications are
    (a) mutually exclusive
    (b) overlapping
    (c) depends on the specific version of CASE
    (d) none of the above

13. Tools, workbenches, and environments are different types of CASE tools, classified according to their
    (a) price
    (b) performance
    (c) popularity
    (d) specific scope and type of software engineering activity they support

14. 'The use of CASE can bring in order(s) of magnitude increases in the development time of a software project.' This statement
    (a) is true
    (b) is false
    (c) is sometimes true, sometimes false
    (d) depends on the situation

15. Which of the following statements about the use of CASE is/are true?
    (a) It is a novel approach in software development.
    (b) It has been around since the past two decades.
    (c) It is limited in commercial software development.
    (d) Both (b) and (c) are true.

16. The essence of CASE, which is applicable to many situations where use of CASE may or may not be appropriate, is use of
    (a) tools
    (b) automation
    (c) workbenches
    (d) environments

Reflective Questions

Reflective Questions require you to think more deeply about some of the ideas and come up with your own interpretations and answers.

1. What do you think is the most unique feature of software engineering that makes it less amenable to automation, as compared to other engineering branches?

2. What do you think is the most laborious aspect of building a software system? Has it been sufficiently automated? If yes, how can such automation be improved? If no, why has it not been automated?

3. Point out some of the drawbacks of the COMP-REF technique. How do you think the technique can be further improved?

4. As we discussed, very simple tasks of software development can also be under the purview of CASE. Take a software development activity where you think you have used CASE and explain how that could have been done without the use of CASE.

5. What do you think is the key difference between tools, workbenches, and environments? Can you give examples of their use in a non-software engineering scenario?

6. Do you think CASE can be used more effectively by experienced software engineers than inexperienced ones? Would automation help new software engineers understand software engineering better?

7. Do CASE and automation share a superset-subset relationship? Can there be automation in software development outside the purview of CASE?

Numerical Problems

Numerical Problems require you to do quick 'back-of-the-envelope' kind of calculations to arrive at answers to some simple but insightful problems.

1. In the application of the COMP-REF technique, can the LP solution be interpreted in any other way? Can you think of a case where the recommendations from the technique will be totally counter-intuitive?

2. Given three Requirements and five components, all LP formulations based on the COMP-REF technique will have a solution. Prove this statement or disprove by a numerical example. State any assumption that you will need to make to formulate the LP.

Programming Problem

Programming Problem requires you to analyse or write a program or a program segment to understand a specific problem.

1. Write a program in Java to automate the COMP-REF technique. How would you implement a linear programming solver? What are the pros and cons of using an open-source LP solver?

REFERENCES

Ciupke, O. (1999) 'Automatic Detection of Design Problems in Object-Oriented Reengineering', In *TOOLS '99: Proceedings of the Technology of Object-Oriented Languages and Systems*, p. 18, IEEE Computer Society, Washington, DC, USA..

Datta, S. (2006) 'Agility Measurement Index: A Metric for the Crossroads of Software Development Methodologies', In *ACM-SE 44: Proceedings of the 44th Annual Southeast Regional Conference*, pp 271–73, ACM Press, New York, NY, US.

Datta, S. and van Engelen, R. (2006) 'Effects of Changing Requirements: A Tracking Mechanism for the Analysis Workflow', In *SAC '06: Proceedings of the 2006 ACM Symposium on Applied Computing*, pp 1739–44, ACM Press, New York, US.

Datta, S. and van Engelen, R. (2008) 'Compref: A Technique to Guide the Delegation of Responsibilities to Components in Software Systems', In *Fundamental Approaches to Software Engineering*, volume 4961 of *LNCS*, pp 332–346, Springer.

Davis, M. (2000) *The Universal Computer: The Road from Leibniz to Turing*, W.W. Norton and Company.

Freeman, P. (1973) 'Automating Software Design', In *DAC '73: Proceedings of the 10th Workshop on Design Automation*, pp 62–67, IEEE Press, Piscataway, NJ, US.

Herge, (2003) *Tintin in America*, Casterman Editions.

Jackson, D. (2006a) 'Dependable Software by Design', http://www.sciam.com/article.cfm?chanID=sa006&colID=1&articleID=00020D0%. last accessed on May 19, 2010. 4-CFD8-146C-8D8D83414B7F0000, The Scientific American, June.

Jackson, D. (2006b) *Software Abstractions: Logic, Language and Analysis*, MIT Press.

Karimi, J. and Konsynski, B. R. (1988) An Automated Software Design Assistant, *IEEE Trans. Softw. Eng.*, 14(2):194–210.

O'Keeffe, M. and Cinneide, M.M.O. (2003) 'A Stochastic Approach to Automated Design Improvement', In *PPPJ '03: Proceedings of the 2nd International Conference on Principles and Practice of Programming in Java*, pp 59–62, Computer Science Press, Inc., New York, US.

Sommerville, I. (2004) *Software Engineering*, 7th ed., Prentice Hall.

PART III

MAKING SOFTWARE

Understanding Software Architecture

Learning Objectives

In this chapter, we discuss why an understanding of software architecture is so essential in building large scale software systems. Specifically, we address:

- Architectural views of software
- Different perspectives and definitions of software architecture
- How architecture differs from Design
- Architectural patterns

11.1 MOTIVATION

You employ stone, wood and concrete, and with these materials you build houses and palaces. That is Construction. Ingenuity is at work. But suddenly you touch my heart, you do me good, I am happy and I say "This is beautiful." That is Architecture.

– Le Corbusier (1923), quoted in *Architecture: From Prehistory to Post-Modernism*

The above quotation captures the spirit of *architecture*. We recognize good architecture when we see it, but it is difficult to define what good architecture is. Seeing the Taj Mahal, one of the greatest architectural wonders of the world, one is struck by its beauty, symmetry, poise, and proportion. But one is also aware, there is something nameless beyond these obvious attributes that endows Taj Mahal with its architectural majesty. This sense is imparted to us by the underlying architecture. Whenever we deal with complex structures, we need to explore architectural ideas to simplify the task as well as make the end-product beautiful and resilient.

One of the first organized attempts at studying software architecture can be traced back to the 1995 November/December issue of the *IEEE Software* magazine (Volume 12, Issue 6). Many of the articles published in that issue introduced foundational notions of software architecture, such as Kruchten's 4+1 view model [Kruchten 1995], which we will discuss in detail later in this chapter. The timing of this issue points to an important juncture in software engineering's evolution—more and more complex software systems were being built to support a host of critical activities. At that time, software engineers were becoming increasingly aware of the need to have a framework for abstracting, understanding, documenting, and communicating key ideas about software development. The conception of software architecture was an important step towards building such a framework.

In this chapter, we will discuss why understanding software architecture is so essential in building large-scale software systems. The next section describes the architectural views of software, followed by a discussion on the essence of software architecture and how architecture differs from Design. We will then talk about architectural patterns and the future of software architecture. The Case Study section illustrates a practical scenario of applying software architecture ideas.

11.2 ARCHITECTURAL VIEWS OF SOFTWARE

As mentioned earlier, Kruchten's article, 'The 4+1 View Model Architecture' [Kruchten 1995] is a seminal work in the study of software architecture. Kruchten uses five concurrent 'views'—distinct perspectives of observing a software system—to describe software architecture. Each of these views addresses a specific set of concerns related to different stakeholders of the system.

Logical View

The *logical* view covers the object model underlying the system's Design (assuming the system is designed on object-oriented principles). For intensely data-driven systems, entity-relationship diagrams can constitute the logical view. The logical view is closely connected with functional Requirements—the facilities the system must provide to its end-users. In Chapter 13 we discuss unified modelling language (UML) in detail. The logical view can be thought to correspond to UML class and sequence diagrams.

Process View

The *process* view illuminates concurrency and synchronization aspects of the system's Design. Concurrency and synchronization are concepts often associated with *threads* and techniques related to utilization of the CPU by several parallel execution paths, each of which has the illusion of owning the whole CPU.

The process view is concerned with some of the non-functional Requirements, such as performance, system availability, and fault tolerance.

Physical View

The *physical* view relates to the mapping between the software components and the hardware sub-stratum. It deals with issues such as deployment of the software system on the underlying hardware. The physical view is also associated with throughput, scalability, etc. The physical view can be thought to correspond to UML deployment diagrams.

Development View

The *development* view delineates the software system's static organization in its development environment. It emphasizes how the modules of a software system combine together to deliver the overall functionality. The development view can be taken to correspond to UML package diagrams.

Use-Case View

All the four views are held together by the *use-case* view, which Kruchten refers to as *scenario*. As the name suggests, the use-case view maps to use-case diagrams in UML. Figure 11.1 depicts the 4+1 views.

The 4+1 view model gives a set of interesting abstractions of a software system, based on different concerns. It places primary focus on the '+1' use-case view to consolidate the users expectations from the system. The centrality of the use-case view in Figure 11.1 is no accident; the main objective of any software system is

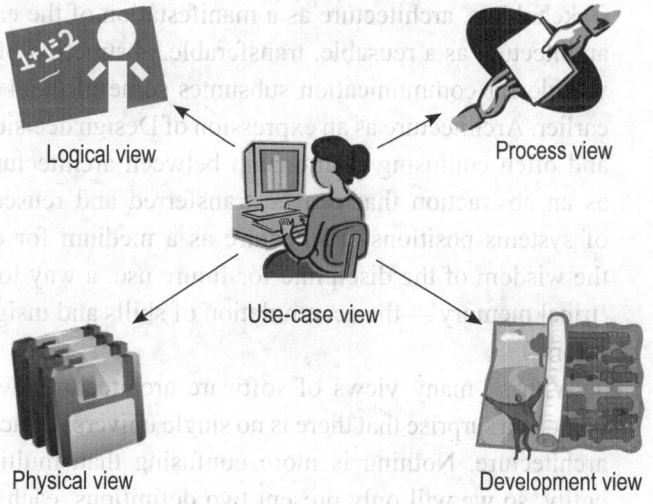

Logical view　　　　　Process view

Use-case view

Physical view　　　　　Development view

Figure 11.1　4+1 views of software architecture

to deliver value to users. It is vital for developers to see the software system as the user sees it and understand all the transactions between the user and the system.

The use of the word *view* in describing the five perspectives mentioned above is very insightful. Just as a vital concern of software architecture is to be able to view the software system as different stakeholders see it, software architecture itself can be viewed at different levels. The 4+1 views proposed in 1995 by no means represent a complete description of views. With the rapid penetration of software into different domains in the subsequent years, the necessity to observe software through other views has naturally arisen. For example, with the widespread use of Web-based systems for supporting critical infrastructure such as banking and finance, security has become a major concern for software developers. Thus specific systems may need to have a separate *security* view to highlight security-related issues. Additionally, the latest trends of outsourcing and offshore development have given rise to a new culture of distributed development, where customers, users, developers, and other stakeholders of software systems are spread out across continents. A *distribution* view may be necessary to specify the issues related to such a development scenario.

In the next section, we will briefly review the different ways software architecture as a whole can be viewed and give some definitions.

11.3 VIEWS AND DEFINITIONS OF SOFTWARE ARCHITECTURE

As stated earlier, like any other interesting concept, software architecture can itself be viewed from different angles. Clements et al. (2001) see software architecture on three different levels: architecture as a vehicle for communication among stakeholders, architecture as a manifestation of the earliest Design decisions, and architecture as a reusable, transferable, abstraction of a system. Architecture as a vehicle for communication subsumes some of the 4+1 views we have discussed earlier. Architecture as an expression of Design decisions brings into focus the vital and often confusing relationship between architecture and Design. Architecture as an abstraction that can be transferred and reused across systems or classes of systems positions architecture as a medium for encapsulating and capturing the wisdom of the discipline for future use: a way to share what Booch calls the 'tribal memory'—the accumulation of skills and insights over countless trials and errors.

With so many views of software architecture swirling around, it should not come as a surprise that there is no single universally accepted definition of software architecture. Nothing is more confusing than multiple definitions of the same entity, so we will only present two definitions, each bringing out a few essential characteristics of software architecture. According to Bass et al. (2003):

'The software architecture of a program or computing system is the structure or structures of the system, which comprise software elements, the externally visible properties of those elements, and the relationships among them.'

Rumbaugh et al. (2005) define software architecture as follows:

'The organizational structure of a system, including its decomposition into parts, their connectivity, interaction mechanisms, and the guiding principles that inform the Design of a system.'

Both these definitions have things in common and those that are not. Both talk about architecture as a way of combining parts of a system to form a cohesive whole, which is indeed the most important criterion of good architecture. When looking at the Taj Mahal, we do not see individual blocks of stone, not even the domes and the minarets by themselves, but we see the monument in totality. Good architecture *seamlessly* connects parts into the whole and does it in such a way that the sum of the parts is greater than the whole. The first definition also mentions 'externally visible properties', which blends with our initial observation that we readily recognize good architecture when we see it. The second definition points to the vital link between architecture and Design, emphasizing that architecture gives us the principles by which a system is designed. The second definition also mentions interaction mechanisms, which is the 'glue' by which a software system is held together. As we will see in the forthcoming chapters, software Design can be regarded as responsibility delegation amongst components, which interact to collectively deliver the functionality of the system within non-functional parameters.

We will discuss how architecture differs from Design a little later. Before that, we need to understand why architecture is essential for large-scale software systems.

11.4 NEED FOR ARCHITECTURE IN LARGE-SCALE SOFTWARE SYSTEMS

Software architecture is not an esoteric concept far removed from the engineering of software. On the contrary, it has grown out of the myriad challenges that software engineering as a discipline has faced over the years. Increasingly, software has been called upon to model complex situations of the real world, solve larger problems, and provide support in more and more critical scenarios. The essence of software architecture can thus be best understood in large-scale systems.

For a typically small software system without real users, say a class project, it matters very little whether we think of architecture. The customary way to proceed with such a project is often called CABTAB (code-a-bit-test-a-bit). We get into coding from the word go, make changes as necessary, and hopefully finish and turn in the project before the deadline. Design decisions in such cases are usually trivial and, more often than not, made for us in the problem statement itself: Write a program in Java that uses recursion to calculate the first ten numbers of the Fibonacci series. These kind of projects help us learn programming better, but they do not reflect the real world. (See Chapter 4 for a discussion of CABTAB.)

In the real world, software engineers are given a problem and asked to solve it within constraints of time and money. As has been proved from time immemorial, the best way to solve a big problem is to break it down into smaller problems. So, the first task when engineering a large-scale software system is to think in terms of sub-systems and/or components that will help us solve the sub-problems. Assuming that we have broken the problem into bits and every bit is being taken care of by some part of the system, a new question arises: How do we combine bits of the solution back into the whole solution? After all, the user is interested in the solution to the whole problem he gave us in the first place. So, there needs be a way for the subsystems to communicate and combine, that is interact, so that each one's contribution fits into the other like a jigsaw puzzle to give us the whole solution.

Whenever we operate under constraints of time and money, it helps to have help, and sometimes we cannot succeed if we do not have help. In the engineering context, this translates into an ability to use wisdom that others have gained in their successes and failures while trying to do something similar. But how do we glean such knowledge from perfect strangers, those we have never met, and will most likely never meet, and those who may literally speak a different language? It is essential that we have a common medium to transmit such ideas from one software engineer to another.

Note how we have been using terms like *subsystems, interactions, parts, whole, reusing wisdom,* etc. in the last few paragraphs. These very words have appeared when we explained and defined software architecture in the earlier sections. In small systems, the stakes are small enough for these factors not to explicitly matter. But for large systems, shying away from these basic decisions at the beginning is a recipe for much pain towards the end. Whether the Taj Mahal would be built using white marble or red bricks was a decision that was taken at the very conception of the monument. Architecture lets us take such decisions from a balanced point of view while incorporating the discipline's best practices. Software architecture was born out of pain and necessity—the pain of failing frequently while building large and complex systems, and the necessity to ensure that we eventually and consistently succeed.

Architecture defines our broad horizons, and sets us up for more detailed thinking, which is the domain of Design. It is very easy to confuse architecture with Design, and the confusion neither helps the practice of architecture nor Design. We take up this topic in the next section.

11.5 HOW ARCHITECTURE DIFFERS FROM DESIGN

There is no definite line that separates architecture and Design. But understanding the difference between the two is vital in the development of large and complex software systems. From the preceding discussion about software architecture, the following points stand out:

- Architecture relates to taking some fundamental decisions about the structure of subsystems, components, and their interactions that together make up the entire system.
- Architecture relates to reusing some of the wisdom that other people have gathered in working with similar systems.

Architecture is not concerned about what goes on inside the subsystems, how individual components deliver their functionality, or how particular algorithms perform specific tasks. And these are precisely what Design is concerned about. As we discuss in the forthcoming chapters, *high-level* and *low-level* Designs respectively involve deciding about Design issues at higher and lower levels of abstraction. High-level Design at its highest level of abstraction may merge into some of the architectural decisions. Yet Design and architecture have very different *intents*.

To illustrate the difference between architecture and Design, let us use an example. We have a credit card company that needs to provide its cardholders the facility of checking their transaction details over the Internet. Based on this single sentence expression of the Requirements (in the real world, initial Requirements may be even more cryptic; they often come as sentence fragments) how do we approach the problem from an architectural as well as Design point of view?

What are the choices of getting the data to the cardholder's computer? One option might be to develop a client application that will be installed on the cardholder's computer, to download transaction data from the company's servers. Another might be to develop a Web-based application giving online access to cardholder's data. Both options have their pros and cons, and this is certainly an architectural decision. Assuming we decide one way, subsequent Design choices would involve selecting a particular programming language, and a database. Both architectural decisions and Design choices involve layers of subtleties. Note how we use the word 'decision' for architecture and 'choice' for Design. 'Decision' differs from 'choice' in scale and scope, and this is an important aspect of the difference between architecture and Design. The Case Study section discusses in detail how these decisions and choices are made in this particular example. With this background, we will now discuss architectural patterns.

11.6 ARCHITECTURAL PATTERNS

Patterns are a fascinating topic in all disciplines that require systematic problem solving, and software engineering is no exception. We are often able to relate to the idea of a pattern in a quick, intuitive way. For example, a widely recommended strategy for taking written examinations is to attempt those questions first whose answers are best known. This is a pattern at its simplest: a general problem, with a suggested solution, in a specific context. A pattern embodies the wisdom of those who have faced a situation before, and devised a useful way of addressing it. Our daily lives are fraught with patterns of behaviour we have picked up from others, developed by ourselves, or have been imposed upon by authority or custom.

The notion of patterns is nothing new. One of the earliest attempts to organize and codify them in the context of the architecture of buildings and cities (civil and structural engineering) was made by Alexander in his book, *The Timeless Way of Building* [Alexander 1979]. Alexander writes:

'Each pattern is a three-part rule, which expresses a relation between a certain context, a problem, and a solution.'

Relation is a key term in the above definition—patterns connect problems, solutions, and contexts through relationships. A pattern may not necessarily suggest a solution to a specific problem (although it may in some cases), it is more likely to address a *class* of problems. Alexander's ideas on patterns have been widely adopted by the software engineering community. There have been proponents of architectural patterns [Fowler 2003], Analysis patterns [Fowler 1996], and Design patterns [Gamma et al. 1995]. There is also criticism on too much preoccupation with patterns, which may lead to sweeping over-generalizations of problems, and overkill in their solutions [Brown et al. 2001].

Patterns can facilitate many architectural decisions and Design choices when used with discretion. The architectural patterns suggested by Fowler are classified under the heads: domain logic patterns, data source architectural patterns, object-relational behavioural patterns, object-relational structural patterns, object-relational metadata mapping patterns, Web presentation patterns, distribution patterns, offline concurrency patterns, session state patterns, and base patterns. Each category includes the discussion of several individual patterns [Fowler 1996].

To get a flavour of the nature and usefulness of patterns, we will now discuss the model-view-controller (MVC) pattern in some detail. MVC was first proposed by Reenskaug in 1978–79. According to Reenskaug, MVC was conceived as a general solution to the problem of users controlling a large and complex data set (http://folk.uio.no/trygver/themes/mvc/mvc-index.html).

The MVC is among the most widely used architectural patterns in Web development, where we have a presentation layer, often called the *graphical user interface* (GUI), a middle layer to handle *business logic* for fetching and manipulating data, and a back-end database for storing data. Figure 11.2 shows a typical MVC scenario. (Notice how one feature of this pattern is that all interactions between view and model are coordinated by the controller.) Deciding on MVC is often easy for a particular system: The tricky part is to map model, view, and controller individually to the underlying sub-systems and components.

Irrespective of the actual Implementation, MVC can express an architectural style with the elements outlined above. We will see a particular mapping of model, view, and controller in the Case Study section. But the most significant point to note here is that an architectural pattern like MVC gives us a way to understand, document, and communicate ideas about architectural styles and idioms across a variety of applications, which may differ to a large extent in their specifics but remain aligned in the broad purpose and method of delivering a solution. The study of architectural styles has been pioneered to a large extent by Shaw [1995].

So, architectural patterns can be taken as generalized strategies for addressing specific situations while building large-scale software systems. In the discussion

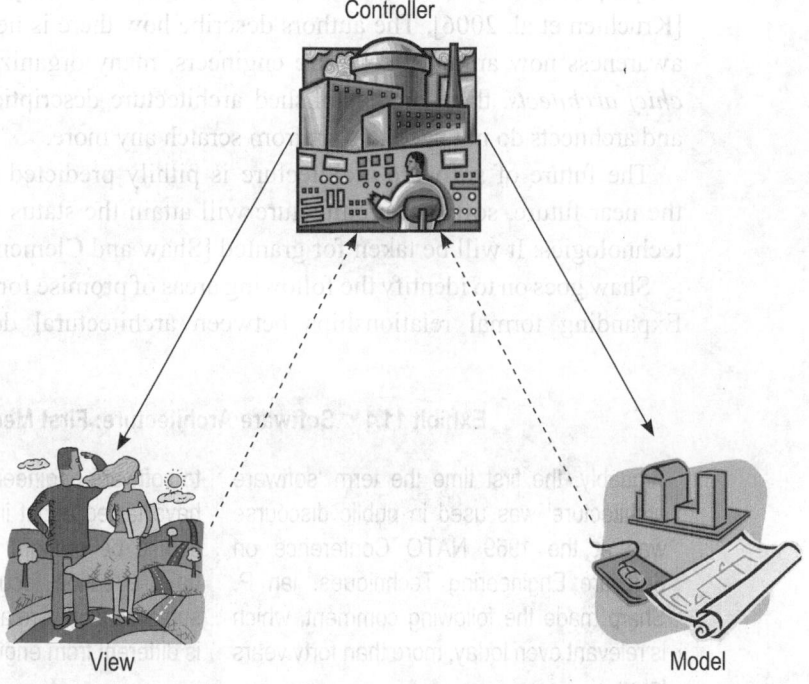

Figure 11.2 Model-view-controller (MVC) pattern

of this chapter, we have usually qualified architecture with 'large-scale software systems'. To reiterate the point we have already made, the concerns of building software that architecture addresses involve an element of scale. It would be of no help, indeed may also harm, if we get too occupied with the architecture of a program of only a few hundred lines of code. As we are able to appreciate architecture in buildings which are notably large, software architecture too, needs to be on a worthwhile scale.

We have so far studied software architecture from a historical perspective. What is the direction software architecture is likely to take in the coming years? In the next section, we briefly review this topic.

11.7 FUTURE OF SOFTWARE ARCHITECTURE

The 2006 March/April issue of *IEEE Software* (Vol 23, Issue 2) was again based on software architecture—as a tenth anniversary tribute to the earlier mentioned 1995 issue which brought many of the architectural ideas into mainstream software engineering. Ten years after introducing the 4+1 model of architectural views, Kruchten et al. describe software architecture as 'the structure and organization by which modern system components and subsystems interact to form systems, and the properties of systems that can best be designed and analysed at the system level' [Kruchten et al. 2006]. The authors describe how there is heightened architectural awareness now amongst software engineers, many organizations have roles like *chief architects*, there are established architecture description languages (ADLs), and architects do not need to start from scratch any more.

The future of software architecture is pithily predicted in Shaw's words, 'In the near future, software architecture will attain the status of all truly successful technologies: It will be taken for granted [Shaw and Clements 2006].'

Shaw goes on to identify the following areas of promise for software architecture: Expanding formal relationships between architectural decisions and quality,

---

Exhibit 11.1   Software Architecture: First Mention

Arguably, the first time the term 'software architecture' was used in public discourse was at the 1969 NATO Conference on Software Engineering Techniques. Ian P. Sharp made the following comment, which is relevant even today, more than forty years later:

'I think that we have something in addition to software engineering: something that we have talked about in small ways but which should be brought out into the open and have attention focused on it. This is the subject of software architecture. Architecture is different from engineering.'

[Kruchten et al. 2006]

finding the language to represent architectures and ways to assure architecture and code are in synchronism, formulating approaches to align software testing with architecture, organizing architectural knowledge so that they can be easily accessed and referenced, developing architectural support for systems that have to adapt to the continual change in resources and user Requirements [Shaw and Clements 2006].

Thus, in spite of the great strides made in understanding and applying software architecture, a lot remains to be done for architectural ideas to permeate software development in the large.

As we just saw the future of software architecture, it is interesting to look at its first mention in Exhibit 11.1.

CASE STUDY

We one again return to Preeti, our protagonist. Preeti is now the architect for a team of software engineers working with a large credit card company that wants to give its cardholders the facility of checking their transaction details online. What would be a suitable architecture for such a system? Developing a suitable architectural style for a system starts primarily with questions, rather than answers. Here are some of the issues that need to be clarified before Preeti can proceed:

- What is the approximate number of cardholders who will be given this new facility?
- What is the average *transaction density* (that is, transactions per unit time) for cardholders? Are there some segments of cardholders that have exceptionally high transaction densities?
- What are the exact online facilities sought to be given to the cardholders? Will they only be able to view their transaction details, or can they dispute a particular transaction also?

The list of such questions is long, and as each one is addressed, new ones come up. And to make life more difficult for software engineers, very often the customer has very little knowledge about these issues. Let us for the time being assume that we have definite answers to these questions. We will now see how they influence some architectural decisions.

From discussions with the customer, Preeti has the following information: There are around half a million cardholders, the online facility needs to be provided in a phased manner, starting with a pilot group of about ten thousand and scaling up to all the cardholders in about three years. The number of cardholders is also expected to increase with a yearly rate of ten percent. Close to ninety percent of the cardholders have less than 500 transactions per month, but the remaining ten percent. The corporate cardholders, have close to 10,000 transactions per month.

Initially, cardholders should be able to check their transaction details for the past three months as well as search and sort the data. Later, facilities for disputing particular transactions, applying for credit line increase and payments of card dues have to be supported.

From the above information, some architectural decisions can be made. Given the nature of the system, three-tier architecture readily suggests itself: a Web-based user interface tier, a middle tier for business logic, and a back-end database tier. Translating this to a J2EE infrastructure on a MVC pattern, Java server pages (JSP) can be used for the user interface (view), servlets and enterprise Java beans (EJB) for the middle tier (controller), and a Java database connectivity (JDBC) supported database in the database tier (model). This architecture will support the needs of those with less than 500 transactions per month, although with such a high number of cardholders, the level of concurrent access will be high, and care needs to be taken so that one cardholder does not see another's data!

But it is hardly if ever the case, one architectural solution fits all. Our case at hand is no exception. A Web-based system is good for quick and convenient access to data, but it is often not suitable for sifting through very large volumes of data, as the speed of such access depends on factors such as network connectivity, etc. For the cardholders with around 10,000 transactions per month, it would be very difficult for a Web-based system to support access to data with the facilities of rapid searching and sorting. One option may be to generate their monthly reports offline and send it via e-mail, another might be to install a piece of software on those cardholders' computers, which will periodically access the server and download their transaction data for frequent local access. To speed up searching and sorting of large volumes of data, it is important to minimize fetching data from the back-end. So the MVC architectural paradigm is *not* the only solution is this case; along with it other architectural concerns have to be addressed. Figure 11.3 highlights the MVC-based solution.

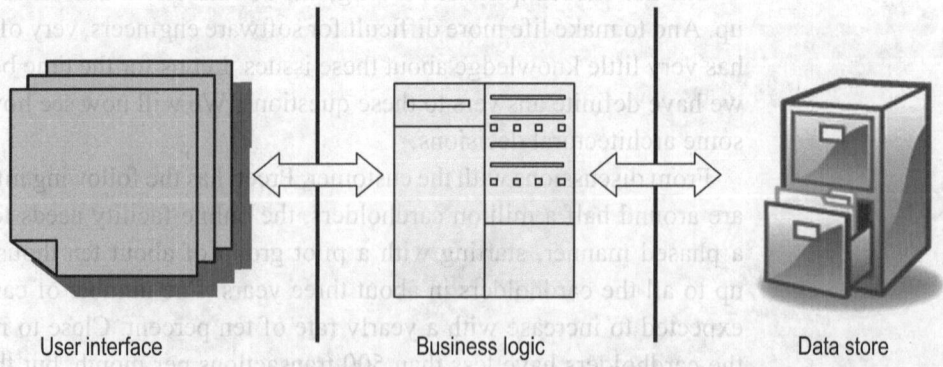

User interface   Business logic   Data store

Figure 11.3   One probable architectural solution for the case study system

Another issue to consider in cases such as this is *integration with legacy applications*. Legacy applications can be described as computer programs (sometimes, whole systems) from another era, written in a language or supported on infrastructure different from the current development paradigms. Legacy applications are almost everywhere, and they are usually doing their jobs very well, and remain remarkably resilient to change. This was discovered to our great surprise during the Y2K confusion. Y2K, a shorthand notation for the year 2000, raised fears that computer programs written decades earlier and supporting vital business applications even in 2000, would misinterpret the year 2000 as 1900 (causing serious errors), since they only had two digits in their internal data representation for the year field. This led to a great amount to resources being invested in correcting this anomaly, and the first major wave of outsourcing of information technology operations came to India. Returning to this particular case, the credit card company must have existing servers and applications that already hold the cardholders' data; how will they be integrated with the new Web-based system? These are some of the open questions that need to be answered as the system matures. As this Case Study shows, architecture does not answer all questions at once, but sets us up to ask the right questions from the beginning.

Summary And Take-aways

In this chapter, we introduced software architecture and reviewed some of its key concepts. Software architecture is an evolving discipline, and its importance is best understood through experience. Given below is a summary of our discussion:

- Software architecture involves decisions about the structure, subsystems, and components of software systems and the mechanisms of interaction between them.
- The 4+1 views of software architecture give a set of perspectives: logical, process, physical, development, and use-case to visualize different stake-holders' interests in the software system. There may be other views of a particular system based on concerns such as security, distributed development etc.
- Software architecture as a discipline can also be seen from different angles: as a strategy for organizing a system's constituents, as a high-level abstraction of a system's Design, as a vehicle for communicating fundamental ideas about a system's development.
- Software architecture is not software Design, although there is no clear line where the former ends and the latter starts. Design relates to choices about how specific functionality is to be delivered, architecture is more about decisions regarding the system's structure.

- Architectural patterns encapsulate tested strategies in a problem-solution-context format. Model-view-controller (MVC) is a widely used architectural pattern.
- In future, with software systems becoming larger and more complex, software architecture ideas will need to address deeper issues of delivering sound solutions to user needs.

WHERE TO LOOK FOR MORE

- Grady Booch, one of the 'three amigos' who formalized the unified modelling language (UML), has started an initiative to compile a *Handbook of Software Architecture* to codify 'the architecture of a large collection of interesting software-intensive systems, presenting them in a manner that exposes their essential patterns and that permits comparisons across domains and architectural styles'. Details of the handbook's development are available at http://www.booch.com/architecture/index.jsp.
- Carnegie Mellon University's Software Engineering Institute (SEI) has interesting information related to software architecture at http://www.sei.cmu.edu/architecture/.

EXERCISES

Review Questions

Review Questions test your understanding of the key concepts presented in this chapter.

1. The 4+1 view of architecture includes
   (a) logical, process, physical, deployment, and use-case views
   (b) logical, process, physical, development, and use-case views
   (c) logical, process, procedural, development, and use-case views
   (d) logical, process, procedural, scenario, and use-case views

2. Software architecture involves
   (a) deciding on the structures and interaction of subsystems
   (b) choosing specific algorithms
   (c) both (a) and (b)
   (d) none of the above

3. Software architecture and software Design are
   (a) the same
   (b) totally different
   (c) sometimes same, sometimes different
   (d) related but different

4. Architectural patterns
   (a) are detailed plans
   (b) are aesthetic features of architecture
   (c) are three-part rules that express a relation between a certain context, a problem, and a solution
   (d) form a new programming language

5. Organized study and application of the ideas of software architecture
   (a) have been around since 1969
   (b) started from the 1990s
   (c) is yet to start
   (d) will start in the near future

6. Model-view-controller (MVC) is
  (a) an architectural pattern
  (b) one among many views of software architecture
  (c) a model for software development
  (d) a technique to control software development

7. Which of the following statements about architectural decisions and Design choices is correct?
  (a) Both are made hand in hand.
  (b) The former should happen before the latter.
  (c) The latter should happen before the former.
  (d) There is no prescribed order of precedence.

8. The idea of patterns came into software from
  (a) electrical engineering
  (b) medicine
  (c) aerodynamics
  (d) civil engineering and architecture

9. Which of the following statements about the selection of a programming language for a software system is true?
  (a) It is an architectural decision.
  (b) It is a Design choice.
  (c) It depends on the scale of the system.
  (d) None of the above is true.

10. 'Architecture is important in large-scale systems'.

  (a) It is true
  (b) It is false
  (c) It is sometimes true, sometimes false
  (d) Architecture is not relevant for large-scale systems

Reflective Questions

Reflective Questions require you to think more deeply about some of the ideas and come up with your own interpretations and answers.

1. Before we knew about software architecture, 'architecture' was associated primarily with physical shapes and forms of buildings. Point out three similarities and three dissimilarities between conventional architecture and software architecture.

2. In the example discussed under Case Study, what are some of the other questions you might want to ask to specify the system's architecture in more detail? How are the answers to such questions likely to affect architectural decision?

3. You are an architect of a project to develop an online result notification system for your university. Students will be able to log in to the system with their unique userid and password to check their examination results and print their transcripts. What are the architectural decisions you will be required to make?

REFERENCES

Alexander, C. (1979), *The Timeless Way of Building,* Oxford University Press.

Bass, L., P. Clements, and R. Kazman (2003), *Software Architecture in Practice*, 2nd ed., Addison-Wesley Professional.

Brown, W.J., R.C. Malveau, H.W.S. McCormick, and T.J. Mowbray (2001), *AntiPatterns:*

Refactoring Software, Architectures, and Projects in Crisis, John Wiley and Sons.

Clements, P., R. Kazman, and M. Klein (2001), *Evaluating Software Architectures: Methods and Case Studies*, Addison-Wesley Professional.

Fowler, M. (1996), *Analysis Patterns: Reusable Object Models*, Addison-Wesley.

Fowler, M. (2003), *Patterns of Enterprise Application Architecture*, Addison-Wesley.

Gamma, E., R. Helm, R. Johnson, and J. Vlissides (1995), *Design Patterns: Elements of Reusable Object-Oriented Software*, Addison-Wesley.

Kruchten, P. (1995), The 4+1 View Model of Architecture, *IEEE Software.*, 12(6): 42–50.

Kruchten, P., H. Obbink, and J. Stafford (2006), The Past, Present, and Future for Software Architecture. *IEEE Software.*, 23(2):22–30.

Rumbaugh, J., I. Jacobson, and G. Booch (2005), *The Unified Modeling Language Reference Manual,* 2nd ed., Addison-Wesley.

Shaw, M. (1995), Comparing Architectural Design Styles, *IEEE Software.*, 12(6):27–41.

Shaw, M. and P. Clements (2006), The Golden Age of Software Architecture, *IEEE Software.*, 23(2):31–39.

Paradigms of Software Development

12.1 MOTIVATION

What do we expect from a house? We expect shelter from sun, snow, rain, and hail, so that our living, working, meeting, etc. are not left to the mercy of the elements. This expectation from a house remains the same across ages, cultures, and places. But how do we go about *building* a house? There are many ways and many things to build with—stone, wood, mud, steel, concrete, and ice. Even though the basic utility of a house, i.e. to provide shelter, as well as the demands it has to meet (from laws of physics, and from architectural and aesthetic needs) do not change, how a particular house will be built gets decided by a number of factors. With experience, one approach comes to be seen as more apt for a certain kind of house—a skyscraper can hardly be built with the same material and techniques as a mud-house.

Similarly, there are many ways to go about building software. We will call these different ways of building software as *paradigms of software development*. 'Paradigm' is a widely used (and often misused!) word. It is formally defined as:

1. A typical example, pattern, or model of something
2. A conceptual model underlying the theories and practice of a scientific subject (http://www.askoxford.com/concise_oed/paradigm?view=uk)

In this chapter, we will use *paradigm* in a sense that draws upon both of the above definitions, as a typical model underlying the theory and practice of how software systems can be conceived and built.

This is predominantly a conceptual chapter. We will introduce some broad ideas that have much influence on how we can create effective, resilient, and beautiful software. The discussion in this chapter is thus rather abstract; and as abstract issues are best understood through concrete examples, we will start with a metaphor.

12.2 A COOKING METAPHOR

After graduating from engineering college, when we enter the real world, reality hits us in many ways. Rather unexpectedly, a common reality hits us not from the demands of hard work or original ideas, but cooking for ourselves. Out into the world, in a new city or even a new country, there is no home or hostel cooking to fall back upon. Many of us try to get by using recipes gleaned from some relative or downloaded from the Internet. A cooking recipe is a set of instructions, a step-by-step guide, starting with some ingredients and ending up with a cooked item. For someone new to cooking, this works reasonably well if the recipe is detailed enough and the food item not too ambitious. Now, let us assume that our new engineer-in-the-world wants to have a party; he or she has invited several friends and wants to cook a multi-course dinner. We also assume there are recipes for every item which the new engineer diligently follows. What are the concerns now for our friend? While cooking for a party he or she has to keep track of many more bowls, pots, and pans. It may be disastrous to pour the ingredient for one menu item into the another, or worse still, mix two finished items in one bowl. Things may be a bit easier with more help, so the new engineer calls friends for collaboration. Even if the friends' cooking expertise is not significant, they can lend a hand in tasks like, cutting vegetables, etc. Hopefully, our new engineer-in-the-world ends up serving a reasonably palatable fare. But surely he or she comes away with an appreciation of how increasing complexity affects even an everyday task like cooking; one can only wonder at how a ten- course dinner for 100 guests ever gets cooked!

But it does, as do more expansive meals for larger number of people. The key is to understand and address the complexity we just mentioned. We will return to this cooking metaphor several times in this chapter.

12.3 CASE FOR SOFTWARE'S COMPLEXITY

Brooks (1995) said in his celebrated essay, *No Silver Bullet*, 'The complexity of software is an essential property, not an accidental one'. *Essential* vis-à-vis *accidental*, in this context, points to the fact that complexity makes software what it is, and is not an auxiliary or adventitious aspect of software.

What is *complexity* in the software context? In common usage, we take 'simple' to mean something we can readily comprehend and/or manipulate, and 'complex' as the opposite of simple. A complex system is made up of simple components but has the significant property of being *far greater than the sum of its parts*. So, merely studying and understanding the behaviour of the parts of a complex system will not guarantee a grasp of the entire system. A ready example of a complex system is the human body—it is made up of cells that combine together to form tissues and organs, which in turn collaborate to create a functioning organism of almost incomprehensible intricacy. Complexity is immanent in many domains; from the structure of plants and animals, to the structure of matter and the structure of social organizations [Booch et al. 2007]. *Complexity theory* is a scientific discipline encompassing many fields [Wessels 2006], [Capra 1997], and [Waldrop 1992].

Not all software systems are complex. Programs that we write when we are first learning to program a computer, or those written by a single programmer for his or her own use, are usually simple. This is not to suggest such programs are not insightful or useful; however, these are seldom used widely, or subjected to widespread modification and reuse. But software that supports large-scale enterprise systems, upon whose success or failure lives and livelihoods depend, are certainly complex. The clear symptom of such complexity is that it is impossible for a single individual to comprehend its Design in full detail and subtlety [Booch et al. 2007]. The paradigms of software development are tools to taming software's inherent complexity.

Software's complexity can be traced to many of its characteristics. We underscore some important points mentioned earlier.

- As reflected upon by Datta (2007), when a building is commissioned as an office, hospital, school, residence or any other purpose, users have a clear idea of what the finished product needs to be, how they will use it, even how it should not be used. This subliminal understanding comes from the fact that human civilization has been using buildings for centuries, if not millennia. So, the *cognitive gap* between what they want engineered and what can be engineered for them, is manageably narrow. But it has only been a few decades, since when people who do not build software themselves have been using software systems widely. So not knowing what they want is a common characteristic of the common software consumer. This leads to frequent and often wide changes in the Requirements of a software system, with users realizing what they *do not* want from a software system only when they start using the system. This situation plays an important role in enhancing the complexity of software.

- Booch et al. (2007) identify one of the fundamental goals of software development as shielding users from the inherent complexity of software, and presenting a relatively simple interface through which users can get their work done [Booch et al. 2007]. This deliberate hiding of complexity from users pushes much of the complexity into the developer's domain. Along with this the explosion in the size of software systems, as well as the number of people building them, have contributed significantly to the complexity of software.

- A curious and unique property of software, that has been called its *plasticity* [Datta, 2007] in the sense of flexibility [Maier and Rechtin 2000] makes it seem very easy to modify software to do something widely different from its original purpose. Changing one line of code in a thousand lines can cause the behaviour of a software system to change drastically. One of the factors that makes software so complex is this inherent plasticity.

- Booch et al. (2007) point to another interesting and often overlooked aspect of software systems that adds to its complexity. A software system consists of many variables and possibly multiple threads of control. This leads to a very large number of possible states. Such 'combinatorial explosion' [Booch et al. 2007] can result in inconceivably many factors influencing the effects of change in one part of the system on other parts. There is often no way to foresee such effects.

Now, given that software is inherently complex, how can we create software systems that work for us with reasonable consistency? This is the central question that every software development paradigm needs to answer.

One of the main reasons we find complexity so difficult to handle is the level of detail—there are just *too many* factors in a complex system that affect the system's behaviour for the human mind to keep track of. Miller (1956) established through researches in psychology that the human mind is capable of keeping track of maximum nine and minimum five (that is seven plus or minus two) distinct pieces of information at any point of time. This limitation of our *channel capacity* [Booch et al. 2007] is linked to the extent of our short-term memory as well as the speed of processing information, which is about five seconds to grasp a new piece of information [Simon 1996]. So, any organized approach to tame complexity must try to work around these constraints. Figure 12.1 highlights the channel capacity of the human mind.

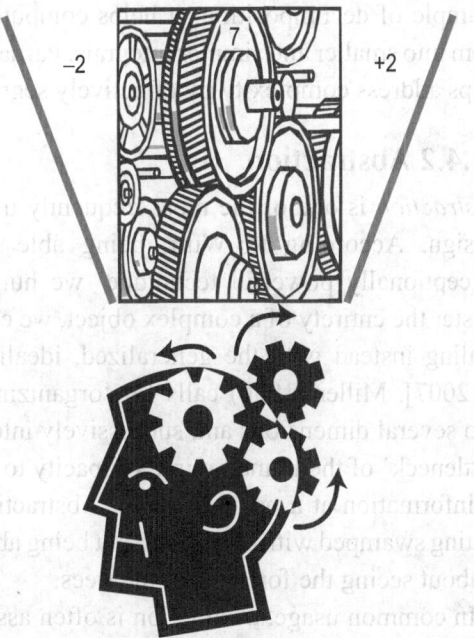

Figure 12.1   Channel capacity of human mind

In the next section, we discuss some of the major strategies for addressing complexity in software systems.

12.4 STRATEGIES FOR ADDRESSING COMPLEXITY IN SOFTWARE SYSTEMS

As the complexity of software cannot be wished away (remember, it is an essential property, not an accidental one!) and we have to build software to serve our needs, we need ways to systematically address software's complexity. Booch et al. (2007) identify three major planks of attack—*decomposition, abstraction,* and *hierarchy.*

12.4.1 Decomposition

Divide and rule is a dictum we have often heard about, especially in the context of India's freedom struggle. *Divide et impera* (divide and rule) is a technique of mastering complexity in vogue since ancient times (Dijkstra 1979). The key idea is to break down a system into smaller and more manageable parts, so that the channel capacity of our comprehension is not breached. Decomposition is something we use very often in our lives, often subconsciously. At the beginning of each school year, the girth of textbooks looks rather daunting; but breaking them down into smaller bits across lessons and periods, makes their conquest easier. The technique of *integration by parts,* which we learnt while studying integral calculus is also an example of decomposition; it helps compute complex integrals by decomposing them into smaller and simpler integrals. Parnas (1985) has noted that decomposition helps address complexity by effectively segregating a system's state space.

12.4.2 Abstraction

Abstraction is one of the most frequently used words in the context of software Design. According to Wulf, being able to abstract from complexity is an 'exceptionally powerful technique' we humans are endowed with. 'Unable to master the entirety of a complex object, we choose to ignore its inessential details, dealing instead with the generalized, idealized model of the object' [Booch et al. 2007]. Miller (1956) calls for 'organizing the stimulus input simultaneously into several dimensions and successively into chunks' to break the 'informational bottleneck' of the human mind's capacity to process seven plus minus two pieces of information at a point of time'. Abstraction is what prevents our minds from getting swamped with details and not being able to make sense of them. Abstraction is about seeing the forest from the trees.

In common usage, abstraction is often associated with generalization. There is indeed an element of generalization in abstraction. A part of abstraction is about making conjectures on larger issues from small observations. But abstraction is more than just generalization. Abstraction is a way of looking at circumstances so that essential ideas are highlighted and inessential ones are hidden. In our lives, we are abstracting all the time about nearly everything. Abstraction is built into the human mind to see coherent patterns out of the constant influx of information through the senses. Abstraction is especially useful when taking decisions. Let us consider an example. When I shop online for flight tickets, my concern about the price is paramount, and then at the airport counter I am usually very bothered about getting an aisle seat. At the gate, just before boarding, my main worry is whether the airplane has four engines, which I think, is the minimum safeguard for a long flight! At each stage of the process from buying tickets to boarding, my primary focus has shifted; some details have blurred to give some other factor main clarity. Abstraction is about this shifting of focus as we tackle a situation in steps [Datta, 2007].

On a grander plane, an example of abstraction would be the *viswaroopdarshan* Krishna endowed Arjuna with just before the *Battle of Kurukshetra* began in the Mahabharata. Viswaroopdarshan does not individually exhibit each aspect of creation, but *abstracts* the origin and destiny of all beings and all circumstances and shows all to be within Krishna's power and knowledge.

How well a software development paradigm facilitates abstraction and how well it helps model the problem domain in terms of the world around us, goes a long way in determining the usefulness of the paradigm.

12.4.3 Hierarchies

Compared to decomposition and abstraction, the idea of hierarchies is slightly more involved. Whenever we examine something new or something that seems to be complex, we try to see what common characteristic it shares with other things we know, or if it is part of something we have seen earlier. We can study a car by studying its wheels, its engine, its passenger compartment etc. separately—this conceptual break down of the car can be said to be according to a *part-of* hierarchy. On the other hand, we can say that the car is an automobile, which is a locomotion device with wheels, engine; or the car's engine is an internal combustion engine or a compression ignition engine. This represents the *is-a* hierarchy perspective. Booch et al. (2007) call the 'is a' hierarchy a *class structure* and the 'part-of' hierarchy an *object structure* (these may not readily map to classes and objects in object-oriented programming).

Recognizing 'is-a' and 'part-of' hierarchies is a very important step in understanding complex systems. It frees us from the drudgery of having to study each individual element by identifying those characteristics the element may share with other elements and how smaller units make up a larger unit of interest. Along with decomposition and abstraction, this offers a mechanism to document and communicate our perspectives about a complex system.

12.5 DIFFERENT SOFTWARE DEVELOPMENT PARADIGMS

So far, we have made a case for the complexity of software and outlined effective strategies for addressing complexity. Now we arrive at the central topic of this chapter—how specific software development paradigms approach software development.

12.5.1 Algorithmic Paradigm

An *algorithm* is defined as 'a process or set of rules used in calculations or other problem-solving operations' (http://www.askoxford.com/concise-oed/algorithm?view=uk). The algorithmic paradigm of software development sees a software system as the realization of an algorithm, 'wherein each module in the system denotes a major step in some overall process' [Booch 2007]. The algorithmic paradigm is referred to by several names, some of which have subtle variations in meanings: *structured, top-down, procedural, imperative, functional*. Instead of worrying about the similarities and dissimilarities of these variants, we will focus on understanding the overall spirit of the algorithmic paradigm.

The value from running a computer program comes primarily from its manipulation of some data of interest to us (the output), based on the program's logic as well as the data we feed into the program (the input). The central idea of the algorithmic paradigm is to devise the steps that the program will execute, to transform the input data to the output data as per our needs. It has to be noted that in the algorithmic paradigm, data and operation are essentially disengaged from one another. In the cooking metaphor we gave earlier, an algorithm is analogous to the recipe for a particular item—a sequence of steps that will help us transform the 'ingredients' (input data) into 'cooked food' (output data). Figure 12.2 highlights the algorithmic paradigm.

Let us see the algorithmic paradigm applied to an example. We will revisit the same example again while discussing the object-oriented paradigm later.

From our knowledge of high school mathematics, we should be familiar with the notion of *complex numbers*. With reference to Figure 12.3, a complex number z can be expressed in the *rectangular* or *Cartesian form*: $z = x + iy$ or the *polar form*: $z = r(\cos \theta + i\sin \theta)$ or $z = re^{iq}$, where $i^2 = -1$, that is, $i = \sqrt{-1}$. The elements of the two forms are related by the expressions $x = r\cos \theta$, $y = r\sin \theta$ and $r = \sqrt{x^2 + y^2}$, $\theta = \arctan(y, x)$. Addition and subtraction of two complex numbers is easily done if they are expressed in the Cartesian form:

$$z = z_1 + z_2 = (a + ib) + (c + id) = (a + c) + i(b + d)$$
and $$z = z_1 - z_2 = (a + ib) - (c + id) = (a - c) + i(b - d) \qquad .$$

But multiplication and division of two complex numbers is more convenient if they are expressed in the polar form:

$$z = z_1 * z_2 = r_1 e^{i\theta_2} = r_1 r_2 e^{i(\theta_1 + \theta_2)}$$
and $$z = \frac{z_1}{z_2} = \frac{r_1 e^{i\theta_1}}{r_2 e^{2\theta_2}} = \frac{r_1}{r_2} e^{i(\theta_1 - \theta_2)} \qquad .$$

Now, we want to write a program that is able to add, subtract, multiply, or divide two complex numbers (z_1 and z_2). Let us call the program *ComplexPlay*. How should we Design it using the algorithmic paradigm? Of course, we need to think of an algorithm—a series of steps which are given below:

1. Accept the x coordinate of the first complex number (z_1) from the user and store it as x_1.
2. Accept the y coordinate of the first complex number (z_1) from the user and store it as y_1.

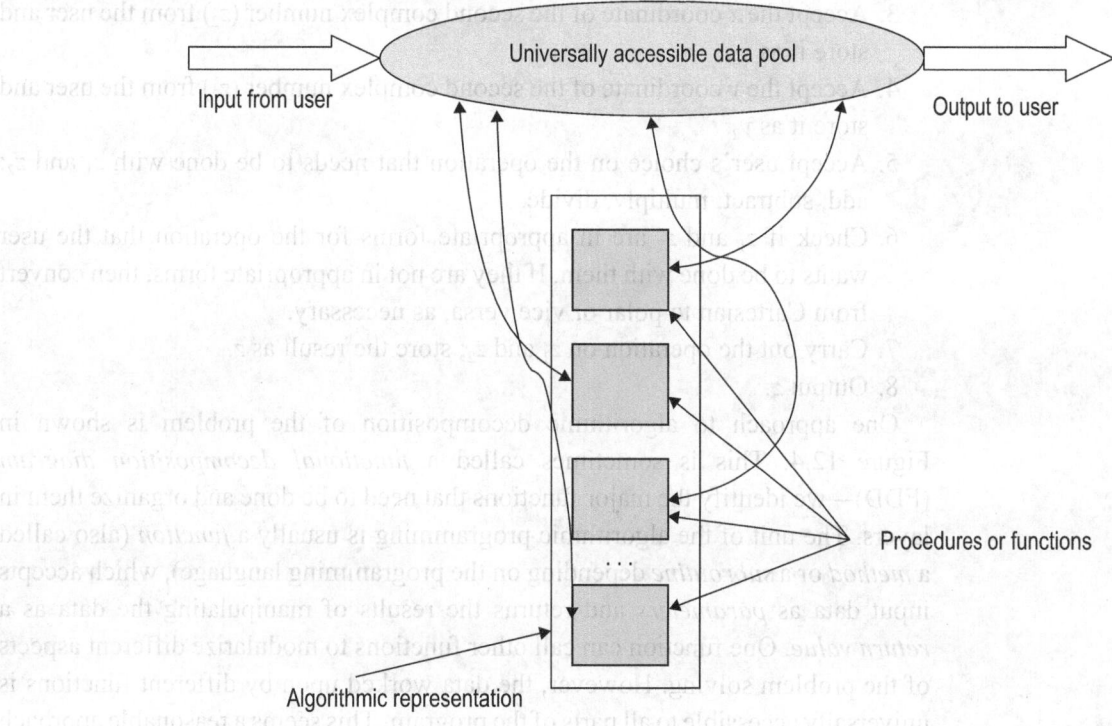

Figure 12.2 Idea of an algorithmic paradigm

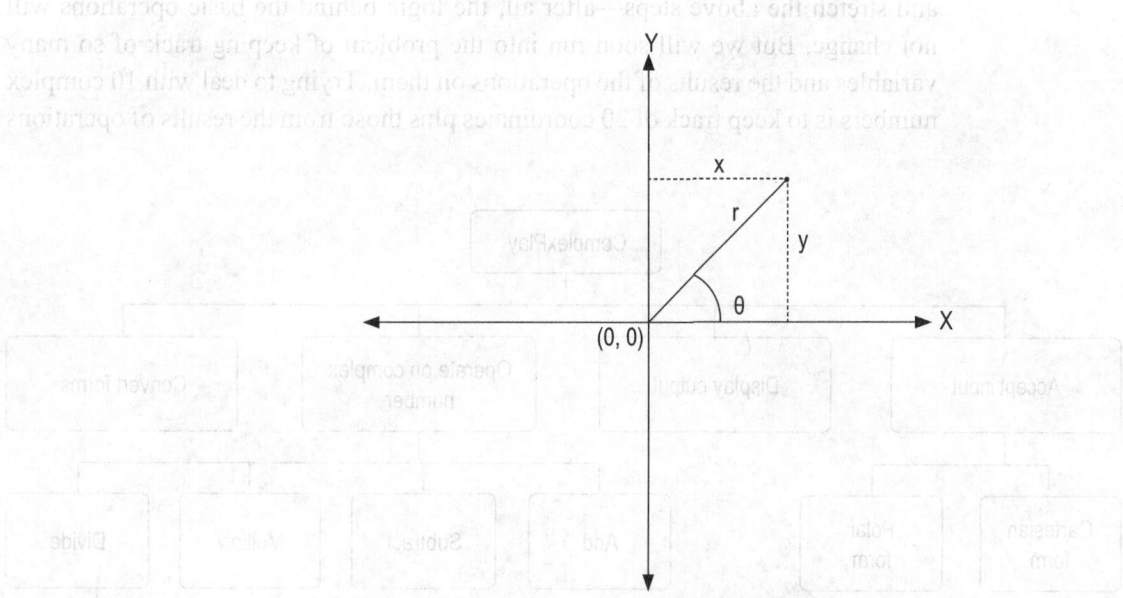

Figure 12.3 Cartesian and polar representations of a complex number

3. Accept the x coordinate of the second complex number (z_2) from the user and store it as x_2.

4. Accept the y coordinate of the second complex number (z_2) from the user and store it as y_2.

5. Accept user's choice on the operation that needs to be done with z_1 and z_2: add, subtract, multiply, divide.

6. Check if z_1 and z_2 are in appropriate forms for the operation that the user wants to be done with them. If they are not in appropriate forms, then convert from Cartesian to polar or vice versa, as necessary.

7. Carry out the operation on z_1 and z_2; store the result as z.

8. Output z.

One approach to algorithmic decomposition of the problem is shown in Figure 12.4. This is sometimes called a *functional decomposition diagram* (FDD)—we identify the major functions that need to be done and organize them in layers. The unit of the algorithmic programming is usually a *function* (also called a *method* or a *subroutine* depending on the programming language), which accepts input data as *parameters* and returns the results of manipulating the data as a *return value*. One function can call other functions to modularize different aspects of the problem solving. However, the data worked upon by different functions is universally accessible to all parts of the program. This seems a reasonable approach when we are dealing with operations on two, or just a *few* complex numbers. Now what happens if we want to play around with *many* complex numbers? We can try and stretch the above steps—after all, the logic behind the basic operations will not change. But we will soon run into the problem of keeping track of so many variables and the results of the operations on them. Trying to deal with 10 complex numbers is to keep track of 20 coordinates plus those from the results of operations

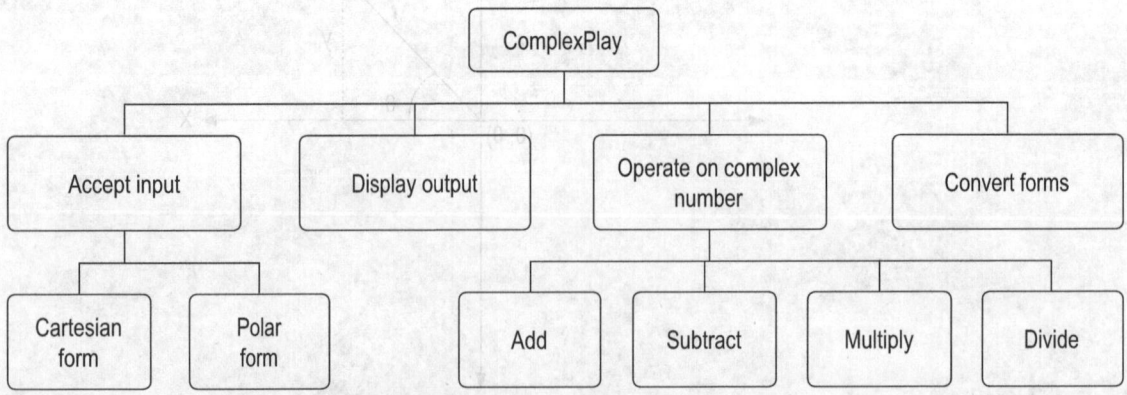

Figure 12.4 One way to algorithmic decomposition of the complex play program

on them. It is a sizeable nuisance, and dealing with 100 complex numbers can be disaster. If such a simple programming task, turns so messy just as we try to scale things up, one can imagine the roadblocks the algorithmic paradigm can run into when we face a complicated problem. This is evidently not something we can resolve by merely being more attentive to details (remember, Miller's 'seven plus minus two' dictum; we can only account for as much in our minds!). We need a new way of thinking.

12.5.2 Object-Oriented Paradigm

In the real world, entities are identified by information—*who they are*—as well as behaviour—*what they do*. A human being has a name, colour of eye, date of birth, gender; he or she can also walk, run, swim, read, write, etc. A bird has a type, colour, size, and it can fly, sing, etc. Object-oriented paradigm models software entities as *objects* on the lines of real-world entities, like a human or a bird, by synthesizing data and behaviour in a single unit of abstraction. In this paradigm, tasks are fulfilled by the *collaboration* between objects. Objects collaborate by exchanging *messages* amongst one another, a message is a request from the calling object to the called object to invoke a particular *method* on it, and a method is an encapsulation of a particular behaviour offered by the object. (A 'method' in object-oriented paradigm is very similar in intent and purpose to the 'function' as described earlier in the algorithmic paradigm. However, a method is an integral part of the object and cannot be seen in isolation from the object's data members, which are sometimes called *attributes* or *fields*.) The data and the behaviour *together* constitute an object, which facilitates the modelling of a corresponding real-world entity (like a man, bird, or car) or a thought construct (like a complex number). Ideally, an object should have the sole authority to manipulate its own data, to ensure accidental corruption of data does not occur. Accidental corruption of data in this context is akin to inadvertently mixing the ingredients of different bowls or pouring one curry onto another in the same pot, in our cooking metaphor.

In Figures 12.5(a) and 12.5(b), we use the object-oriented paradigm to visualize *ComplexPlay*. Note how the complex numbers are now *ComplexNumber* objects with the attributes *x, y, r, theta* and the methods *add*, *subtract*, *multiply*, *divide*, *cartesianToPolar, polarToCartesian*. Another object, which we call *Coordinator*, creates two complex numbers, *complex1* and *complex2* and calls methods on *complex1*.

The unified modelling language (UML) can be used as the notation for the object-oriented paradigm. Class diagram and sequence diagram [Figure 12.5(a) and 12.5(b)] are two of the modelling constructs of the UML. We will study these in more details in Chapters 13 and 14. Class diagram is relatively easy to read, the lowermost section of the rectangle represents the methods, and the one over it represents the attributes or data members. The positive (+) and negative (–) signs for the methods and data members respectively denote whether they are 'public' or 'private'; that is whether access to them is restricted to the owning class itself, or open to other classes. The arrows of the sequence diagrams represent methods called by one object to another; time flows vertically from top to bottom; so the method marked with the number '1' is called before the one marked '2' and so forth.

One of the areas in which we found the algorithmic paradigm inadequate was when we wanted to start playing with many complex numbers. In the object-oriented paradigm, we have seen how an object is used to represent a particular complex number, complete with its coordinate attributes and methods. What do we need if we want to create as many *ComplexNumber* objects as we want? We need something like a cast, something that is used to make many statues that are replicas. A *class* is such a template for creating objects. 'A class is the descriptor for a set of objects with similar structure, behaviour, and relationships' [Rumbaugh et al. 2005]. Described in terms of a class, an object 'is a concrete manifestation of an abstraction; an entity with a well-defined boundary and identity that encapsulates state and behaviour; an instance of a class' [Booch et al. 2005]. In the above definition, *state* refers to the values stored in the data members of an object and *instance* means realization of a class through an object. The ideas of class and object and their implications on one another remain at the very foundation of the

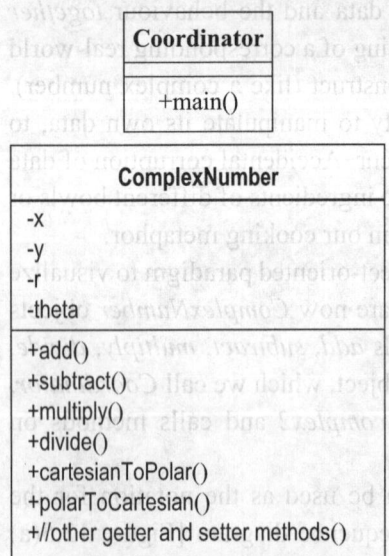

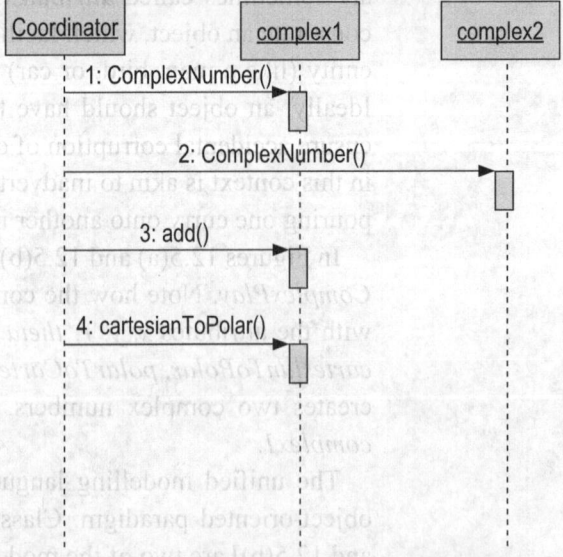

Figure 12.5(a) Class diagram representing *ComplexPlay*

Figure 12.5(b) Sequence Diagram Representing *ComplexPlay*

object-oriented paradigm. Booch et al. (2007) recognize an object as any of the following: a visible or tangible thing, something which we can create or comprehend intellectually, or 'something towards which thought or action is directed'; an object has state, behaviour, and identity. As against the *concrete* existence of an object, either in the world or in our minds, a class is an abstraction, the 'essence' of an object [Booch et al. 2007].

The key theme of the object-oriented paradigm is the cooperation of objects to deliver a system's overall functionality. There are other important concepts, like *inheritance*—which allow the propagation of characteristics of a parent class to its children (in the same sense we inherit qualities from our parents). But the central idea that makes the object-oriented paradigm so effective relates to the collaboration of objects. Returning to our cooking metaphor, an object-oriented way to organize a large dinner would be to have a number of chefs who specialize in cooking different items (specific behaviour) and have been given all the ingredients they need (specific data) to operate—maybe under the coordination of a head chef. The object-oriented paradigm is very appropriate for building large and complex systems. However, in such systems, there is often a replication of functionality in different parts of the system. For example, chopping of vegetables may be needed for a number of items in our hypothetical dinner. Chopping the same vegetables separately for two separate items is not the best utilization of chopping resources and may also lead to waste. Something similar to such replicated functionality is recording the running of a computer program at several execution points. It is a common activity in different contexts. Returning to our *ComplexPlay*, what if we want every step in running of the program logged? The easiest and the clumsiest way to do this is to put in *print* statements everywhere in the code. Is there was a way to encapsulate this kind of *crosscutting* functionality through some abstraction? (Crosscutting functionality means functionality that cuts across different parts of the system.) The object-oriented paradigm does not have an easy way to abstract this kind of functionality. Thus we need to look at the situation from another *aspect*.

12.5.3 Aspect-Oriented Paradigm

A relatively new addition to the pantheon of software development paradigms, aspect-oriented development is based on a wholly different abstraction. An *aspect* is defined as 'a programming unit designed to capture a functionality that crosscuts an application' [Pawlak et al. 2005]. As an idea, aspects closely relate to what the word 'aspect' means: a way of looking at things or how something appears when observed. In the software context, that translates to looking at the functionality of a system for common behaviour that can be isolated. The aspect-oriented paradigm is concerned with formalizing ways of discovering, understanding, and using aspects as a software development artefact. Aspects are implemented through tools and frameworks works as a cohesive unit. *AspectJ* – (http://www.eclipse.org/aspectj/) is one such tool.

The aspect-oriented paradigm complements the object-oriented paradigm by offering a new way of encapsulating functionality that spreads across several classes. Lopes (2002) highlights this positioning of aspects vis-à-vis objects as follows: 'Aspects are software concerns that affect what happens in the Objects (sic) but that are more concise, intelligible and manageable when written as separate chapters of the imaginary book that describes the application' (Lopes 2002). ('Application' is taken to mean the entirety of a software system.) In Figure 12.6, we illustrate how an aspect can gather and encapsulate crosscutting functionality. The vertical rectangles are three different classes. The shaded horizontal bands represent some crosscutting functionality like logging. The shaded horizontal rectangle represents an aspect that encapsulates all of the crosscutting functionality from the classes.

One way to view aspects is to recognize that whereas the object-oriented paradigm (as well as the algorithmic paradigm) takes a predominantly *vertical* view of abstraction, the aspect-oriented paradigm has more of a *horizontal* perspective, as it tries to collect in aspects, functionality that stretches across multiple units of code. We have seen earlier that the execution of a program's steps and collaboration

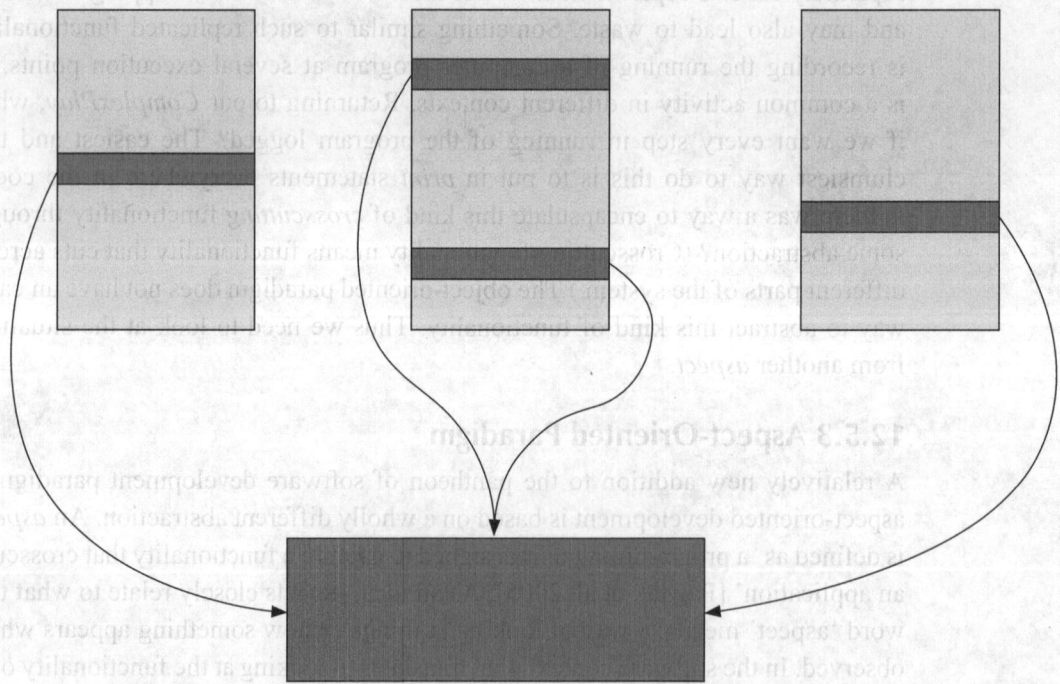

Figure 12.6   Encapsulation of Crosscutting Functionality in an Aspect (inspired by Pawlak et al., 2005)

of objects are the key ideas respectively of how the algorithmic and object-oriented paradigms function. The corresponding idea of the aspect-oriented paradigm is enacapsulation of crosscutting functionality.

An aspect-oriented system contains classes as well as aspects. An operation, known as *aspect weaving,* takes them as inputs and produces an application that integrates the functionalities of the classes and aspects. A *joinpoint* is a point in the control flow of a program where one or several aspects apply; a *pointcut* is a set of joinpoints where an aspect applies; *advice code* is the definition of the behaviour of an aspect [Pawlak et al. 2005]. Given these features of the aspect-oriented paradigm, a key question still remains; how does one decide whether a functionality is best modelled as a class or an aspect?

Crosscutting score is a metric that can help guide this decision [Datta 2006]. While calculating the metric, the components that help a given component fulfill its primary and secondary responsibilities, as well as those that influence any of its conditional behaviour are identified. The crosscutting score metric helps discern the lines of functionality that crosscut an application versus those that are more localized in nature. For a detailed discussion of the crosscutting score metric, refer to *http://www.dattas.net/misc/ccs.pdf*.

Returning to our cooking metaphor once again, the aspect-oriented paradigm addresses common tasks that stretch across—crosscut, that is, the preparation of many of the items in a multi-course dinner. As we mentioned earlier, chopping vegetables is one such task; instead of doing it separately for every individual item, it is often more convenient to collect them all together and assign the common task to one or more expert vegetable cutters, this is aspectualizing crosscutting functionality.

The aspect-oriented paradigm is not a revolutionary paradigm, but an evolutionary one, drawing on many of the ideas of the object-oriented paradigm and viewing them from a fresh perspective. There is lack of consensus yet on whether a system can be conceived and constructed exclusively based on the aspect-oriented paradigm, or how useful such an abstraction would be. But the enthusiasm and ongoing advances surrounding this paradigm underscore the fact that the world of software development paradigms is in ferment. Old ways of addressing software complexity are constantly examined and improved upon, and new ways complement the old.

12.6 PARADIGMS, PERSPECTIVES, AND PROGRAMMING

There is more than one way to approach the complexity of software, as represented by the multiple paradigms discussed in this chapter. The obvious question that arises is that which paradigm is *better*? There is no universally valid answer to such a question.

The paradigms represent different ways of looking at a problem and modelling the solution. The algorithmic paradigm is certainly easier to implement for problems of limited scope. It is also easier to grasp when one first learns to program a computer. A step-by-step plan for solving a problem is at the heart of every computer program; thus the algorithmic paradigm goes to the heart of the computing problem directly. But as highlighted before, difficulties arise when too many data items have to be updated and kept track of it. To address this problem, the object-oriented paradigm consolidates data and behaviour into one conceptual entity; a class—which is instantiated in objects that collaborate to deliver the system's functionality. However, sometimes multiple objects may share similar functionality in different contexts leading to replication of code. The aspect-oriented paradigm offers a way to collect and encapsulate crosscutting functionality.

The key to effectively using a paradigm is to recognize where and how it fits into the overall need to abstract a particular problem into a software solution. A paradigm can only suggest a way of approaching a problem; it cannot enforce details of the solution. That is the concern of the languages of software development, which we take up in the next chapter.

How apt is it to talk about software engineering exclusively in terms of a particular development paradigm? We discuss this issue in the next section.

12.7 A HOLISTIC VIEW

One often hears phrases like 'classical or procedural software engineering', 'object-oriented software engineering', or 'aspect-oriented software engineering'. Can we qualify software engineering with a particular development paradigm? To address this question, let us consider electrical engineering. Ohm's Law is expressed by the relation $V = IR$, where V is the voltage that induces a current of I to flow across a conductor of resistance R. This is a foundational principle of electrical engineering. The law holds irrespective of whether the conductor is made up of copper, silver, or any other material. The quest for materials with better conducting properties is ongoing (and yield significant benefits from time to time), but no matter what the results are, it will not change how Ohm's Law guides the Design of electrical circuits. Although software engineering lacks the surety of something similar to Ohm's Law, its basic principles of understanding user Requirements, followed by analysing, designing, implementing, and Testing a software solution, are not fundamentally altered by the development paradigm used. Development paradigms offer new ways to look at old problems; they do not change the nature of problems being looked at. One of the basic objectives of engineering is problem solving, and software engineering is no exception. So software engineering should qualify the paradigms of software development, not the other way around.

CASE STUDY

Preeti has to prove a point to the customer that using a particular programming language does not necessarily ensure that a particular programming paradigm is being followed. As an example, she implements the same operations on complex numbers, using the same programming language, Java, but using two different development paradigms, algorithmic and object-oriented. Figures 12.7 and 12.9 gives the code for the two approaches and Figures 12.8 and 12.10 list the respective outputs when the two programs are compiled and run.

(The code given in these tables, though not complete in the sense of implementing *all* the methods, nonetheless have been compiled, tested, and run. However, there is no promise stated or implied on whether the code follows all recommended coding conventions for Java. Detailed discussion of Java programming is beyond the scope of this book.)

By comparing Figures 12.8 and 12.10, it is apparent that the outputs of the two programs are exactly the same. The programs operate on the same input data. The formulas operating on complex numbers are also the same. But by studying the code, it is apparent there is a paradigmatic difference between the ways the two programs perform the same task. Relating the actual code to the earlier discussion of algorithmic and object-oriented programming, the difference in the spirit of the two paradigms is evident. So the same programming language, Java, (a predominantly object-oriented language) can be used to express very different programming idioms. This is an interesting observation; which will be helpful, as we study the languages of software development in the next chapter.

```java
import java.math.*;

public class ComplexPlay {

    static double x = 0;
    static double y = 0;
    static double r = 0;
    static double theta = 0;

    public static void main(String args[])
    {
    // Define complex numbers 2 + i3
    double x1 = 2;
    double y1 = 3;

    // Define complex numbers 4 + i5
    double x2 = 4;
```

```
            double y2 = 5;
            // Add the complex numbers 2 + i3 and 4 + i5
            add(x1,y1,x2,y2);
            // Convert the complex number 2 + i3 to its polar form
            cartesianToPolar(2,3);
    }

        public static void add(double x1, double y1, double x2,
    double y2)
        {
            x = x1 + x2;
            y = y1 + y2;
            System.out.println('The sum of the given complex
    numbers is: ' + x + '+'+ 'i' + y);
        }
        public static void subtract(double x1, double y1, double x2,
    double y2)
        {
            // Code for subtraction
        }
        public static void multiply(double x1, double y1, double x2,
    double y2)
        {
            // Code for multiplication
        }
        public static void divide(double x1, double y1, double x2,
    double y2)
        {
            // Code for division
        }
        public static void cartesianToPolar(double xCoordinate,
    double yCoordinate)
        {
            r = Math.sqrt((Math.pow(xCoordinate, 2) + Math.
    pow(yCoordinate, 2)));
            theta = Math.atan2(yCoordinate, xCoordinate);
            System.out.println('The polar coordinates of the given
    complex numbers are: ' + 'radius = ' + r + ' and angle(in radians)
    = ' + theta);
```

```
        }
        public static void polarToCartesian(double radius, double
angle)
        {
            // Code for converting polar to Cartesian coordinates
        }
    }
```

Figure 12.7   Implementing complex play using the algorithmic paradigm

```
The sum of the given complex numbers is: 6.0+i8.0
The polar coordinates of the given complex numbers are: radius =
3.605551275463989 and angle(in radians) = 0.982793723247329
```

Figure 12.8   Output of the code listed in Figure 12.7

```
    import java.math.*;

    public class Coordinator {

        public static void main(String args[])
        {
            // Define a complex number by creating the complex1
object of ComplexNumber class
            ComplexNumber complex1 = new ComplexNumber();

            // Set the x and y coordinates of complex1
            complex1.setX(2);
            complex1.setY(3);
    // Define a complex number by creating the complex2 object of
ComplexNumber class
            ComplexNumber complex2 = new ComplexNumber();

            // Set the x and y coordinates of complex2
            complex2.setX(4);
            complex2.setY(5);
```

```
                    // Add complex number complex1 with complex2
                    ComplexNumber addResult = complex1.add(complex2);
                    System.out.println('The sum of the given complex
        numbers is: ' + addResult.getX() + ' ' + '+ 'i' + addResult.
        getY());

                    // Convert the complex number 2 + i3 to its polar
        form
                    ComplexNumber       polarForm       =       complex1.
        cartesianToPolar();
                    System.out.println('The polar coordinates of the
        given complex numbers are: ' + 'radius = ' + polarForm.getR() + '
        and angle(in radians) = ' + polarForm.getTheta());


                }


            }


        class ComplexNumber { // a class abstracting the idea of a
        complex number


                private double x = 0;
                private double y = 0;
                private double r = 0;
                private double theta = 0;

                public void setX(double a)
                {
                        x = a;
                }


                public void setY(double b)
                {
                        y = b;
                }
```

```
public double getX()
{
        return x;
}

public double getY()
{
        return y;
}

public void setR(double c)
{
        r = c;
}

public void setTheta(double d)
{
        theta = d;
}

public double getR()
{
        return r;
}

public double getTheta()
{
        return theta;
}

public ComplexNumber add(ComplexNumber complex)
{
        ComplexNumber result = new ComplexNumber();
        result.setX(complex.getX() + this.x);
        result.setY(complex.getY() + this.y);
        return result;
}
```

```
                public ComplexNumber subtract(ComplexNumber complex)
        {

                ComplexNumber result = new ComplexNumber();
        // Code for subtraction
                return result;

        }

        public ComplexNumber multiply(ComplexNumber complex)

        {

                ComplexNumber result = new ComplexNumber();
        // Code for multiplication
                return result;

        }

        public ComplexNumber divide(ComplexNumber complex)

        {

                ComplexNumber result = new ComplexNumber();
        // Code for division
                return result;

        }

        public ComplexNumber cartesianToPolar()

        {

                ComplexNumber result = new ComplexNumber();
        r = Math.sqrt((Math.pow(this.x, 2) + Math.pow(this.y,
        2)));
        theta = Math.atan2(this.y, this.x);
        result.setR(r);
        result.setTheta(theta);
        return result;

        }

        public ComplexNumber polarToCartesian()

        {

                ComplexNumber result = new ComplexNumber();
        // Code for converting polar to Cartesian coordinates
                return result;

        }
}
```

Figure 12.9 Implementing complex play using the object-oriented paradigm

```
The sum of the given complex numbers is: 6.0 + i8.0
The polar coordinates of the given complex numbers are: radius =
3.605551275463989 and angle(in radians) = 0.982793723247329
```

Figure 12.10   Output of the code listed in Figure 12.9

SUMMARY AND TAKE-AWAYS

In this chapter, we have introduced some of the paradigms of software development. Our discussion can be summarized as:

- Software is inherently complex.
- A key characteristic of a complex system is that it is far greater than the sum of its parts.
- Decomposition, abstraction, and hierarchies are among the common strategies for addressing complexity.
- The algorithmic paradigm of software development, sometimes referred to as the procedural paradigm, looks to deliver a system's functionality through a step-by-step execution of tasks.
- The central idea of the object-oriented paradigm is that objects collaborate via messages to deliver a system's functionality.
- Aspects provide a mechanism to encapsulate cross-cutting functionality across programming units.
- Different paradigms offer different perspectives on the fundamental issues of software engineering.
- Implementing a particular paradigm is an issue of programming style, rather than choice of a programming language.

WHERE TO LOOK FOR MORE

- Object Management Group (OMG) provides useful information about evolution and state of the art of object technology at http://www.omg.org/
- Aspect-Oriented Software Development Community and Conference maintains a comprehensive source of information about the aspect-oriented paradigm of software development at http://aosd.net/.

EXERCISES

Review Questions

Review questions test your understanding of the key concepts presented in this chapter.

1. A software development paradigm is similar to
   (a) building a house using wood
   (b) building a house using stone
   (c) building an igloo
   (d) a general approach to house-building, irrespective of the material used

2. A key feature of a complex system is that
   (a) it is equal to the sum of its parts
   (b) it is less than the sum of its parts
   (c) it is greater than the sum of its parts
   (d) there is no correlation between the system and its parts

3. According to Brooks, complexity is
   (a) an accidental property of software
   (b) an essential property of software
   (c) sometimes accidental, sometimes essential property of software
   (d) none of the above

4. One of the reasons for the inherent complexity of software is its
   (a) plasticity
   (b) rigidness
   (c) mutability
   (d) supportability

5. The combinatorial explosion that adds to the complexity of a software system may arise from
   (a) changing Requirements
   (b) large number of users
   (c) states of its variables as well as multiple threads of control
   (d) all of the above

6. According to Miller, human mind can simultaneously comprehend

   (a) maximum nine pieces of information
   (b) minimum five pieces of information
   (c) only seven pieces of information
   (d) (a) and (b)

7. Divide-and-rule relates to which of the following approaches towards addressing complexity
   (a) abstraction
   (b) decomposition
   (c) hierarchies
   (d) all of the above

8. Abstraction involves
   (a) looking at every possible detail
   (b) looking at selected details
   (c) dealing with the generalized, idealized model while ignoring details
   (d) comparing details with the generalized, idealized model

9. 'Is-a' and 'part-of' hierarchies, respectively called class structure and object structure
   (a) map to classes and objects in object-oriented programming
   (b) are software development paradigms
   (c) are ways to implement abstractions
   (d) are abstractions to examine and understand complex systems

10. The algorithmic paradigm of software development involves
    (a) devising a step-by-step solution to the task at hand
    (b) combining data with operations
    (c) looking at cross-cutting concerns
    (d) all of the above

11. In the algorithmic paradigm, data and operation,
    (a) depend on one another
    (b) the former influences the latter

(c) are disengaged from one another

(d) none of the above

12. Common problem(s) with the algorithmic paradigm when building large and complex systems is/are

(a) complexity of the algorithm

(b) accidental data corruption

(c) bloated programs

(d) all of the above

13. In the object-oriented paradigm,

(a) classes collaborate by sending messages to one another

(b) classes collaborate by sending messages to objects

(c) objects collaborate by sending messages to classes

(d) objects collaborate by sending messages to one another

14. Objects combine, in one unit of abstraction

(a) security and performance

(b) data and Design

(c) data and operation

(d) operation and reliability

15. A class is a template

(a) for creating methods

(b) for creating attributes

(c) for creating fields

(d) for creating objects

16. An aspect is a

(a) type of object

(b) an algorithm

(c) a unit of code that captures crosscutting functionality

(d) all of the above

17. The aspect-oriented paradigm and the object-oriented paradigm

(a) contradict one another

(b) complement one another

(c) supplement one another

(d) support one another

18. The crosscutting score metric

(a) helps separation of concerns

(b) helps devise optimal algorithms

(c) helps decide which piece of functionality is best modelled as a class and which as an aspect

(d) none of the above

19. Which of the following is true about the software development paradigms?

(a) They offer perspectives on addressing software complexity.

(b) They enforce specific programming styles.

(c) They provide mechanisms for software testing.

(d) None of the above is true.

20. 'Software engineering should be qualified by specific development paradigms.'

(a) True

(b) False

(c) True depending on what development paradigm is used

(d) True depending on what software engineering problem is being addressed

Reflective Questions

Reflective Questions require you to think more deeply about some of the ideas and come up with your own interpretations and answers.

1. Social systems have been mentioned as examples of complex systems. Point out one social system that you think is complex. Where does the complexity of this system originate from?

2. Complexity is a property which many interesting systems seem to have. What do you think are some of the characteristics of a complex system vis-à-vis a simple one?

3. An example of a simple *continuous* system is tossing a ball into the air; it will come

down to the earth according to established principles of physics. On the other hand, a software program executed on a digital computer is a *discrete* system [Booch 2007]. Which of these two types of system do you think is more likely to be a complex system? Based on your answer, discuss whether all continuous systems are complex or all discrete systems are complex.

4. Can you think of any other reasons for software's complexity in addition to the ones discussed in this chapter?

5. We have reflected on a channel capacity of the human mind, in the sense it can comprehend seven plus minus two distinct pieces of information at a point of time. We have a software development team of say, 10 software engineers. Would the channel capacity of the team be 10 times that of each individual, or more/less than that? Justify your answer.

6. We have observed that one of the most important characteristics of a complex system is that it is greater than the sum of its parts. But decomposition as a way to address complexity prescribes breaking down a system into smaller and simpler parts that are easier to understand. Do you see any contradiction between the very nature of complex systems and this approach towards understanding complexity? How can such contradiction be resolved?

7. The human body is a complex system. To understand its complexity, identify some 'is-a' and 'part-of' hierarchies.

8. An incandescent light bulb is a familiar item of utility. However it is a fairly complex system in terms of the material and specifications of its filament, the need to have a vacuum around it, the amount of light that it emits, compared to the energy it consumes, etc. If you are asked to devise an abstraction for the light bulb, which of its features will you include and which will you ignore. Justify your choices.

9. For someone who is just learning to program, which paradigm—algorithmic, object-oriented, or aspect-oriented—do you think will be the easiest to grasp? Justify your answer.

10. We observed that the object-oriented paradigm is more closely aligned to the real world than the algorithmic paradigm. Does this mean all real world problems would be easier to abstract through the object-oriented paradigm? Can you think of a problem which can be better addressed through the algorithmic paradigm?

11. What do think is the most basic difference as well as similarity between the algorithmic and object-oriented paradigms?

12. Can the aspect-oriented paradigm complement the algorithmic paradigm and vice versa? If so, how?

13. Are some types of systems more prone to have crosscutting concerns than others? What kind of functionality *can not* be modelled as crosscutting functionality?

14. Historically, the order of the paradigms has been: algorithmic, object-oriented, aspect oriented. What do you think is the reason behind this ordering? Could we have had aspect-oriented paradigm arrive before the algorithmic paradigm?

15. The function, class, and aspect can be said to be the fundamental units of the three development paradigms. Is there any correlation between these three? Can you express the same idea through these three different constructs?

Programming Examples

Programming Examples require you to analyse or write a program or a program segment to understand a specific problem.

1. Write a program in Java that calculates the factorial of a given number, using the algorithmic paradigm as well as the object-oriented paradigm. Compare the sizes of the two programs (in lines of code), the time taken to develop each of them, and the number of bugs detected in the first round of Testing. Which paradigm do you think is more useful for this particular problem?

2. Complete the programs (the methods of subtract, multiply etc.) of Figures 12.7 and 12.9 compile and run them. Now extend each program so that we can multiply three complex numbers. Which program took more time to write and debug? How much more time? Which program has more lines of code?

REFERENCES

Booch, G. (2007), 'Speaking truth to power', *IEEE Software*, 24(2): 12–13.

Booch, G., Maksimchuk, R.A., Engel, M.W., Young, B.J., Conallen, J., and Houston, K.A. (2007), *Object-Oriented Analysis and Design with Applications*, 3rd edition, Addison-Wesley Professional.

Booch, G., Rumbaugh, J., and Jacobson, I. (2005), *The Unified Modeling Language User Guide,* Second Edition, Addison-Wesley.

Brooks, F.P. (1995), *The Mythical Man-Month: Essays on Software Engineering,* 20th Anniversary Edition, Addison-Wesley.

Capra, F. (1997), *The Web of Life: A New Scientific Understanding of Living System*, Anchor.

Datta, S. (2006), 'Crosscutting score: an indicator metric for aspect orientation', In *ACM-SE 44: Proceedings of the 44th annual southeast regional conference*, pp. 204–208, ACM Press, New York.

Datta, S. (2007), *Metrics-Driven Enterprise Software Development: Effectively Meeting Evolving Business Needs*, J. Ross Publishing.

Lopes, C. (2002), Aspect-oriented Programming: An Historical Perspective.

Maier, M.W., and Rechtin, E. (2000), *The Art of Systems Architecting,* Second Edition, CRC Press.

Miller, G. (1956), The magical number seven, plus or minus two: Some limits on our capacity for processing information, The Psychological Review.

Pawlak, R., Retaille, J.P., and Seinturier, L. (2005), *Foundations of AOP for J2EE Development*, 3rd edition, Apress.

Rumbaugh, J., Jacobson, I., and Booch, G. (2005), *The Unified Modeling Language Reference Manual,* Second Edition, Addison-Wesley.

Simon, H.A. (1996), *The Sciences of the Artificial*, 3rd edition, The MIT Press.

Waldrop, M.M. (1992), *Complexity: The Emerging Science at the Edge of Order and Chaos*, Simon and Schuster.

Wessels, T. (2006), *The Myth of Progress: Toward a Sustainable Future*, University of Vermont Press.

Languages of Software Development

13.1 MOTIVATION

This is one of the central chapters of this book. The material presented here puts
into context much of the discussion of preceding and succeeding chapters.

This is a chapter about the languages of software development. Over the last 12
chapters we have discussed software engineering in English, and will continue to
do so in the next ten chapters. What is so special about the languages of software
development, and why and how are they different from the language this book is
written in? These are the questions we are concerned with.

The online Oxford dictionary gives the following definitions of 'language'
(http://www.askoxford.com/concise_oed/language?view=uk):

1. The method of human communication, either spoken or written, consisting of the use of words in a structured and conventional way.
2. The system of communication used by a particular community or country.
3. The phraseology and vocabulary of a particular group: *legal language*.
4. The manner or style of a piece of writing or speech.
5. Computing: a system of symbols and rules for writing programs or algorithms.

It is very tempting to pick out the last definition – a system of symbols and rules for writing programs or algorithms – to be the most apt in our context. Temptation would lead us astray here (as it does frequently). The whole point why all five definitions were enumerated above is that not one definition fully fits the notion of 'language' in the software engineering context. A *language* for us would be a combination of all the ideas each of the definitions capture. The easiest way to know more about the thought and action of any group is to be able to decipher their language. This is the common strategy archaeologists adopt when they come across the ruins of a long lost civilization, or the anthropologists do when they study a new ethnic group. Language not only serves as a vehicle of communication, disseminating, and documenting thoughts and deeds. It is also a window into what all thoughts might be thought and what all deeds done.

As we shall see in this chapter, languages of software development operate at several levels, such as facilitating expectations from a software system to be shared between stakeholders (specification languages), capturing Analysis and Design artefacts through a common set of idioms and notation (modelling languages), and implementing the system in executable code (programming languages). Each of these levels represents a particular *layer* of abstraction; and there is no indication stated or implied as to which level is higher or lower. However, in the process of software building, there is certainly a sequence in which these languages are used. Looking outside-in into software development, a software system is conceived as an idea, and then realized as a set of executing programs. From this view, we need the services of specification, modelling, and programming languages, in that order. But this is not the view a software engineer has, especially those joining the industry fresh from college. Looking inside-out, to them the world revolves around programming—modelling and specification seem niches far up in the clouds—and programming languages are their bread and butter. And as successful software engineers know, learning programming languages well is the first step towards using modelling and specification languages effectively. On the lines of what we have been discussing so far, Figure 13.1 reflects on the interplay of the three different types of languages and their areas of relevance. In the sequence software development life cycle (SDLC) activities, specification languages come first, followed by modelling languages, and then programming languages. On the other hand, programming languages exude highest familiarity, followed by modelling, and specification languages (for example, many more people know Java than

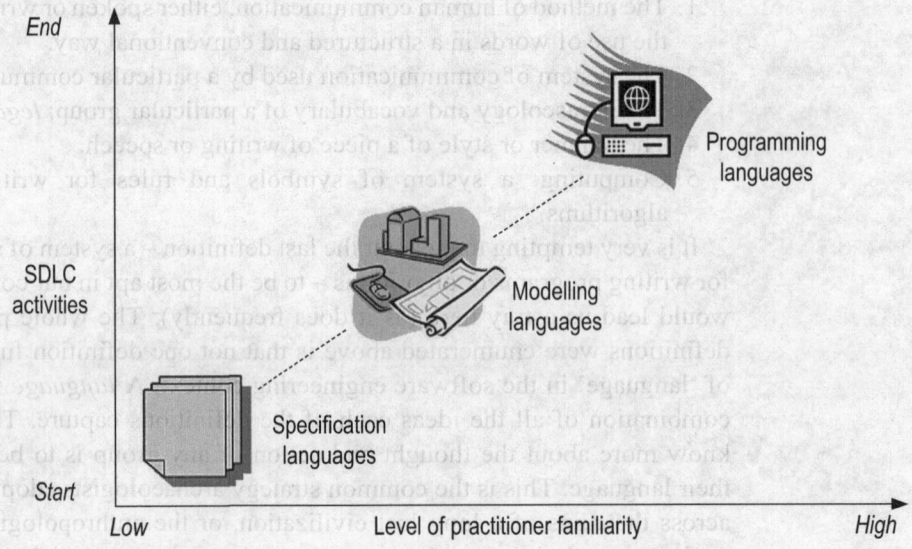

Figure 13.1   Spectrum of languages in software development

those who know UML, and many more people have a familiarity with UML than those with a smattering of Z—a specification language we will discuss later in this chapter). In our discussion of this chapter, we will address programming languages first and then move on to languages at the other levels.

Before going deeper, it is important to reflect on the thinking behind the collection of the topics in this chapter. We will cover a wide swath of ideas, notation, and idioms here. The common thread binding them together is that they all relate to the languages of software development. Merely learning the alphabets or the rules of grammar do not ensure learning a language. The real power comes from a language when we are able to express ideas which we could not even conceive if we did not know the language. This chapter gives an overview of the languages software engineers need to know to practice their profession, to have a better understanding of the principles that underlie software engineering, and to be able to disseminate their knowledge to software engineers of the future.

This chapter has three main threads of discussion. In the order of presentation they are: programming languages, modelling languages, and specification languages. Each topic is treated differently. Assuming that readers of this book have sufficient exposure to programming, we do not emphasize on how to program a computer while talking about programming languages. Instead, we highlight the journey and classifications of programming languages. On the other hand, a software engineering course is often the first opportunity for students to know about modelling languages. So our discussion of modelling languages—centring on the Unified Modelling Language (UML)—is aimed at giving students an overview of UML (illustrated in more detail in the next chapter) and enough leads to learn the subtler aspects of the language later. Specification Languages are something even seasoned software engineers confess to having seldom heard about. Thus our discussion of specification languages—centring on Z—will be mostly introductory.

13.2 INCREMENTAL APPROACH TO LEARN LANGUAGES

At several places in this book we have underscored how an incremental approach is helpful for building large and complex software systems. The essence of this approach lies in building in small increments, reviewing what has been built, and building for the future upon what has been built in the past. The incremental approach is also very suitable for learning languages. When we first start learning a language, we do not take a dictionary and memorize all the words, or take a grammar book and learn all the rules and their combinations. Instead, we start with a small set of words and rules that have strong intuitive appeal for us, learn to use them when apt, and then learn new words and rules in terms of what we already know. (As Helen Keller has described in her autobiography, her first appreciation of language came with the tactile sensation of water and the spelling of W-A-T-E-R on her palm. Many of us can empathize with that electric moment, even though we did not have to face Keller's great challenges.) A dictionary is *circular*; each word is defined in terms of other words contained in the same dictionary. So if we do not start small by learning few words at a time whose meanings we already know from intuition or experience, we can not move forward. We will adopt a similar incremental approach while learning the languages of software development. This chapter will try to give just enough to get you started on learning these languages. As you learn more you can build on the basic understanding imparted here.

13.3 PROGRAMMING LANGUAGES

In this section, we will discuss programming languages. The fact that you have come to this point in this software engineering book can be taken to indicate that you are familiar with at least one programming language. Chances are high that you know several programming languages and are proficient with at least one. Here we will pesent the genesis of programming languages, to help you put the languages you use and those you do not, in perspective.

13.3.1 Journey of Programming Languages: Milestones

In the 1940s it was customary to refer to computers as 'electronic computers'. Electronics was very different at that time—without transistors, vacuum tubes held sway. Computers were energy guzzling and fragile behemoths, they tripped every time a tube burned, which was frequently. These computers were programmed in binary machine code, sometimes called *machine language*, first by switches and then by 'punched cards'. Code written by such means was *not reusable or relocatable*, that is, the code could only be run on the machine it was meant for. To address the lack of reusability and relocatability, *assembly languages* were introduced. They introduced a layer of abstraction over the bare-bones machine code, letting machine operations be expressed in easy to remember 'mnemonic' abbreviations. This facilitated the writing of reusable and relocatable programs on a larger scale. But the machine understood only machine code, so there had to be someone who would translate the easy to remember abbreviations that humans wrote to the code that machine could understand. Enter the *assembler*—the specialized program which bridged between assembly and machine codes.

The mid 1950s saw the advent of FORTRAN, which is often cited as the first *high-level programming language*. 'High-level programming language' is an interesting phrase; used widely and loosely. To have understanding of 'high-level' in the programming language, context, see Exhibit 13.1. The primary need for computers in the 1950s was to solve scientific problems using numerical methods and FORTRAN (its name derived from the phrase FORmula TRANslator) was specially designed with that view. The syntax of FORTRAN closely mimics the mathematical notation for writing algorithms. After FORTRAN, a slew of other high-level languages appeared: Algol 58 as an improvement on FORTRAN, COBOL—Common Business-Oriented Language, Lisp—the name derived from 'List Processing Language'—for artificial intelligence, BASIC—Beginner's All-purpose Symbolic Instruction Code, C for system's programming etc. The thinking was, if you have a specific need for computer programming, learn and use a specific language. (One of the Reflective Questions in the exercises section explore, whether there has been a change in this line of thinking.)

The 1980s saw the coming of object-oriented programming. Object-oriented programming is both an old and a new phenomenon. The idea of object orientation partly comes from the Simula 67 language from the 1960s, which had the notion of data type abstraction (from which the concept of a 'class' was derived) but did not allow inheritance. Investment of thought and resources into object-oriented programming increased in the 1980s and made the technology more robust, reliable, and useful [Sebesta 2007]. (For a detailed discussion of object orientation and its place in software engineering, refer to Chapter 12.)

Exhibit 13.1 High and Low of Programming Languages

In our discussion so far we have mentioned 'high-level programming languages' several times. Although we have not explicitly said so, recognizing *high-level* languages seems to automatically lead to some languages being seen as *low-level*. What sets high-level languages apart from their low-level cousins?

A quick answer would be: our vanity. Human beings tend to regard themselves as highest in the hierarchy of life. Thus computer languages that pretend to get closer to the so-called *natural* languages of humans are regarded as high-level ones. This is true in an ironic sort of way (we will soon reveal why), but then, not entirely true. What a high-level computer language does is hide particulars of the underlying machine by allowing programmers to write code in a language that is closer to the language they use in their everyday lives. So, *further* a language removes the programmer from worrying about how the computer hardware would *actually* perform the tasks it is being asked to, the *higher* the language comes to be regarded as. But here comes the irony.

As we all know from our first lessons in electronics, computer hardware only understands 0s and 1s. This seems pretty adequate until you try to have the hardware do interesting things for you. Matters get very complicated in trying to communicate our ideas to the machine merely through 0s and 1s. So, we take recourse in leveraging some of the expressiveness (and our familiarity) of natural languages to tell the computer what to do. The job of translating our language into the primordial 0s and 1s is left to another computer program (specific to the language being used), broadly referred to as the *compiler* (see Exhibit 13.2). The irony lies in the fact that in our perception of computer languages, 'high-level' tries to get closer to the incredibly complex structures of natural languages, away from the simplicity of 0s and 1s, whereas in common understanding, a *high-level view* of a situation is expected to hide the complexity and give you the essence.

In Figure 13.4, we illustrate what gives 'height' to a programming language.

High-level languages have provoked the explosion in the number of computer programmers, fuelled large and complex software systems, and engendered a burgeoning software engineering industry. There was a time not long ago when programming was viewed as an esoteric art, which only those with rare talents could learn and practice. Programming is now as easily learnt as riding a bicycle; perhaps more. That has been primarily due to high-level languages.

The quest for higher levels of programming languages is ongoing. Lopes et al., in their book [Lopes et al. 2003], look beyond Aspect Oriented Programming—one of the more recent programming paradigms (see Chapter 12) towards 'naturalistic programming' which observes the fact that '... humans have been thinking and describing complex systems for thousand of years using natural languages' and try to use some of the elements of natural languages to make programming languages even more familiar for humans.

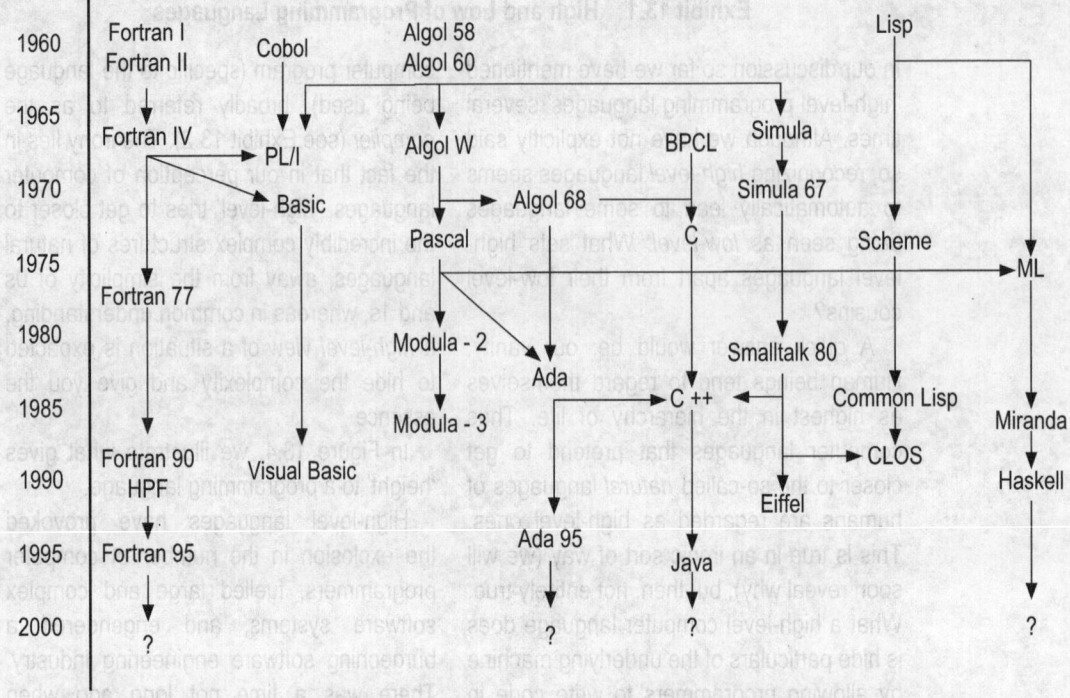

Figure 13.2   Family tree of programming languages

Figure 13.2 shows a family tree of programming languages. This is by no means a complete representation. Some existing languages and scripting languages (like Perl, Python, Ruby etc.) may not have not been shown. A key point worth noting is that unlike the usual family tree where each offspring has exactly two parents, computer languages can derive from one or more parent language(s), sometimes as many as three, (like C++). As indicated by the question marks at the bottom, the tree will continue to grow.

13.3.2 Profusion of Programming Languages

From the above discussion, it is clear that there are *many* programming languages, in fact *very* many. The questions that easily come to mind are why there are so many of them, and why some languages become popular, while others wither. The first question can be addressed in a number of ways. On one hand, evolutionary trends influence what is regarded as *good* versus *bad* programming constructs. For example, the notorious GO-TO based control flow gave way to while-loops and case statements in the 1970s. So as newer concerns emerged or old concerns were sought to be addressed in newer ways, newer programming languages came into existence. Additionally, many languages were designed for specific purposes, e.g. Java was conceived to be specially suited for Internet based programming. Personal preferences also play a very important role in the proliferation of languages. Programming is often as much a matter of individual style as, dress and food habits. And just as it is impractical to expect there will be a universal dress code or a universal menu across countries and cultures, it is equally unlikely there will be one programming language loved or even accepted by all.

Yet, it is apparent that some languages are used more than others, have a larger fan following, and a deeper penetration into the programmer's mind space. Why is that? The following are some of the reasons that account for the popularity of one language over another.

- *Power of expression*: Although all languages are meant to be powerful, a language's specific features influence a programmer's interaction with it for reading, writing, enhancing, modifying, and debugging code. The abstraction mechanisms offered by a language also increase its expressive power.
- *Ease of learning by beginners*: A relatively flat learning curve goes a long way in popularizing a language. Very often, the ease with which we can code, debug, and run our first programs warm us up to a language, and that warmth helps us build a stronger and longer relationship with that language.
- *Ease of operation:* Languages like Java that run on a wide range of architectures, and offer platform independence, have a bigger chance of being popular.
- *Tool support*: Powerful compilers and versatile tool support, such as Integrated Development Environments, go a long way in making a language popular.
- *Miscellaneous factors*: Many languages succeed over the heads of better counterparts due to strong financial and political backing. Another important factor—as evidenced during the Y2K scramble to modify long running 'legacy' code—is that older languages continue to run in spite of newer and better languages having come to the fore. This often happens as critical business functionalities remain 'locked in' old systems which are not touched due to unavailability of know-how and unacceptability of replacement downtime.

13.3.3 Classification of Programming Languages

Programming languages are sometimes broadly categorized into four groups, *imperative*, *functional*, *logic*, and *object oriented* [Sebesta 2007]. It should be noted, however, that this classification is not exclusive and languages overlap and share features across these boundaries.

- Imperative languages are designed around the *von Neumann* architecture for computers. In von Neumann computers, data and programs reside in the same memory, whereas the central processing unit, which executes the program

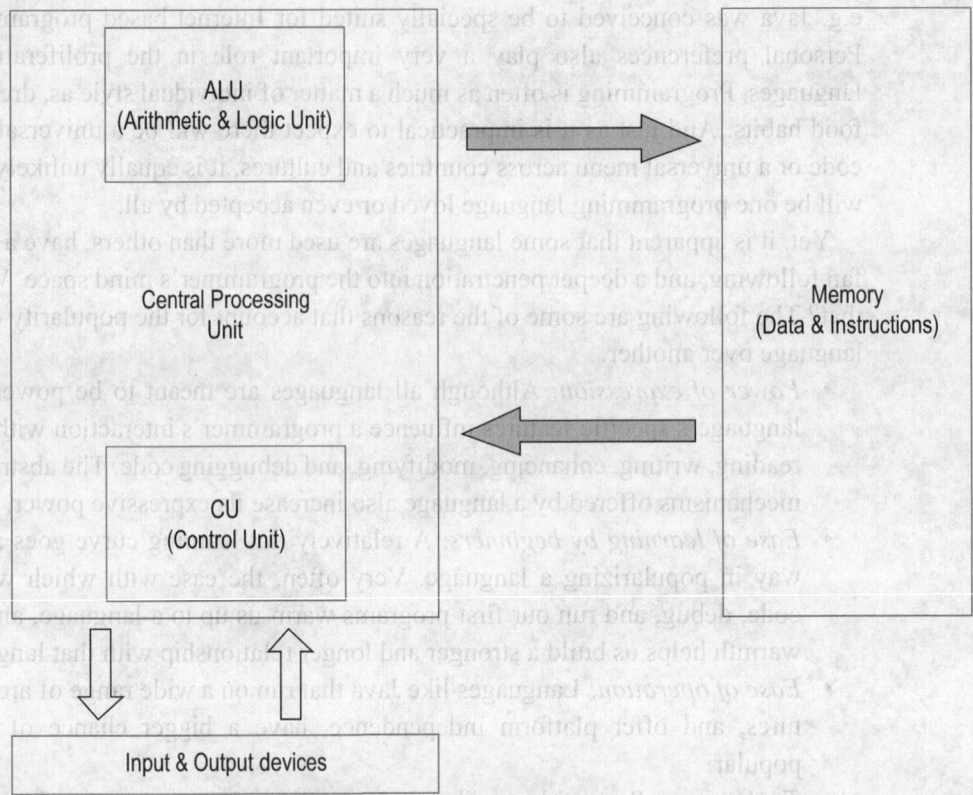

Figure 13.3   Von Neumann architecture of a computer

instructions, is housed separate from the memory. Thus data and program instructions need to be transmitted from memory to the CPU, and the results of the operations returned to the memory. Since the 1940s, a large majority of (digital) computers are based on the von Neumann architecture (Figure 13.3). As they are based on the von Neumann architecture, imperative languages have *variables* modelling memory locations, *assignment statements* to model the transmittal of information between the memory and CPU, and *iteration* for repetitive tasks.

- In functional languages, the computing paradigm is to pass parameters to *functions*—chunks of code that perform a specific task. Functional languages, allow us to program without the use of variables, assignment statements, and iteration, unlike imperative languages. When a task needs to be done repetitively, the function calls itself, that is, performs a *recursion*.

- Logic programming languages are rule-based. Compared to an imperative language, where every step of an algorithm as well as the sequence of their execution need to be specified in great detail, rule-based languages operate

Table 13.1 Generations of Programming Languages [Booch et al. 2007]

Generations	Languages	Characteristics
First-generation languages (1954 – 1958)	FORTRAN I, ALGOL 58, Flowmatic, IPL V	Mainly used for mathematical calculations; consists only of global data and sub-programs.
Second-generation languages (1959 – 1961)	FORTRAN II, ALGOL 60, COBOL, Lisp	Use extended to business applications; artificial intelligence; subroutines, block structure, data types introduced.
Third-generation languages (1962 – 1970)	PL/1, ALGOL 68, Pascal, Simula	Use extended to wider applications; ideas of modules and data abstraction introduced.
The generation gap (1970 – 1980)	C, FORTRAN 77	Many languages invented with few surviving; small executables, thrust towards standardization.
Enhanced popularity of object-orientated languages (1980 – 1990)	Smalltalk 80, C++, Ada83, Eiffel	Languages derived from previous ones; the idea of a class as a basic unit of abstraction.
Emergence of frameworks (1990 – present)	Visual Basic, Java, Python, J2EE, .NET, Visual C++, Visual Basic .NET	Widespread use of integrated development environments (IDE); focus on Web-based systems.

differently, with the language Implementation system deciding which order the rules need to be executed. Logic programming represents a way of thinking very different from that underlying other types of languages.

• Object-oriented languages offer the abstraction of a *class* to model real-world entities. *Objects* are specific instances of classes, which collaborate amongst themselves via messages to fulfill a system's functionality. For a detailed discussion of the object-oriented way of programming, refer to Chapter 12.

Another way of classifying programming languages is to look at their *generations*. Drawing on Figure 13.2 (family tree of programming languages), Table 13.1 presents the generations of programming languages, and highlights important characteristics of each generation.

13.3.4 Choice of a Programming Language

Whatever we have discussed so far has not touched upon an important consideration: From the myriad programming languages, how does one select a language for a specific purpose? Choice of a programming language usually gets decided by the specific domain the software system is meant for. As stated earlier, the first high-level language FORTRAN was meant for scientific applications.

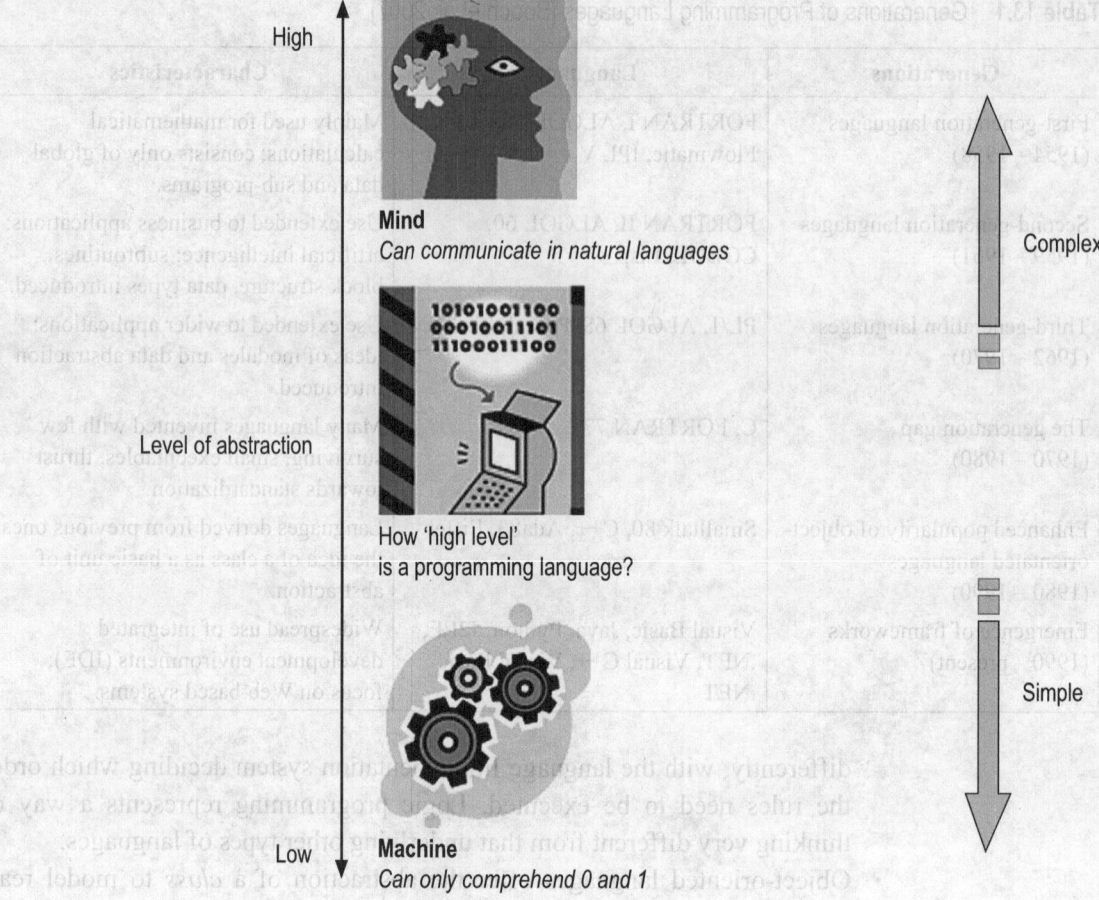

Figure 13.4 Height of a programming language

On the other hand, COBOL was meant to support widespread use of computer systems for business applications. Artificial intelligence needs specific languages like Lisp and Prolog to carry out rule-based reasoning. C is still widely used for systems programming even as other languages have taken its place for large-scale user applications. Scripting languages starting with Unix shell scripts such as *sh*, to *JavaScript* and *Perl* are strictly not programming languages, but none the less do help in programming tasks. So the choice of a programming language ultimately depends on what you want to do, and how you want to do it. There is almost always a specific language suited to the kind of system that needs to be built and it is in the best interest of the system to choose such a language, instead of falling for some passing fad.

Exhibit 13.2   Black Magic of Compilation

As we have discussed in this chapter, high-level languages make our task of communicating with the computer significantly easy. But the computer understands nothing but 0s and 1s. So something must be happening between what we say and what the computer hears, to keep both sides happy. This 'something happening between' is *compilation*. Conceptually, compilation can be said to be the process of translating source code into binary target code; 'target' since the process is aimed at giving the CPU something it can execute. So, a compiler is a program that may run on the same platform as the target code, or on a different platform, in which case the process is known as cross compilation. Figure 13.5 gives the conceptual schema of a compiler. A clear distinction between compiler and interpreter is often elusive. An *interpreter* can be viewed as a *virtual machine*, and the CPU is taken to an Implementation of the virtual machine in hardware. There are languages which cannot be purely compiled into machine code, and these need *assemblers*. The objective of compilers is to be able to make as many decisions as possible at compile time instead of deferring them till

runtime. Compilers enhance performance through a number of ways, primary among which are the allocation of variables without lookup at runtime, code optimization to utlize specific hardware features etc. The integrated development environments (IDEs) we have mentioned earlier combine various programming tools together in a transparent and seamless way; editors, compilers, interpreters, assemblers etc. Figure 13.6 shows the different phases of the compilation process.

In this context, we may note that a *context-free grammar* defines the syntax of a programming language, specifying syntactic categories such as statements, expressions, and declarations. Categories are subdivided into more detailed categories; for example, a statement can be a *for statement*, an *if statement*, an *assignment* etc. Context-free grammars are a fascinating area—on the basis of a few simple rules they allow the rich complexity of a programming language's structure to be expressed. As a part of the compilation process, the context-free grammar structure of a programming language's syntax is parsed.

13.4 MODELLING LANGUAGES

In this section, we will explore the role of modelling languages in software development. Modelling is one of our basic approaches towards addressing complexity. Starting with a discussion of the essence of a model, we see how modelling languages help software development.

13.4.1 Essence of a Model

What do you see in Figure 13.7? When this question is posed to a class of undergraduate students, the answers are ready and obvious: 'It is an electric circuit'

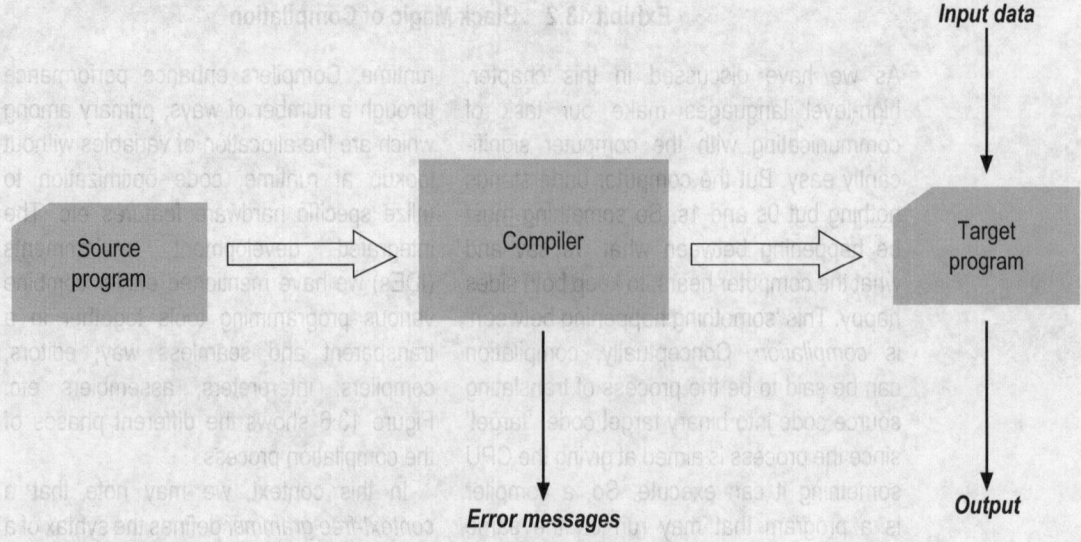

Figure 13.5 Conceptual schema of a compiler

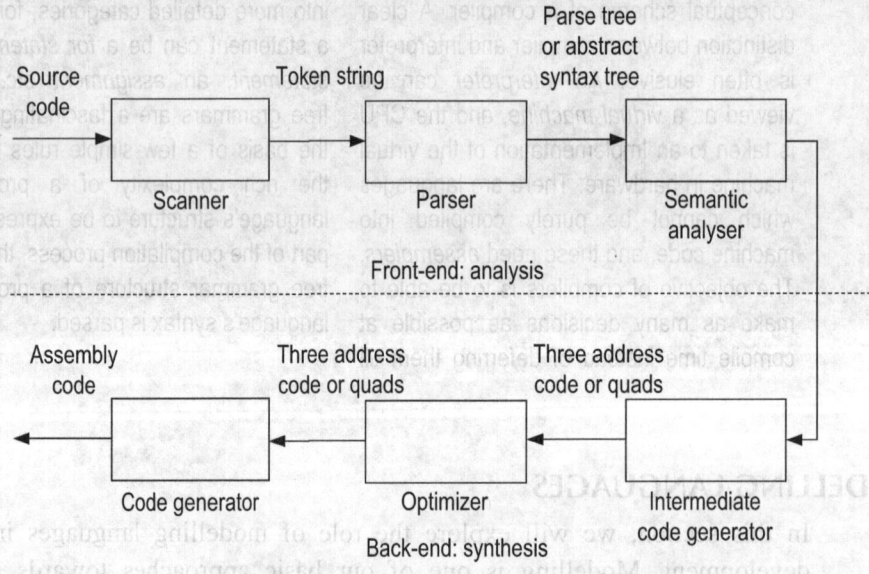

Figure 13.6 Phases of a compiler

or, 'on the left is an electric circuit with only a resistor and a battery, on the right we have a circuit with a resister, inductor, capacitor, and a source of alternating current.' But, choosing to be argumentative, I may say what I see are only lines

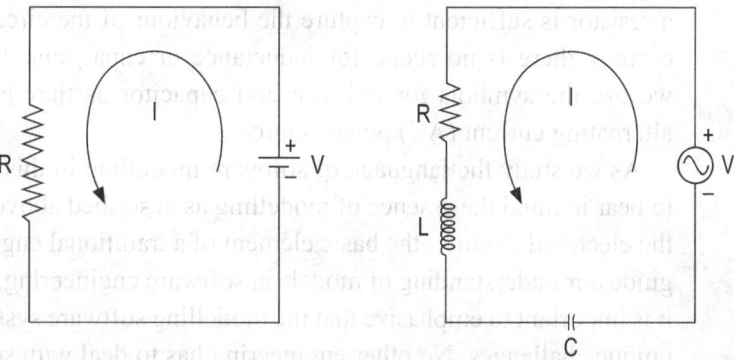

Figure 13.7 A model

on a piece of paper, some parts straight, some parts crooked, some broken with horizontal bars (straight and bent), with seemingly arbitrary alphabets beside some stretches of the line. Why are we seeing the same thing differently? Who is correct, the class of clever students, or me?

Both are correct; I am seeing what the picture *shows*, the students are seeing what it *represents*. Fundamentally, a model is a *representation*: a view of reality from a particular point of interest. To be able to give that view, some aspects of reality need to be highlighted and some hidden, based on certain assumptions. Table 13.2 presents the essence of the electrical circuit's model shown in Figure 13.7.

A model consists of notation, such as the standard symbols for the circuit's components is shown in Figure 13.7. The notation is associated with the *syntax*— the structure of representing something—as well as the *semantics*—what that representation denotes. We do not have to use the entire range of notation available to us; for example in the direct current (DC) circuit on the left of Figure 13.7 only

Table 13.2 Essence of a model

Reality	Model	Highlighted	Hidden	Assumptions
An electrical circuit, consisting of physical elements.	Straight and crooked lines on a piece of paper.	Circuit components and their specific properties (resistance, inductance, capacitance, battery or alternating current power source).	Physical details such as length and gauge of wire, number of turns of the inductor, distance between capacitor plates, the materials of the elements etc.	1. The resistance, inductance, and capacitance are individually *lumped* at specific points in the circuit. 2. The resistor offers resistance only and does not have inductance or capacitance; similarly for the inductor and the capacitor. 3. All other points in the circuit do not offer any resistance, inductance, or capacitance.

a resistor is sufficient to capture the behaviour of the circuit (without alternating current, there is no scope for inductance or capacitance), whereas on the right we use the symbols for inductor and capacitor as they become relevant for an alternating current (AC) power source.

As we study the language of software modelling in this section, it is important to bear in mind the essence of modelling as described above. We deliberately used the electrical circuit—the basic element of a traditional engineering discipline—to guide our understanding of models in software engineering. However, at this point it is important to emphasize that the modelling software systems does present some unique challenges. No other engineering has to deal with something even close to the *invisibility* and *unvisualizability* of software that Brooks pointed out decades ago [Brooks 1995]. We can see and feel an electrical circuit (the latter perception, if we are careless, can even fetch a sharp 'shock'), and this helps us get started with modelling it. But software offers no such facility.

Fortunately, in comparison with programming languages, there are far few modelling languages in software engineering. In fact, there is one—the Unified Modelling Language—that is *almost* universally accepted. I said 'almost' since there are some dissenting voices which question the primacy of UML (as we discuss later). Let us discuss the broad outlines of UML.

13.4.2 Unified Modelling Language

'A modelling language is a language whose vocabulary and rules focus on the conceptual and physical representation of a system' [Booch et al. 2005]. In this context, UML can be said to be a language for visualizing, specifying, constructing and documenting the artefacts of a software system [Rumbaugh et al. 2005]. Visualization, specification, and documentation are the uses of UML which are most readily utilized. Construction of a software system, in the sense of generating *executable* code from UML artefacts, needs sophisticated tool support that is not very common yet.

The view of UML that is more down to earth is that of a collection of diagrams that *together* lets us *see* the statics and dynamics of a software system. There are a number of different UML diagrams, nearly half the number of alphabets in the English language. Why are their so many diagrams?

There is a famous poem by John G. Saxe where several visually challenged men are examining an elephant (http://bygosh.com/Features/092001/blindmen.htm). Since they can not see, to each one of them the elephant is only what he touchs, either the tail or the legs or the trunk. Only if they had recognized and shared each other's experience, the men could have had a more holistic view of the elephant.

The poem perhaps is a metaphor for the ways all of us are limited from seeing the whole. It highlights the need to acknowledge that *one* view will not give us *all* that we need to see. No single UML diagram is sufficient to capture all that we are interested in a software system, all diagrams may not even be relevant for a given system, but the combination of diagrams can give us the complete view. When live telecasts of cricket matches first started in India in the late 1970s and early 1980s, it was customary to have one camera capture the central action, usually kept near one of the sight screens. As the bowling changed ends after each over, TV broadcasts either showed the bowler running in to bowl to the batsman, or the back of the batsman who was facing the bowler. Compare that to the number of views we have now of the action on a cricket ground, some of which are necessary for the third umpire to take a decision. Not all views from all cameras capture something interesting all the time, but when taken together, they do give a detailed picture of the cricketing action.

Before, we start discussing the UML diagrams in detail, it is imperative to keep in mind that the '... only place a system is conceived is the mind of the designer' and the act of drawing a diagram is merely capturing that conception [Booch et al. 2007]. In short, UML is neither Analysis nor Design nor any other activity of the software development life cycle. It is merely a notation.

Figure 13.8 shows the taxonomy of UML. Although diagrams are by far the most visible component of UML, the vocabulary of UML actually consists of three kinds of building blocks: *things*, *relationships,* and *diagrams*. Things represent the basic abstractions—structural things, behavioural things, grouping things, and annotational things. Things are tied together by the relationships—dependency, association, generalization, and realization. Diagrams—both *structural* and *behavioural*—are ways to group interesting collections of things. Structure diagrams are essentially *snapshots*, they capture some static structure of the system, frozen at a given point of time. Behaviour diagrams capture the dynamic aspects of a system—what happens to the system, or how the system reacts to external stimuli.

We will now briefly describe each of the different diagrams of UML [Booch et al. 2005]. No particular importance needs to be attached to the order in which they are presented here.

- A *class diagram* depicts a set of classes, interfaces, and collaborations and their relationships. Class diagrams are used most frequently while modelling object-oriented systems.
- A set of object and their relationships are shown in an *object diagram*, which is a snapshot of instances of classes represented in class diagrams.
- A *component diagram* represents an encapsulated class and its interfaces, ports, and internal structure which may consist of nested structures and connectors. Differences are sometimes made between a *composite structure*

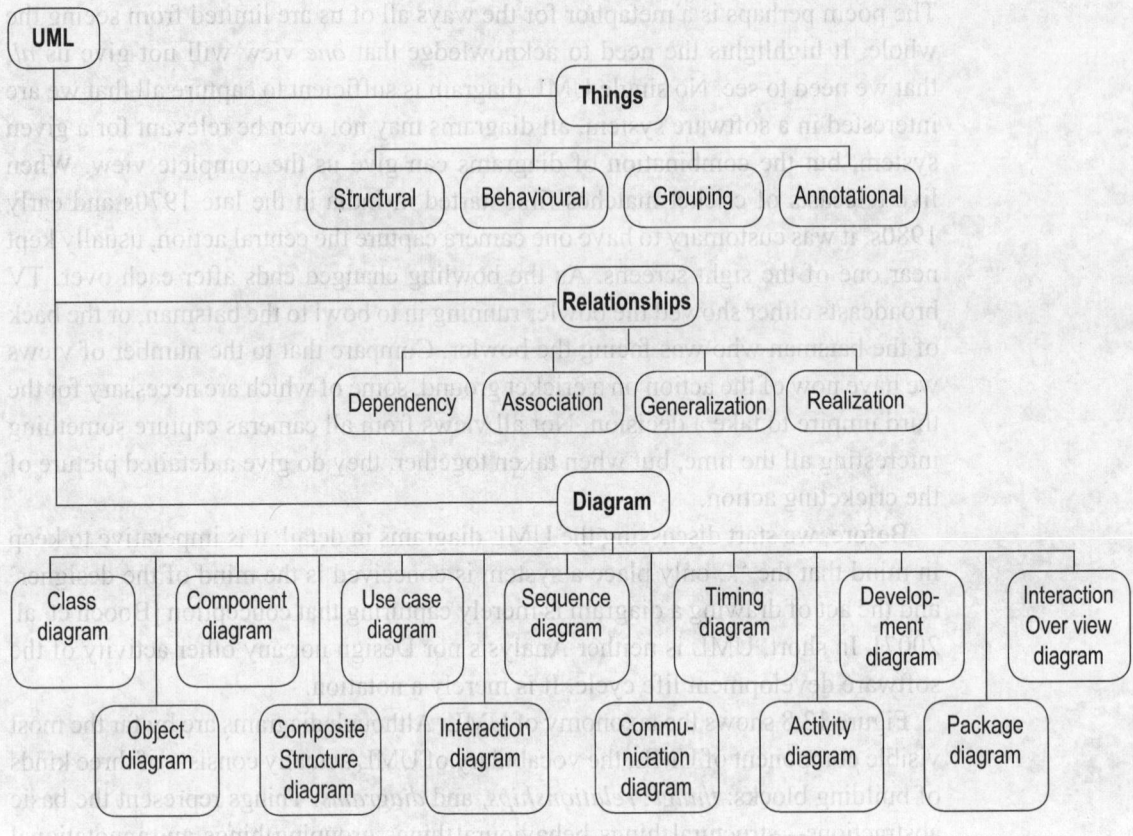

Figure 13.8 Taxonomy of UML

diagram applicable to any class and a component diagram, but we can ignore such details for the time being.

- A *use-case diagram* depicts a set of use-cases and actors and their relationships.
- An *interaction diagram* shows the interaction between objects or roles, via sending of messages back and forth between them.
- A *sequence diagram* is an interaction diagram that highlights the order in which the messages are exchanged.
- A *communication diagram* is an interaction diagram that highlights the structural organization of the entities that exchange messages.
- A *timing diagram* shows the actual times at which the messages are exchanged between objects or roles vis-à-vis their relative sequences.
- An *activity diagram* is UML's closest counterpart to the *flow chart* of the structural programming world; it shows the structure of a process or other computation with a step-by-step flow of control and data.

- A *deployment diagram* represents the configuration of run-time processing nodes and components that run on them.
- A *package diagram* depicts the decomposition of the model into organizational units and how they are dependent amongst one another.
- An *interaction overview diagram* represents a hybrid of an activity diagram and a sequence diagram.

This is not an exhaustive list of diagrams; UML provides the facility to define and use other types of diagrams. But as we will illustrate in a later chapter, the degree of usage of these diagrams varies to a large extent across projects.

The story of UML's has been chronicled in several books and articles: It gives a rare insight into the combination of diverse ideas into a standard expected to be universally applicable. The drafting of the initial version of the UML—as well as its subsequent enhancements and modification—in a way represents, the coming of age of the software industry. Subsequent to the creation of UML, Booch, Rumbaugh, and Jacobson came to be known as the *three amigos*. As the three amigos have said on many occasions, they claim no exclusivity as the creators of UML, the language unifies the ideas from many sources, named and unnamed. (See Exhibit 13.3.)

What does the 'unified' of the Unified Modelling Language signify? It means: *Across historical methods and notations, across the development lifecycle, across application domains, across Implementation languages and platforms, across development processes,* and *across internal concepts* [Rumbaugh et al. 2005]. Thus UML seeks to subsume the many ways software is conceived and constructed, and present a set of uniform idioms to developers.

Exhibit 13.3   Making of UML

With the advent (or rebirth) of object-oriented technologies in the 1980s, object-oriented software development methodologies began to be popular. A number of methodologies, with some overlapping and some distinct features were used by their own groups of followers: such as the object-modelling technique (OMT), the Booch method, object-oriented software engineering (OOSE) etc. Gradually, a need was felt to integrate the best practices from these methodologies. In the mid-1990s, Grady Booch, James Rumbaugh, and Ivar Jacobson joined hands to collect all the threads together and created what was to become the Unified Modelling Language. UML 1.1 was accepted by the Object Management Group (OMG)— a consortium in charge of proposing, reviewing, and maintaining standards—in November 1997. After minor revisions (versions 1.3 and 1.5) in the intervening years, a major revision, UML 2.0 was adopted in 2004, which serves as the current standard.

It is important to note that UML *does not* enjoy uniform favour among academics and practitioners. A common criticism of UML is that it is a product of *Design by committee*. The maxim 'a camel is a horse designed by committee' is quoted to impugn the 'unifying' approach of UML. Additionally, there are voices which deprecate the over-enthusiasm surrounding UML and model-driven development. Thomas [Thomas 2004] warns 'used in moderation and where appropriate, UML ... code generators are useful tools, although not the panaceas that some would have us believe.' Bell goes a step further. In an article ominously titled *Death by UML Fever* [Bell 2004], the author begins by saying 'A potentially deadly illness, clinically referred to as UML (Unified Modelling Language) fever, is plaguing many software-engineering efforts today'. In spite of some of its drawbacks, UML remains a widely used system of notation for describing software development artefacts.

13.5 SPECIFICATION LANGUAGES

We have remarked earlier in the chapter that programming languages are used most and specification languages the least, in the usual course of software development. This is a curious statement. It is in a sense contradictory too, as all of us, whether or not we have studied software engineering, know and use at least one specification language. In this book we have been using a specification language—English. Even a natural language is a specification language when we use it to express or 'specify' our ideas. The great power of natural languages comes from their expressivity, and their biggest pitfall lies in their potential for ambiguity. When building an engineering artefact, we want to be able to communicate *exactly* what the user wants, through the specifications, without the scope or need to read someone's lips or read between the lines. The quest for exactitude is what sets apart the specification languages for software development from other natural languages.

Sometimes the phrases 'formal methods' or 'formal specifications' are used in the broad sense of specification languages. According to Sommerville, formal methods refer to any activities that rely on 'mathematical representations of software' and a formal software specification '.. is a specification expressed in a language whose vocabulary, syntax, and semantics 'are formally defined' [Sommerville 2004]. Formalism in this context is the use of mathematics, more specifically, the concepts and tools of set theory, logic, and algebra drawn from discrete mathematics. Two basic approaches, differing in their use of mathematical constructs in detailed description of software systems, have been identified [Sommerville 2004]:

1. *Algebraic approach*: Operations and relationships are used to describe the system.

2. *Model-based approach*: The system is modeled using sets and sequences and system operations are defined to bring about the changes to the system state.

Use of formal methods centring on specification languages has had a checkered history. In the 1980s they were thought to have much promise, and predictions were made that the future of software engineering lay in the widespread adoption of formal methods. The optimism was not unfounded, as formal methods brought into software engineering something that the conventional engineerings have been successfully following for many years: the use of mathematical techniques in the engineering process. But formal methods have not made the expected inroads into the mainstream of commercial software development. The main reason comes out of a combination of a characteristic of formal methods, and a market trend for software products.

With reference to Figure 13.9, we observe that the use of formal methods pushes the cost of development *upfront*. Conventional processes result in validation costs that are about 1.5 times Design and Implementation costs, while the latter is double the specification costs. But formal methods lead to almost equal specification and development costs, and comparatively modest validation costs. While the system is being specified with great care to ensure a higher quality product, there is nothing 'running' to show the customer. This goes against a common expectation from certain software products, where *time to market* is taken to be of higher concern than attention to *quality*. From the point of view of engineering integrity, this demand is by no means justifiable. But regrettably, this is how the industry has run so far. A very large software product company has seemingly made a virtue out of this vice: Release bug-ridden software rapidly and follow up with an infinite series of 'patches' to clean things up later. Evidently, the careful use of formal methods is not at all amenable to this quick-and-dirty approach, and hence it has so far been under-utilized in the industry.

It is very heartening that through recent works [Hinchey et al. 2008], [Jackson 2006a] and [Jackson 2006b], formal methods are generating renewed interest amongst practitioners. Before we move to an example of a specification language in the next section, it would not be out of place to go over the so-called *Ten Commandments* that guide the choice of formal methods for a project.

13.5.1 Ten Commandments of Formal Methods

In 1995, Bowen and Hinchey postulated the *Ten Commandments of formal methods* in an eponymous paper [Bowen and Hinchey 1995]. The quaint phrasing of each commandment is reproduced as in the original. We outline them in brief below:

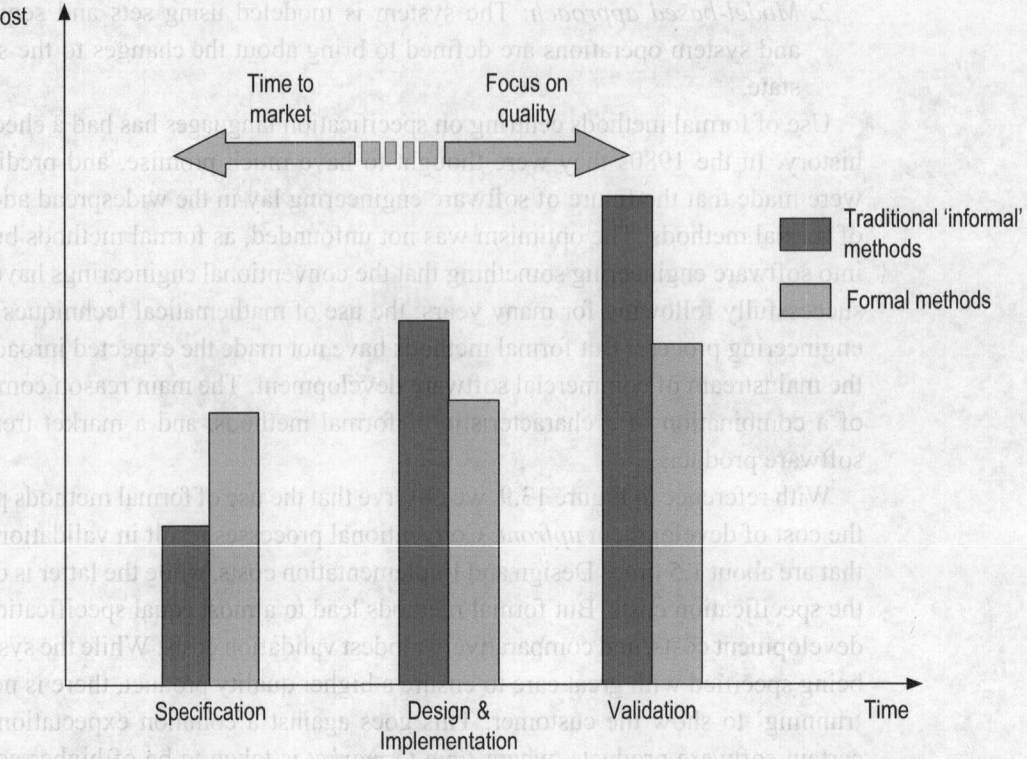

Figure 13.9 Cost versus time of formal and traditional methods (inspired by Sommerville 2004)

1. *Thou shalt choose the appropriate notation*: Choice of notation should depend on the type of the software systems as well as the domains the formal methods will be applied to.

2. *Thou shalt formalize but not overformalize*: Formal methods very often get their bad name from overkill. Just because formal methods are being used does not mean they have to be used for *every* aspect of a system. For example, it is reasonable to use formal methods for specifying the most critical components of a system only.

3. *Thou shalt estimate costs*: As we mentioned earlier, use of formal methods invariably increases cost early in the development cycle. Whether this is something acceptable to the paying customer, has to be clearly understood upfront.

4. *Thou shalt have a formal methods guru on call*: Skill and experience in using formal methods need a level of preparation every software engineer does not always have. It is impractical to expect that everybody in a development team will be formal method experts. So the team should have ready access to someone who can provide specific guidance on formal methods.

5. *Thou shalt not abandon thy traditional development methods*: Again, adopting formal methods should not obligate a team to adopt them *exclusively* and forget all about traditional, non-formal methods. The objective is not to prove that formal methods are better than traditional ones (or vice versa), but to build better software systems. The trick lies in combining the two approaches so that their respective strengths are amplified.

6. *Thou shalt document sufficiently*: It is easy to be led astray by the elegance of the specification idiom offered by formal methods and decide to forsake natural language documentation altogether. This is a mistake. The more concise a formal description is, chances are higher it may have missed out on an important aspect due to oversight or lack of understanding. Complementing these gaps with natural language documentation is essential.

7. *Thou shalt not compromise quality standards*: Formal methods do not represent a magic wand whose mere touch can do away with all quality issues. Even as formal methods are being followed, standard quality assurance procedures need to be in place and followed.

8. *Thou shalt not be dogmatic*: Formalism is not synonymous with dogma or obduracy. The mere fact that formal methods have been employed is not a guarantee of correctness. Users of formal methods need to have the flexibility to recognize that errors and omissions are still possible, and work towards fixing them.

9. *Thou shalt test, test, and test again*: With reference to Figure 13.9, note that while using formal methods, validation cost goes down, but it is not zero. The primacy of Testing (as established in Chapter 17) still holds, and it needs to be upheld by technical and managerial mandate and consistent efforts.

10. *Thou shalt reuse*: If software development has to leverage economies of scale in the long run, there has to be planned and deliberate effort to facilitate reuse. Formal methods can encourage reuse through unambiguous specifications.

A decade after proposing these commandments, the same authors revisited their views [Bowen and Hinchey 2005]. They concede that the integration of formal methods into mainstream software development has not happened to the extent they thought it would, and reflect with great insight, that formal methods are, '... certainly not completely reliable since humans, as well as mathematics, are involved and the logical models must relate to the real world in an informal leap of faith ...' [Bowen and Hinchey 2005]. Bowen and Hinchey see the future of formal methods in the industrial strength tools they have been lacking so far.

To give a flavour of a formal specification language, we will illustrate a brief example using Z. There are many formal techniques, such as Anna—a formal specification language for Ada [Luckham and Von Henke 1985]. The Vienna definition method (VDM) [Jones 1990] can be used for specification, as well as Design and Implementation. The *Communicating Sequential Processes* language [Hoare 1985] is a specification language based on the idea of viewing specifications as an event sequence, where each event is a simple action or some message that transfers data into or out of the system [Schach 2005]. Z is chosen for our illustration here as it is one of the most widely used specification languages.

13.5.2 Simple Example Using Z

We have highlighted the formalism of specification languages in our discussion so far. While it is beyond the scope of this book to describe a specification language in detail, we will show through a simple example, how 'specific' the specification through such a language can be.

The formal specification language Z was invented by Jean-Raymond Abrial [Abrial 1980] and named after the mathematician Ernst Friedrich Ferdinand Zermelo. It is pronounced as 'zed' as the British pronounce the last alphabet of the English language and not the Americanized 'zee' [Schach 2005]. As with any powerful language, Z can be used to express many complex ideas, and accordingly, its constructs can get complicated. Our objective here is not to parade such complexity but demonstrate the use of Z through a simple example.

Let us consider an elevator's operation. This is something which should seem pretty straightforward, based on our regular experience with elevators. We can think of the elevator as a *chamber* that goes up and down a *shaft*, and is controlled by a set of *buttons*, some of which lie inside the chamber and some on the *levels* the elevator can stop at, as depicted by Figure 13.10. Now, how do we specify the elevator's behaviour using Z?

While using Z, a system is taken to be an abstract state and a sequence of operations are defined on this set. First some basic states are introduced, followed by the definition of an *abstract state*, in terms of sets, relations, functions, sequences, etc. An *initial state* is then defined which represents the system just after it starts (say, turning the electricity on in the elevator). Change of state can be caused by operations on the system, signifying a *before* state and an *after* state. What the operation does is defined by the number of predicates, as well as inputs and outputs. Operations are taken to be *atomic*, that is, they can not abort in between, either the operation goes all the way through, or it does not happen at all. This is certainly not the entire gamut of Z vocabulary, but only a small part of it. Once the system has been specified and the Design formulated using Z, theorems can be stated and proven about the system that helps ensure mistakes are caught

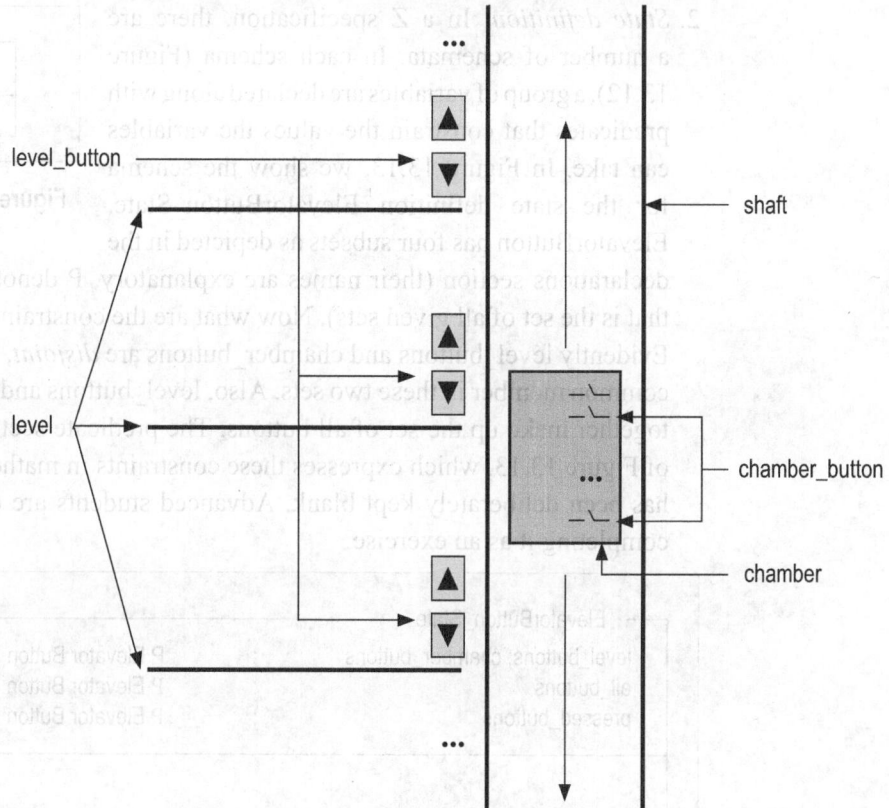

Figure 13.10   An elevator

early. This is an involved process, and very laborious if done manually, but helps impart a reasonably complete understanding of the system and its Design, before committing the Design to Implementation [Bowen 1996].

The following discussion is drawn from the description of a similar case study [Schach 2005]. In our example we shall consider the simplest form of a Z specification, which consists of four sections:

1. *Given sets, data types, and constants:* Given sets are sets that are taken as given and need not be specified further. In the elevator problem, we call the given set as ElevatorButton (depicted in Figure 13.11); it consists of the set of all buttons.

[ElevatorButton]

Figure 13.11   ElevatorButton given set

2. *State definition*: In a Z specification, there are a number of schemata. In each schema (Figure 13.12), a group of variables are declared along with predicates that constrain the values the variables can take. In Figure 13.13, we show the schema for the state definition ElevatorButton_State. ElevatorButton has four subsets as depicted in the

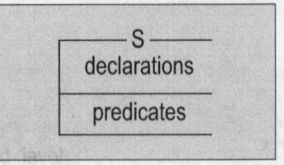

Figure 13.12 Z schema

declarations section (their names are explanatory, P denotes the power set, that is the set of all given sets). Now what are the constraints of this schema? Evidently level_buttons and chamber_buttons are *disjoint*, that is, there is no common member in these two sets. Also, level_buttons and chamber_buttons together make up the set of all buttons. The predicate section in the schema of Figure 13.13, which expresses these constraints in mathematical notation, has been deliberately kept blank. Advanced students are encouraged to try completing it as an exercise.

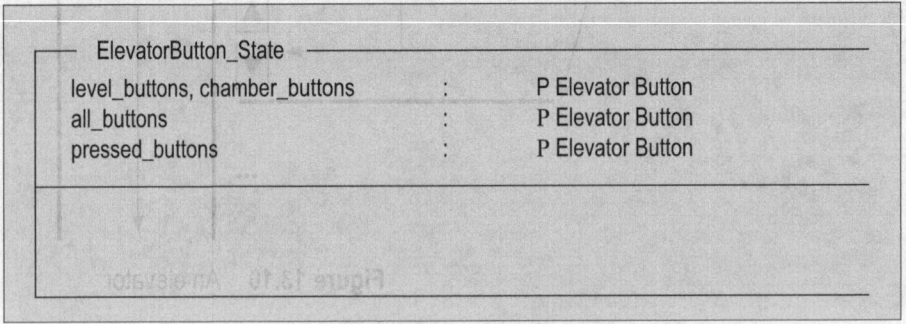

Figure 13.13 Schema for the state definition ElevatorButton_State

3. *Initial state*: The abstract initial state for the elevator is shown in Figure 13.14. This specifies that when the elevator is first turned on, there are no pressed buttons, thus the pressed_buttons set is empty. Let us not worry too much about all the symbols and their meanings; we will explain whatever is necessary.

ElevatorButton_Init $\triangleq$ [ElevatorButton_State $'$ | pressed_button $'$ = $\triangle$]

Figure 13.14 Abstract initial state for the elevator

4. *Operations*: Figure 13.15 defines the operation Press_Button. When a button is pressed for the first time, it is added to the set pressed_buttons.

ΔElevatorButton_State in the first line of the declarations signify that this operation changes the state of the ElevatorButton_State. button? is an input variable. In Z, '?' denotes input whereas, '!' denotes output. For an operation, the predicate part consists of *preconditions* that must exist before the operation starts, and the *postconditions* that must hold once the operation is complete. What would happen if the operation occurs without the preconditions being satisfied is not specified.

```
┌─ Press_Button ──────────────────────────────────────────
│ Δ ElevatorButton_State
│ button ?            :           ElevatorButton
├─────────────────────────────────────────────────────────
│ (button? ∈ all_buttons) ∧
│ (((button? ∉ pressed) ∧ (pressed' = pressed ∪ {button?})) ∨
│ ((button? ∈ pressed) ∧ (pressed' = pressed )))
└─────────────────────────────────────────────────────────
```

Figure 13.15 Definition of the press_button operation

In Figure 13.15 we check that button? is a member of all_buttons, this is the first precondition. The second precondition, checks whether button? is *not* pressed, and if that is so, then the set of pressed buttons is updated to include button? (A new value of a variable in Z is denoted by a prime, as in pressed'.) Thus the post-condition says that after operation Press_Button has been performed button? must be added to the set of pressed buttons. What about the possibility of an already pressed button being pressed again? Although nothing happens in this case, as indicated by the last statement in the predicate—pressed' = pressed—it is important this be specified as otherwise this very common occurrence (of pressing an already pressed button) would lead to an unspecified state.

The above discussion covers a very simplified view of the elevator's functioning. We did not consider another plausible situation when a button changes from a pressed to an unpressed state (elevator chamber arriving at a level)—this is left as an exercise—and many other situations (for example, what happens if a chamber_button as well as a level_button are pressed at the same time?) every real life elevator encounters frequently. But this example reflects on the level of detail and clarity which a formal specification language like Z extracts. Such detail and clarity can surely facilitate—within the limitations described earlier—development of quality software.

SUMMARY AND TAKE-AWAYS

- In this chapter, we have discussed programming languages, modelling languages, and specification languages. These represent three distinct but related facets of software development.

- As with natural languages, languages of software development also offer many layers of details. Out of these, only those relevant to the problem at hand need to be considered in a given situation.

- The journey of programming languages started from the 1940s with the advent of FORTRAN and is ongoing. From the 1980s, object-oriented languages have started being used extensively.

- Some of the factors that lead to the popularity of a programming language are: power of expression, ease of learning by beginners, ease of running on a wide range of architectures, tool support etc.

- Programming languages can be broadly categorized into four groups, imperative, functional, logic, and object oriented; and into the first, second, third, and fourth generations.

- The choice of a particular programming language for developing a system depends on the kind of system it is. Each programming language has a definite focus area, such as, C for systems programming etc.

- How high-level a programming language is depends on the degree to which it hides the underlying complexity of the physical machine that executes the instructions given in the language by the human programmer.

- Compilation is the process of translating source code into binary target code. High-level languages need compilers to translate programs written by humans into machine intelligible format. A context-free grammar defines the syntax of a programming language; compilers essentially parse the context-free grammar structure of a programming language's syntax.

- A model is based on some assumptions; it highlights some aspects of interest in a system, while hiding others.

- Challenges to modelling software arise to large extent from its inherent 'invisibility' and 'unvisualizability'.

- Unified Modelling Language (UML) is a language for visualizing, specifying, constructing, and documenting the artefacts of a software system.

- UML consists of three kinds of building blocks: things, relationships, and diagrams.

- UML diagrams are classified as structural and behavioural—the former captures the static aspects of a system while the latter captures the dynamic aspects.

- Although very popular, UML is not universally accepted in the software engineering community. A common criticism is that it is a product of Design by committee.
- Specification languages such as Z broadly relate to the formal methods of software development, using mathematical concepts and constructs to describe the behaviour of a software system.
- Use of formal methods leads to an increase in time and cost earlier in the development life cycle, vis-à-vis the use of traditional, non-formal methods. This is one of the reasons why formal methods are not used widely for commercial software development, where time to market is often given higher importance than quality.
- The quest for exactitude and completeness is paramount in a specification using a language such as Z.
- Specification languages and formal methods, when used in combination with traditional approaches can facilitate the Construction of quality software.

WHERE TO LOOK FOR MORE

- http://www.cs.waikato.ac.nz/~marku/languages.html – Presents the breadth of the programming languages spectrum.
- http://www.uml.org/ – The definitive source for UML on the Web.
- http://www.comlab.ox.ac.uk/archive/z.html – A central repository of Z related information, as well as the general ideas surrounding the use of formal methods.

EXERCISES

Review Questions

Review Questions test your understanding of the key concepts presented in this chapter.

1. The languages of software development can be classified into
   (a) four levels
   (b) two levels
   (c) three levels
   (d) unspecified number of levels
2. A language of software development is essentially a programming language.
   (a) This is a true statement.

   (b) This is a false statement.
   (c) Whether it is true or false depends on the language.
   (d) None of the above.
3. The so-called incremental approach is
   (a) helpful for building large and complex software systems
   (b) helpful for learning natural languages
   (c) helpful for learning languages of software development
   (d) all of the above

4. Comparing machine language and assembly language, we can say
   (a) the former is at a higher level of abstraction than the latter
   (b) the latter is at a higher level of abstraction than the former
   (c) both are at the same level of abstraction
   (d) the levels of abstraction represented by them can not be compared

5. The fact that there are so many programming languages can be explained by
   (a) evolutionary trends in the understanding and Implementation of effective programming constructs
   (b) backing of large corporations
   (c) programmer preferences and coding styles
   (d) (a) and (c)

6. FORTRAN can be said to be the first
   (a) programming language
   (b) high-level programming language
   (c) low-level language
   (d) assembly language

7. The idea of object orientation first appeared in
   (a) Java
   (b) C++
   (c) Simula
   (d) C

8. Profusion of programming languages has been influenced by
   (a) increase in the number of computer programmers
   (b) increase in the number of personal computers
   (c) widespread computer literacy
   (d) evolutionary changes in the perception of good versus bad programming constructs

9. The use of GO-TO, which is regarded as an undesirable programming construct, can be replaced by
   (a) a set of variables
   (b) an if-else statement
   (c) an efficient compiler
   (d) all of the above

10. Among the following, which is *not* a reason contributing to the popularity of one programming language over another?
   (a) Power of expression
   (b) Ease of learning by beginners
   (c) Tool support
   (d) None of the above

11. Imperative, functional, logic, object oriented are classes of
   (a) programs
   (b) programmers
   (c) programming constructs
   (d) programming languages

12. C and FORTRAN 77 are
   (a) first-generation languages
   (b) second-generation languages
   (c) third-generation languages
   (d) languages of the so-called generation gap

13. Choice of a programming language for a particular system should primarily be driven by
   (a) the application domain
   (b) programmer skill
   (c) productivity needs
   (b) all of the above

14. Whether a programming language can be called high-level depends on
   (a) how high it is in the family tree of programming languages

(b) how high it is in the popularity rating of programming languages

(c) the level of abstraction it introduces to hide the machine details from the programmer

(d) none of the above

15. A compiler is a

(a) a program that enhances performance

(b) a program that translates source code written in a programming language to CPU-executable target code

(c) a human being who compiles computer programs

(d) a type of assembler

16. Syntax of programming languages are based on

(a) conceptual grammar

(b) context-free grammar

(c) context-sensitive grammar

(d) natural language grammar

17. The electrical circuit shown in Figure 13.7 is an example of

(a) notation

(b) diagram

(c) model

(d) all of the above

18. 'A model is a representation of reality, usually simpler than the reality itself'.

(a) True

(b) False

(c) Depends on the reality being modeled

(d) Depends on the model

19. Every model has assumptions. Which of the following is a plausible assumption for a model of a software system that only uses class and sequence diagrams?

(a) The system does not exhibit significant changes of state.

(b) The timing of messages between objects is not important.

(c) (a) and (b).

(d) None of the above.

20. The need for multiple diagrams in UML can be appreciated in the light of

(a) the invisibility and unvisualizability of software, as per Brooks

(b) the multiplicity of programming languages

(c) the inherent complexity of software

(d) (a) and (b)

21. UML vocabulary consists of

(a) names, places, things

(b) things, actions, positions

(c) things, relationships, diagrams

(d) relationships, diagrams, constructs

22. UML diagrams can be classified into

(a) structural and functional

(b) structural and behavioural

(c) functional and behavioural

(d) positional and behavioural

23. A class diagram captures which of the following aspect of a system?

(a) Static

(b) Dynamic

(c) Configurational

(d) Implentational

24. The 'unified' in Unified Modelling Language means,

(a) unity of its founders

(b) universal acceptance of the language

(c) the vision of the language to be relevant across methods, development life cycles, languages, platforms etc

(d) all of the above

25. Formalism in the context of specification languages usually means the use of
    (a) engineering principles
    (b) mathematical constructs
    (c) ideas from physics
    (d) structured programming techniques

26. Algebraic approach and model-based approach are two ways of
    (a) writing computer programs
    (b) Testing software systems
    (c) describing software systems in detail
    (d) none of the above

27. What do you think is the main reason for formal methods not making expected inroads into the mainstream of commercial software development?
    (a) They are too difficult for software engineers to learn.
    (b) They are not always accurate.
    (c) They can not handle all types of software systems.
    (d) Time to market is often of bigger interest to stakeholders than quality.

28. While using formal methods, it is recommended that
    (a) they be used in conjunction with traditional development methods
    (b) they be used in the exclusion of traditional development methods
    (c) all team members be trained as experts in formal methods
    (d) all documentation be done in formal specification languages only

29. The use of formal methods ensure
    (a) quality standards are automatically met
    (b) there is no need for separate Testing
    (c) code components are reusable
    (d) none of the above

30. Z is
    (a) a formal specification language for Ada
    (b) a formal specification language invented by Jean-Raymond Abrial
    (c) a formal specification language invented by Ernst Friedrich Ferdinand Zermelo
    (d) none of the above

31. English can be taken as a specification language.
    (a) True
    (b) False
    (c) Depends on what is being specified
    (d) Depends on the level of English skill

Reflective Questions

Reflective Questions require you to think more deeply about some of the ideas and come up with your own interpretations and answers.

1. In our discussion of the journey of programming languages, we have remarked that during the early era of high-level languages, languages were defined for specific purposes such as scientific programming, business use, systems programming etc. Do you think this line of thinking is still relevant today? To be able to operate across different domains, how many languages do you feel software engineers of today need to know?

2. We have traced the journey of programming languages from the 1950s. What do you think has been the latest 'big hit' in programming languages? How long do you expect it to remain a hit?

3. Figure 13.2 depicts a family tree of programming language. What criteria do you think is used to decide on the parentage of a language? Why do you think a language like

C++ has three parents (Ada, C, and Smalltalk 80) whereas Java has only one (C++)?

4. You are asked to Design an improved version of Java. What features of the current version will you retain? What new features will you introduce?

5. Usually, whenever we want to model a phenomenon, pictures are more helpful than words. Do you agree with this statement? If so, do you think pictures have more expressive power than words? If not, why?

6. Among the three basic elements of UML, which do you think is most important and why?

7. Do you think the use of formal methods and specification languages needs to be increased in the software industry? If so, how can that be done, given the most important reason why they are not widely applied in the first place?

8. The predicate section of the Z schema of Figure 13.13 has been kept blank. Can you fill it up using the description of the schema given wihile discussing Figure 13.13. (All you need is basic knowledge of set theory.)

9. We have reflected how a natural language can be used as a specification language. Do you think a programming language can also serve as a specification language? What will be its strengths and limitations?

Programming Examples ———————

Programming Examples require you to analyse or write a program or a program segment to understand a specific problem.

1. Arvind is an all-rounder. He is a member of the school's football team, debating society,

in addition to being a Boy Scout. Using Java, create an AllRounder class. How would you ensure it has the properties of a member of the football team, debating society, as well as Boy Scout? Could you implement the solution similarly in C++? Do you feel Java imposes some specific constraints vis-à-vis C++ while trying to implement multiple inheritances? Why do you think such constraints are imposed? How do you get around that constraint?

2. One difference between imperative and functional ways of programming is how a repetitive task is handled. The former uses iteration, whereas the latter uses recursion. Write a program to calculate the factorial of a number iteratively. Now write another program to do the same task recursively. Did you use two different programming languages or the same one? Are the langauge(s) you used imperative or functional?

3. Going seamlessly and transparently from specifications to code remains a holy-grail of software engineering. Write a Java program of your choice within 25 lines of code. Write a specification for it using any notation (words, flowchart, UML diagrams) you like. Write five sentences on the Design of a system that would automatically translate the specification into code, without human intervention.

4. You are asked to meet a user, understand and document his or her needs, and write a program in Java. How many different languages will you need to complete the entire exercise?

REFERENCES

Abrial, J.R. (1980), 'The specification language z: Syntax and semantics', Programming Research Group, Oxford University.

Bell, A.E. (2004), 'Death by uml fever', *Queue*, 2(1): 72–80.

Booch, G., Maksimchuk, R.A., Engel, M.W., Young, B.J., Conallen, J., and Houston, K.A. (2007), *Object-Oriented Analysis and Design with Applications*, 3rd edition, Addison-Wesley Professional.

Booch, G., Rumbaugh, J., and Jacobson, I. (2005), *The Unified Modelling Language User Guide,* Second Edition, Addison-Wesley.

Bowen, J. (1996), *Formal Specification and Documentation Using Z: A Case Study Approach*, Intl Thomson Computer Pr.

Bowen, J.P. and Hinchey, M.G. (1995), 'Ten commandments of formal methods', *Computer*, 28(4): 56–63.

Bowen, J.P. and Hinchey, M.G. (2005), 'Ten commandments revisited: a ten-year perspective on the industrial application of formal methods', In *FMICS '05: Proceedings of the 10th international workshop on Formal methods for industrial critical systems*, pp. 8–16, ACM, New York.

Brooks, F.P. (1995), *The Mythical Man-Month: Essays on Software Engineering,* 20th Anniversary Edition, Addison-Wesley.

Hinchey, M., Jackson, M., Cousot, P., Cook, B., Bowen, J.P., and Margaria, T. (2008), 'Software engineering and formal methods', *Commun. ACM*, 51(9): 54–59.

Hoare, C. (1985), *Communicating Sequential Processes*, Prentice Hall.

Jackson, D. (2006a), 'Dependable software by Design', http://www.sciam.com/article.cfm?chanID=sa006&colID=1&articleID=00020D0%, 4-CFD8-146C-8D8D83414B7F0000, *The Scientific American*, June 2006, last accessed on May 1, 2010.

Jackson, D. (2006b), *Software Abstractions: Logic, Language and Analysis*, MIT Press.

Jones, C.B. (1990), *Systematic software development using VDM,* 2nd edition, Prentice-Hall, Inc., Upper Saddle River, NJ.

Lopes, C.V., Dourish, P., Lorenz, D.H., and Lieberherr, K. (2003), Beyond aop: toward naturalistic programming. *SIGPLAN Not.*, 38(12): 34–43.

Luckham, D. and Von Henke, F. (1985), 'An overview of anna, a specification language for ada', *Software, IEEE*, 2(2): 9–22.

Rumbaugh, J., Jacobson, I., and Booch, G. (2005), *The Unified Modelling Language Reference Manual,* Second Edition, Addison-Wesley.

Schach, S. (2005), *Object-oriented and Classical Software Development,* Sixth Edition, McGraw-Hill International Edition.

Sebesta, R.W. (2007), *Concepts of Programming Languages*, eighth edition, Addison-Wesley.

Sommerville, I. (2004), *Software Engineering*, 7th edition, Prentice Hall.

Thomas, D. (2004), 'Mda: Revenge of the modelers or uml utopia?', *IEEE Software*, 21(3): 15–17.

Software Development across Workflows and Phases

14.1 MOTIVATION

In the last chapter, we learnt about the languages of software development. Learning a language is always more fun when one has something interesting to say with it. As we saw, there is much that is interesting and much that needs to be said with the languages of software development. Based on that discussion, this chapter outlines the contours of how a software system is actually developed.

In this book we have been talking about how to build good software. To be more precise, we have been talking about how to construct software that fits user needs better. In real terms, that translates to planning, tracking, and delivering across the software development life cycle. This involves a complex mix of technological, human, and managerial factors, not all of which we have the scope to go into detail in this book. This is a book about software engineering—essentially, the technical aspects of software development. Like all other engineers, software engineers too need to operate in a space defined by developer perceptions, user needs, and business concerns. We desire not merely to be software engineers, but also *good* software engineers. An important criterion of being good at anything is to look beyond the immediate, and see the larger context.

To see the larger context, let us look at Figure 14.1. A developer sees the software world *bottom up*, starting with specifications, going on to the model, then to the program, and finally on to the system. For a user, the view is essentially *top down*; for him or her value and utility derived from the system are paramount. Anything beneath the utility layer is opaque to the user. The expectation of deriving this utility is what drives users to pay for the software system to be built. Managers have to see sideways to be able to make sure software that developers build translates into value for the user. This translation involves messy details, like keeping up to schedule and budget, and some leaps of faith, like helping users know what they want. But in the end, for the development effort to succeed, the software system must ensure user satisfaction. The *workflows* and *phases* we present in this chapter are components of the software engineer's strategy to make software development succeed.

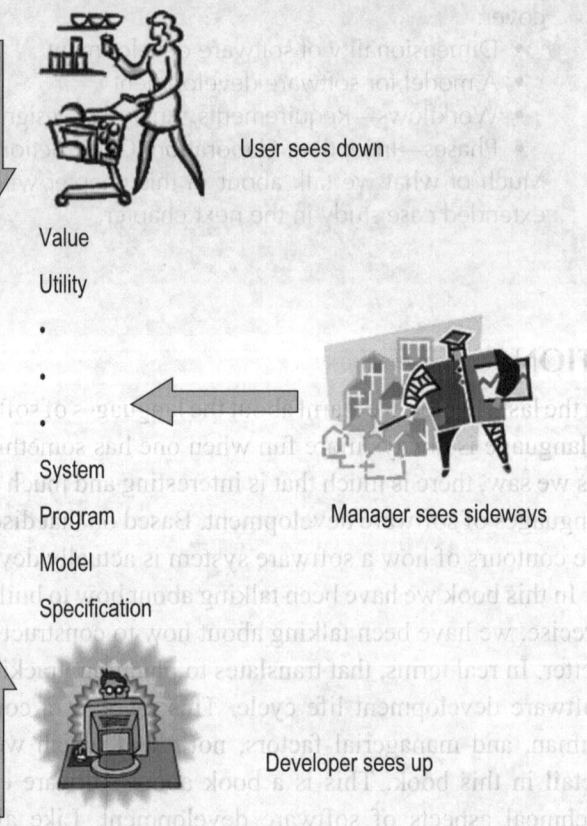

User sees down

Value

Utility

.

.

.

System

Program   Manager sees sideways

Model

Specification

Developer sees up

Figure 14.1 Layers of perception

As this is a chapter about how a software system is built; it represents a confluence of many ideas discussed throughout the book. The ideas intersect closely with one another and we will cross-reference related discussion as we go along. Figures 14.2 and 14.3 puts into perspective how different chapters of this book relate to the discussions around workflows and phases presented in this chapter.

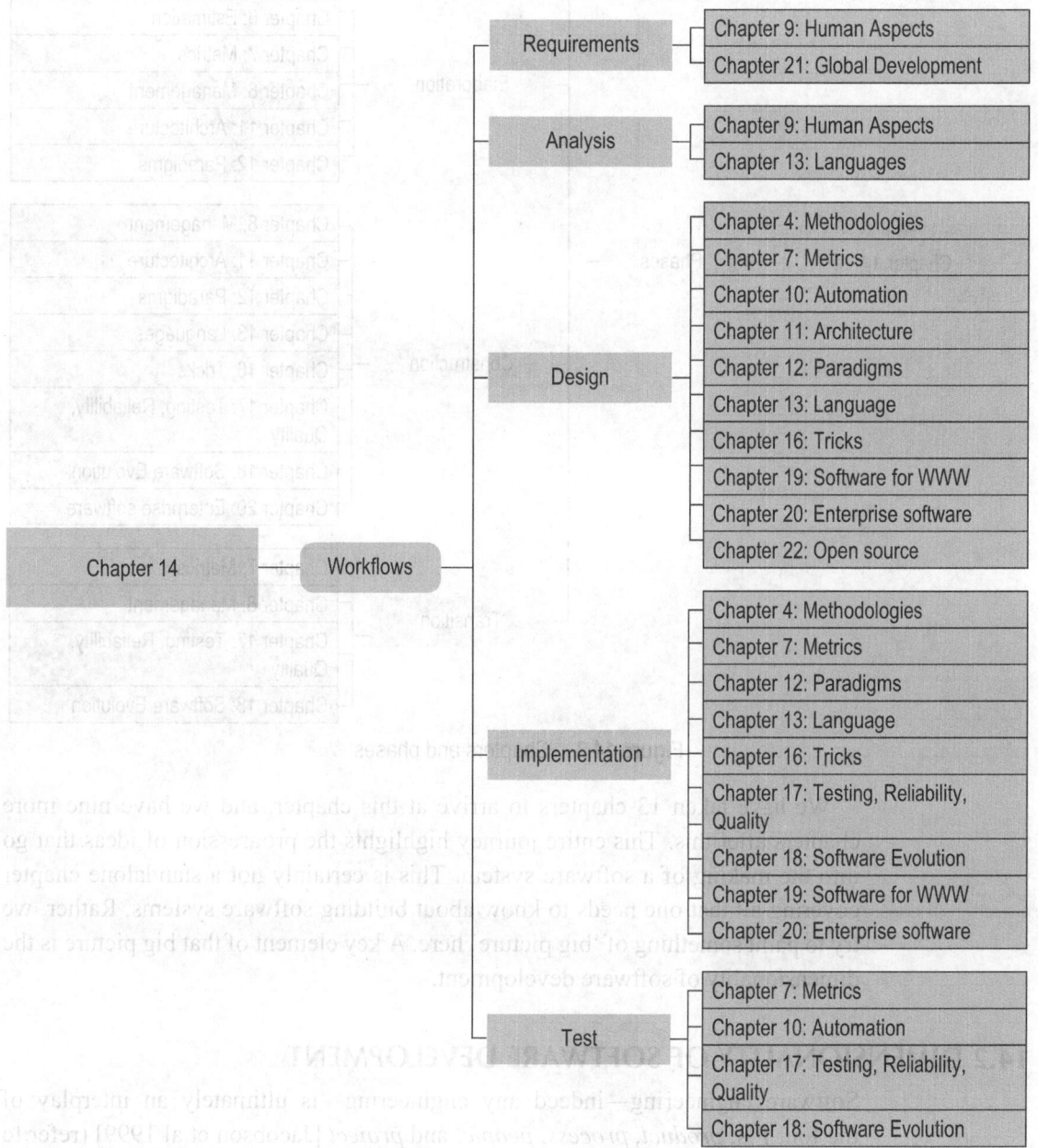

Figure 14.2   Chapters and workflows

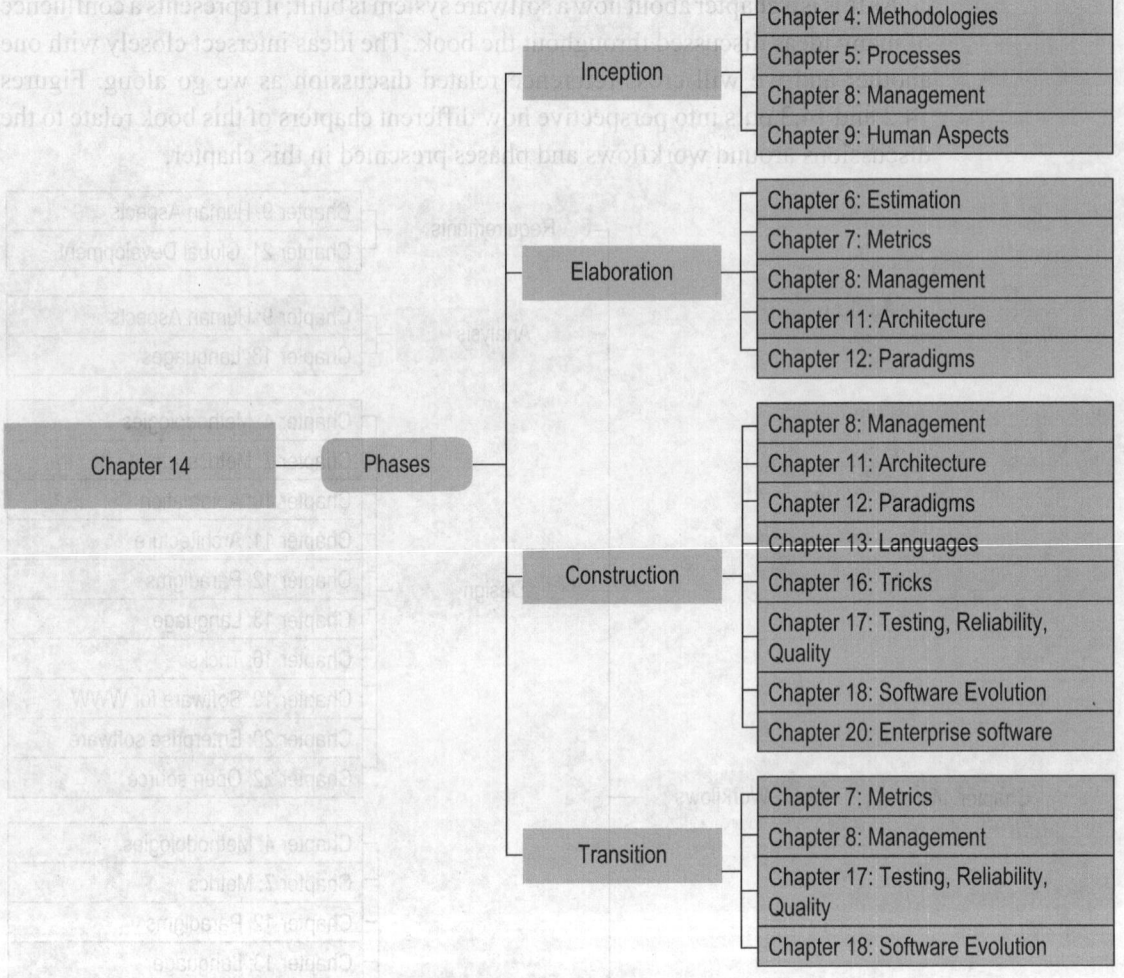

Figure 14.3 Chapters and phases

We have taken 13 chapters to arrive at this chapter, and we have nine more chapters after this. This entire journey highlights the progression of ideas that go into the making of a software system. This is certainly not a standalone chapter covering all that one needs to know about building software systems. Rather, we try to paint something of 'big picture' here. A key element of that big picture is the dimensionality of software development.

14.2 DIMENSIONALITY OF SOFTWARE DEVELOPMENT

Software engineering—indeed any engineering—is ultimately an interplay of the four P's: *product, process, people,* and *project* [Jacobson et al 1999] (refer to Chapter 8). Very simply we may say that the 4 P's represent four basic questions any software engineering endeavor has to address: *what, how, who,* and *when*. The product is all about 'what' needs to be built, the process is 'how' it will be built, the people are those 'who' will build it, and the project is by 'when' it needs to be built. Each of the four P's is an agent in a network of interactions, influencing, and being influenced by others. The objective of software development is to harness the power of this interaction, and confront its challenges. With reference to Figure 14.4, we observe that product (the 'what') inspires process (the 'how') which in turn guides people (the 'who') who drive the project (the 'when'), which in turn feeds into product. This is not to say there are no other inter-connections. For example, the product is influenced by the people who build as well as use it, and the process certainly impacts the project. In Figure 14.4, we show one cyclic flow of the four P's. Now let us cut across the plane in two ways. Product and process together represent the major technological concerns; whereas people and project are conventionally under the purview of management (see Chapters 8 and 9). On the other hand, project and product are the focus of the phases, whereas process and the people concern the workflows. And workflows and phases are our

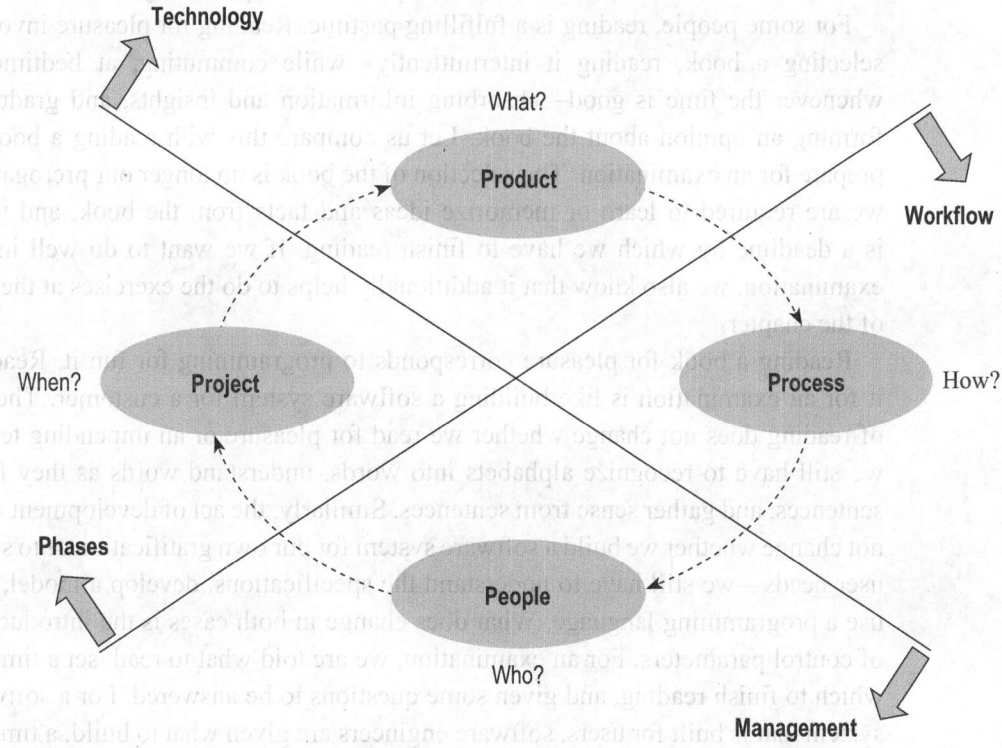

Figure 14.4   Demarcation of the software development plane

area of interest in this chapter. The way we have cut up the development plane may not be unique; there may be other ways to group each of the four P's. The point Figure 14.4 tries to make is that phases and workflows are born out of an interaction of several factors, and they address the interplay of technological and managerial concerns in software development. This leads us to the dimensionality of software development.

The dimensionality of software development arises from the intersection of the technological and the managerial aspects. In an engineering enterprise, technology and management go hand-in-hand. It is the task of management to ensure that technology fulfills business aspirations, whereas technology guides the way business aspirations are managed. As we shall see in this and the next chapter, software development is two-dimensional in nature. It is a matrix of phases and workflows. Phases predicate the development life cycle so that all stakeholders know where the development started from, where it is now, and which way it is heading. Workflows are concerned with how the software gets conceived, specified, designed, built, and deployed. Why do we need to see software development in two dimensions? Let us first address this question metaphorically.

For some people, reading is a fulfilling pastime. Reading for pleasure involves selecting a book, reading it intermittently—while commuting, at bedtime or whenever the time is good—absorbing information and insights, and gradually forming an opinion about the book. Let us compare this with reading a book to prepare for an examination. The selection of the book is no longer our prerogative, we are required to learn or memorize ideas and facts from the book, and there is a deadline by which we have to finish reading. If we want to do well in the examination, we also know that it additionally helps to do the exercises at the end of the chapter.

Reading a book for pleasure corresponds to programming for fun it. Reading it for an examination is like building a software system for a customer. The act of reading does not change whether we read for pleasure or an impending test— we still have to recognize alphabets into words, understand words as they form sentences, and gather sense from sentences. Similarly, the act of development does not change whether we build a software system for our own gratification or to serve user needs—we still have to understand the specifications, develop a model, and use a programming language. What does change in both cases is the introduction of control parameters. For an examination, we are told what to read, set a time by which to finish reading, and given some questions to be answered. For a software system that is built for users, software engineers are given what to build, a time by which to finish building and quality attributes that must be met. (But interestingly, unlike the pre-defined syllabus of an examination, user expectations from a software system are prone to change. See Chapter 18 for further discussion.)

Now, what specifically are the control parameters for software development? Evidently, time is a control parameter. But more importantly, a tracking mechanism for progress is needed whenever we need to finish something on time, and meet certain acceptability criteria. While preparing for an examination, the end-chapter exercises serve as parts of this tracking mechanism, they help evaluate what has been learnt so far and how well we are prepared for what comes next. The act of software building involves a number of sub-acts—understanding Requirements, analysing, designing, implementing, and Testing. These are the so called 'workflows'. The 'phases'—*Inception*, *Elaboration*, *Construction*, and *Transition*—constitute the control mechanism; helping us understand where we are in the development life cycle. A phase is defined as 'the span of time between two major milestones of a development process' [Jacobson et al. 1999], whereas a workflow is a sequence of activities performed with a view to completing one software development process. This two-dimensionality of software development—the act of development and the structure to monitor it—ties in closely with the concept of iterations and increments as discussed earlier. As the system is developed through increments, each iteration addresses concerns across phases and workflows. Figure 14.5 highlights the key questions across workflows and phases.

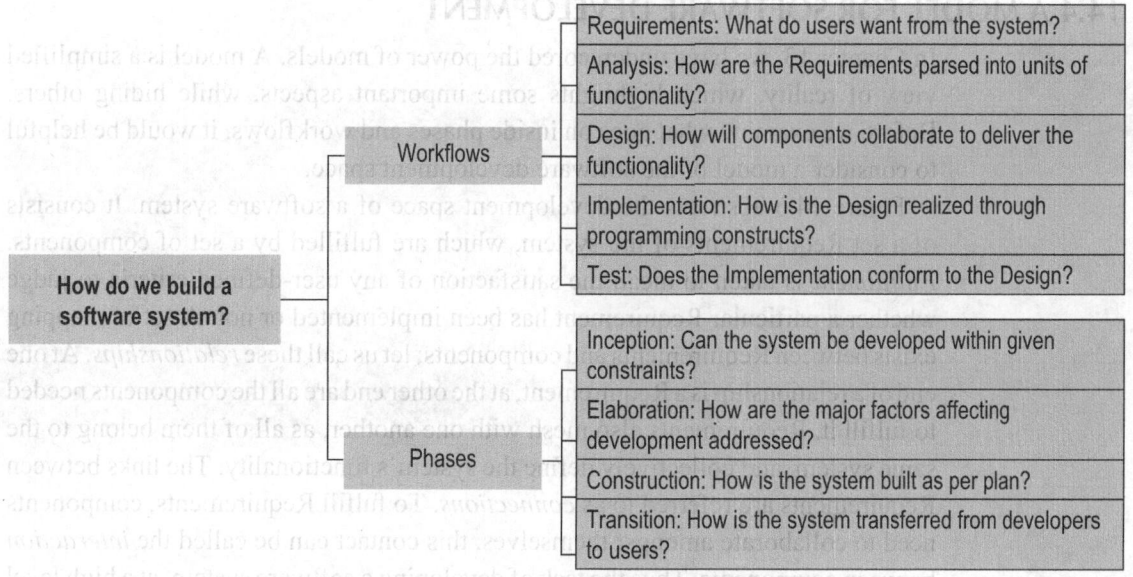

Figure 14.5 Questions of software development

14.3 PHASES AND WORKFLOWS IN PERSPECTIVE

The development of a software system often starts from a casual conversation, or fragment of an e-mail, or chance encounter at the canteen. 'Can you build a system that would help us get ahead of the competition?', or if (fortunately) more specific, 'Can you help us generate reports faster and have them available online?'. How does a software engineer proceed from such a start? The answer lies in being able to ask the right questions.

What do we do when faced with a problem? The time-tested divide-and-conquer strategy tells us to break down a large problem into smaller, more manageable problems, solve them seperately, and then synthesize the solutions. The goal of breaking down software development into phases and workflows is also to divide and conquer. Starting with the basic question—*how do we build a software system?*—each workflow and phase addresses a sub-question (Figure 14.5).

Our discussion of the phases and workflows in the following sections is drawn heavily from the book *The Unified Software Development Process* [Jacobson et al. 1999]. While discussing the Unified Process in Chapter 5, we highlighted its scope. Software development presents many layers of complexity. The Unified Process addresses such complexity through a number of powerful mechanisms, many of which are beyond the scope of this book. In this chapter, we discuss the key themes of each workflow and phase.

14.4 A MODEL FOR SOFTWARE DEVELOPMENT

In Chapter 13, we have underscored the power of models. A model is a simplified view of reality, which highlights some important aspects, while hiding others. Before we examine what goes on inside phases and workflows, it would be helpful to consider a model of the software development space.

Figure 14.6 abstracts the development space of a software system. It consists of a set Requirements of the system, which are fulfilled by a set of components. *Fulfillment* is taken to mean the satisfaction of any user-defined criteria to judge whether a particular Requirement has been implemented or not. A set of mapping exists between Requirements and components; let us call these *relationships*. At one end of a relationship is a Requirement, at the other end are all the components needed to fulfill it. Requirements also mesh with one another, as all of them belong to the same system, and collectively define the system's functionality. The links between Requirements are referred to as *connections*. To fulfill Requirements, components need to collaborate amongst themselves, this contact can be called the *interaction* between components. Thus the task of developing a software system, at a high level of abstraction, can be viewed as: Given a set of connected Requirements, devise

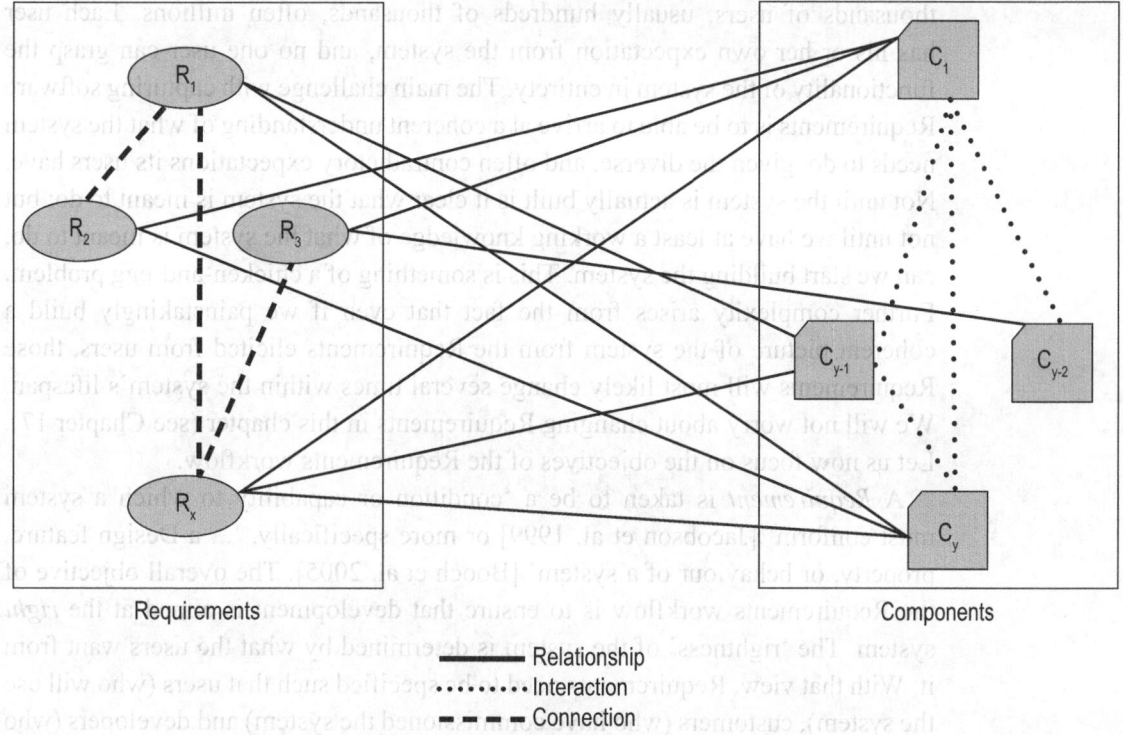

Figure 14.6 Model for software development

a set of interacting components, such that the system's functionality is delivered within given constraints. Evidently, as Requirements change, their relationships with components also change concomitantly, the components themselves and their interaction must also change, to fulfill the changed Requirements. Thus Figure 14.6 is a *snapshot* of a particular instance of the development life cycle. As we study workflows and phases, it would help to keep this model in perspective.

14.5 WORKFLOWS

Workflows are the *activities* of the software development life cycle: gathering Requirements, Analysing, Designing, Implementing, and Testing. They are the sequence of actions that guide a software system's journey from conception to completion.

14.5.1 Requirements

A typical software system is built by tens or hundreds of developers. Before the coming of the Web, it was customary to have a software system used by hundreds or maybe thousands of users. Now-a-days, a Web-based system minimally has thousands of users, usually hundreds of thousands, often millions. Each user has his or her own expectation from the system, and no one user can grasp the functionality of the system in entirety. The main challenge with capturing software Requirements is to be able to arrive at a coherent understanding of what the system needs to do, given the diverse, and often contradictory expectations its users have. Not until the system is actually built is it clear what the system is meant to do; but not until we have at least a working knowledge of what the system is meant to do, can we start building the system. This is something of a chicken-and-egg problem. Further complexity arises from the fact that even if we painstakingly build a coherent picture of the system from the Requirements elicited from users, those Requirements will most likely change several times within the system's lifespan. We will not worry about changing Requirements in this chapter (see Chapter 17). Let us now focus on the objectives of the Requirements workflow.

A *Requirement* is taken to be a 'condition or capability to which a system must conform' [Jacobson et al. 1999] or more specifically, '... a Design feature, property, or behaviour of a system' [Booch et al. 2005]. The overall objective of the Requirements workflow is to ensure that development is aimed at the *right* system. The 'rightness' of the system is determined by what the users want from it. With that view, Requirements need to be specified such that users (who will use the system), customers (who have commissioned the system) and developers (who will build that system) can agree on what the system *should* as well as *should not* do. The closer users can relate to the Requirements that are captured and documented by the developers, the easier it will be to reach an agreement about the system's intent, rather than end up building a system user's do not want.

What do we do in the Requirements workflow? Those activities are discussed next [Jacobson et al. 1999].

14.5.1.1 List Candidate Requirements

A 'wish list' is something that enumerates what we do not have now, but would like to have in future. Our wishes usually have priorities. For a software system that is just being conceived, a *feature list* is something that corresponds to a wish list. It is the first cut at identifying Requirements. Some of the features may graduate to full-blown Requirements, some of the features may be shelved for the time being, and some may be discarded as being too wishful. The candidate Requirements for development come from the feature list. Each feature has a name and a brief description. In addition, each feature in the list can be annotated with *planning values* such as *status* (whether accepted as a Requirement), *estimated Implementation cost*, *priority* (such as vital, essential, desirable), and *risk* (associated with implementing the feature). The planning values come out of consultation between stakeholders, and at this point in the development life cycle, they are merely educated guesses.

14.5.1.2 Understand System Context

When the development life cycle starts, developers may not have a working knowledge of the context in which the system must function. *Domain modelling* and *business modelling* are activities towards acquainting developers with the system's context. A domain model describes the important entities of the system and their relationships. It helps developers build a glossary of key terms and their definitions. For example, in a banking application, the domain model would consist of notions such as *account, deposit, withdrawal, account balance*, as well as the information on how they are related. A business model, on the other hand, aims at describing the business processes that the system must support. A business model can be said to contain the domain model, as the domain is defined by the business context. The business model for the banking example would contain details such as how an account is opened, how money is transferred from one account to another, the rules and conditions for dishonoring a check drawn on an account etc. Between the business model and the domain model, developers seek to capture all the contextual issues that influence the development of the system.

14.5.1.3 Capture Functional Requirements

Equipped with the feature list and an understanding of the system's context, the system's functional Requirements are next explored. Very simply, functional Requirements relate to the users' *transaction* with the system. As we will see in the next chapter's case study, functional Requirements are best distilled through *use cases*, which are the various ways users can use the system. As the name suggests, functional Requirements describe how a system *functions*—what it does to be of value to the user. Although use cases represent functional Requirements and some of their relationships, there is no clear guideline for translating functional Requirements into use cases. Use cases are part of the developer's idiom. In the Requirements workflow, Requirements need to be described in the language of the end-user. Thus they are usually expressed as a natural language description of how the system functions.

14.5.1.4 Capture Non-functional Requirements

Merely working according to functional Requirements is not all that a software system is expected to do. It also needs to conform to some non-functional parameters affecting user experience. These non-functional parameters are generally classified as *usability, reliability, performance, supportability* [Grady 1992]. Specific systems may have some other non-functional Requirements like security, ease of navigation, etc. To understand the significance of non-functional Requirements, let us go back to our banking example. Suppose we are building a Web-based system for the bank's account holders to check their account balances online. There will surely be a screen which lets users enter their user ID and password and the system will allow entry only if these credentials are successfully verified. The functional Requirement is for a user to be allowed to enter if the user ID and password combination is correct. But what would happen if the system takes an unreasonable amount of time—say more than ten minutes—to decide whether a given user ID/password combination is correct? Most likely, users will lose patience. So, the implicit non-functional Requirement here is for the system to judge the credentials *within an acceptable* amount of time. Non-functional Requirements are often implicit and thus prone to be ignored initially. It is very important they are identified, documented, and addressed from the start of the Requirements workflow.

14.5.1.5 Requirements in Summary

As we shall see in the next chapter's example, use cases play a vital role in driving the development process from the Requirements workflow onwards. Intuitively, a use case is a particular way in which the system can be exercised by a user, such that the user derives some value from it. When we describe a use-case, the system is essentially viewed as a black-box. We remain oblivious of its inner workings. User interaction with the system is essentially viewed as a sequence of actions (on the user's part) and responses (on the system's part). We will illustrate the Transition from Requirements to use cases in the next chapter.

Figure 14.7 depicts what goes into the Requirements workflow and what comes out of it. The output of the Requirements workflow is the input to the Analysis workflow. As we move into the next workflow, it is important to underscore that a key activity of the Requirements workflow is communication between stakeholder's of the system. Developers should assume nothing about the system, and users should be free to say whatever they want from the system. Developers should not think like users, or users like developers. The communication can happen through talking or writing, but it must happen in a language, and dialect users are most comfortable with. Requirements elicitation is *not* the time for judging the value of, or exploring interconnections amongst Requirements. For that and much more, we have the Analysis workflow.

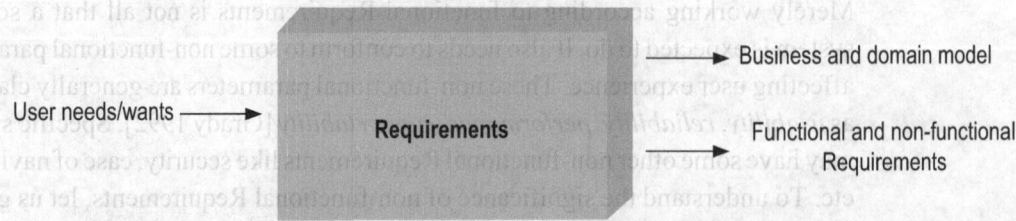

User needs/wants → **Requirements** → Business and domain model

→ Functional and non-functional Requirements

Figure 14.7 Requirements: input and outputs

14.5.2 Analysis

Let us assume four friends have gone to a restaurant for dinner. The menu card lists many items, each equally delicious. After much indecision, the order is placed. To sample the variety of the restaurant's fare, each of the four diners order different dishes, as widely apart as they can be. As they wait eagerly for the food to arrive, what do you think is happening in the kitchen? There can be separate sets of ingredients for each of the menu items and separate chefs to prepare them. But this is hardly feasible, if the restaurant has to make profit as well as satisfy its customers. No matter how *independent* each patron's order is, back in the kitchen, the items are prepared from a common pool of ingredients and by the same set of cooks. True, a particular item can have a special ingredient—like saffron in *biriyani*—and one chef may be an expert in a particular item. But on the whole, no matter how different the items of an order are, they are made from common resources.

When Requirements are captured, they come in the language of the user (the menu card) and they are taken to be independent (each friend orders different items). However, in the developer's domain (the kitchen), the system has to be built using the common set of modelling and coding constructs (same ingredients) by the same developers (same chefs) so that it fulfills all the Requirements of the system (serve the entire order to the four friends). The key to doing this is to identify where and how the Requirements interact and how such interaction can influence how the system is developed. The dinner analogy should not be stretched too far. But the point is, like preparing dinner, developing a software system also needs to find commonality between the consumer's diverse needs and leverage that commonality towards a solution. A key theme of the Analysis workflow is to understand how and where Requirements interact and what it means for the system.

With Analysis, we are getting into the reflective aspects of software development. The act of analysing involves breaking down large, coarse-grained observations and concepts to smaller, fine-grained ones so that their inter-relations can be studied more carefully. Although there are some common strategies to get started, there is no step by step recipe for Analysis, as every system or situation is unique. The more experienced we get at analysing, the better we are at discerning trends and patterns early.

The Analysis workflow takes as input the artefacts created during the Requirements workflow. Starting with the description of Requirements—both functional and non-functional—and the analyst's insight, the Analysis workflow involves the activities [Jacobson et al. 1999] described in the following subsections.

14.5.2.1 Specify Requirements in Developer's Language

As we mentioned earlier, Requirements need to be described in the language of the users, which is most commonly natural language. However, as discussed in Chapter 13, natural language is far removed from the ultimate medium of software development i.e. programming languages. For large and complex systems, there is no ready translation mechanism from natural language to programming language, to go from the system's Requirements to the system itself. It is a step-by-step process; at every step essential ideas need to be discerned and inessential ones discarded. Analysis is the first step in this translation process. During Analysis developers internalize the system's description (as expressed in the user's language), examine it in the light of their experience and skill, and come up with the system representation in the developer's language. The developer's language is not merely another set of vocabulary and grammar rules, it is a way of thinking and viewing different from the user's perspective.

14.5.2.2 Identify Requirement Interconnections

A central factor in seeing the system from the developer's point of view is to identify how Requirements are interconnected. Requirements come from users as seemingly unrelated chunks of functionality. However, the fact that all of a system's Requirements ultimately relate to the same business domain, leads to Requirements being inevitably interlinked. During the Requirements workflow, the focus is to know all that the user wants. But during Analysis, we need to look for interconnections. For a given set of n Requirements (related to the same business domain), one integrated software system needs to be built, not n different systems each catering to one Requirement. How does a single system address a set of diverse but related needs? The answer lies in collaboration. Referring to Section 14.4, we recognize that the software system will consist of components, each of which delivers specific responsibilities and cooperates with other components to collectively fulfill the system's Requirements. To be able to assign responsibilities to components, as well as determine how they collaborate, we first need to know how the Requirements are interconnected. During Analysis, developers have to discover these connections amongst Requirements, so that they can get started on thinking who will do what amongst the system's components. Note however, Analysis is not the time to think about *how* components will fulfill their tasks. As we shall soon see, that is the realm of Design.

To be able to start considering who will do what in the system, we need to know exactly what needs to get done.

14.5.2.3 Detect Ambiguities and Inconsistencies in Requirement Descriptions

Describing Requirements in natural languages is good for stakeholder consensus, but bad for precision in Requirement specification. This is due to the scope natural languages offer for implications, innuendoes, and reading between the lines. Thus unwittingly, ambiguities and inconsistencies creep into Requirement descriptions. During Analysis, the aim is to find out exactly what the system needs to do, as well as what it does not need to do. Each Requirement is thus parsed closely to make sure it means what it means, and nothing but what it means. As we saw during the discussion of the specification language Z in Chapter 13, how excruciatingly detailed specifications need to be, if they are to be exact. Not all software systems need that degree of exactitude, but Analysis needs to make sure major confusions related to a Requirement's scope are weeded out.

With the connections amongst Requirements identified, and the ambiguities removed the *internal* view of the system can now be developed.

14.5.2.4 Develop an Internal View of the System

While understanding Requirements, the system is viewed as a black-box. It is seen to respond to user actions, but we do not worry about what is going on inside the system, or how the system behaves the way it does. The system is essentially viewed from the outside. But at the time of Analysis, developers need to look inside the system. To develop the internal view of the system, the first step is to think in terms of placeholders for chunks of functionality. These are the so-called *Analysis classes*.

14.5.2.5 Identify Analysis Classes and their Collaborations

Analysis classes are not classes in the strict object-oriented sense of the term. But as we shall see later, the 'actual' classes can grow out of Analysis classes during Design. Analysis aims at a conceptualization of the system's components, in terms of assigning responsibilities. The focus here is entirely on abstraction; without Design or Implementation concerns. Analysis classes are of three types: *boundary*, *entity*, and *control*.

Let us try to visualize how a software system serves users. First, the system should offer some way to accept user input and exhibit system output back to the user. Second, the system must be able to store and process the information users have supplied, or has been generated within the system, or gathered from other systems. Third, based on user inputs and the stored or processed information, the system must be able to take decisions towards fulfilling user needs. These three aspects map to the boundary, entity, and control classes respectively. Boundary classes facilitate information transfer in and out of the system. Entity classes help in the storage and retrieval of information within the system. Control classes govern the processing of information, and the system's decision making. To reiterate an important point, Analysis classes are the *preliminary* placeholders of functionality, and *not* the final components to be implemented in code. It is quite likely that one boundary or entity or control class identified during Analysis spawns multiple components or becomes redundant during Design.

As we saw in section 14.4, a software system, at its highest level of abstraction is a set of interacting components each of which have specific responsibilities and collaborates with other components to collectively fulfill the system's Requirements. Identifying the Analysis classes is the first step towards developing these components. In addition to being placeholders for functionality, they also help understand how components need to collaborate. Components collaborate by passing messages between themselves.

How do we identify Analysis classes and their collaborations? To get started, there is a simple rule of thumb. Given the description of a particular Requirement or use case in natural language, identify the nouns (or noun phrases). These are candidates for Analysis classes. Of course, not all nouns are equally important in the system's overall context. Which noun is important and which is not can be judged on the basis of the business model, the domain model, and the glossary compiled during the Requirements workflow. Out of the nouns picked as Analysis classes, some will be demoted to being merely attributes (data members) of classes during Design. Once the significant nouns are selected, the attention is turned to the verbs. Verbs associating one noun with another indicate collaboration between the Analysis classes. For example, let us analyse the statement, 'an **account-holder** can **transfer money** between connected **accounts**'. The significant nouns are 'account-holder', 'money', and 'account'. These are candidates for Analysis classes, perhaps entity classes, since they are associated with some information uniquely identifying them. During Design, we will decide whether or not to keep 'money' as a separate class or make it an attribute of the 'account' class. 'Transfer' is the verb which connects 'money' with 'account' and 'account holder'. The act of transferring money between accounts by an account-holder suggests collaboration between these entities, which is realized by the sending of messages, which translates to method calls between components. So we can think of a `transfer()` method called on the source account by an account-holder, passing as parameter the amount of money to be transferred and the destination account identifier. But we are still in the Analysis workflow. So, we should not get too much into the details of the collaboration such as the exact signature or body of the `transfer()` method. At this point it is enough that we have recognized the Analysis classes—

account-holder, account and money—and one probable plank of collaboration, transfer() between them.

There is an interesting artefact, *Class-Responsibility-Collaboration Cards* or *CRC Cards* (http://c2.com/doc/oopsla89/paper.html) which can be used during Analysis. Each CRC Card lists the name of the Analysis class, its responsibilities, and the other classes it collaborates with. CRC Cards can help the development team do a role-playing exercise. Each team member pretends to be an Analysis class, scrutinizes what it has to do and the help it needs from other classes. The case study in the next chapter illustrates the use of CRC cards.

14.5.2.6 Analysis in Summary

In a nutshell, Analysis bridges the user world with the developer world. We start Analysis usually with natural language text as Requirement descriptions (and use case flows). We end Analysis with a set of tentative components—the Analysis classes—and how these collaborate to deliver the system's Requirements. In the meantime, we have drilled down into what Requirements mean and how they are related to one another. It is important to understand that although Analysis and Design are distinct, they are closely related. Like day and night, it is difficult to pinpoint when one ends and the other begins. Little bit of Design forethought is helpful during Analysis; but getting bogged down in Design details is not. If the Requirements workflow concerns with the *what* and Design with the *how*, Analysis definitely needs to lean more towards clarifying the *what*.

At the end of Analysis, we have the set of Requirements that have come from the user. We also have a set of tentative components—Analysis classes—that developers have come up with. With reference to Figure 14.8 Analysis helps us detect connections between Requirements, and the tentative links between Requirements and components. Design will help determine how best the components should interact. Analysis is primarily about detection, Design is about determination. Figure 14.8 summarizes what goes into the Analysis workflow and what comes out of it.

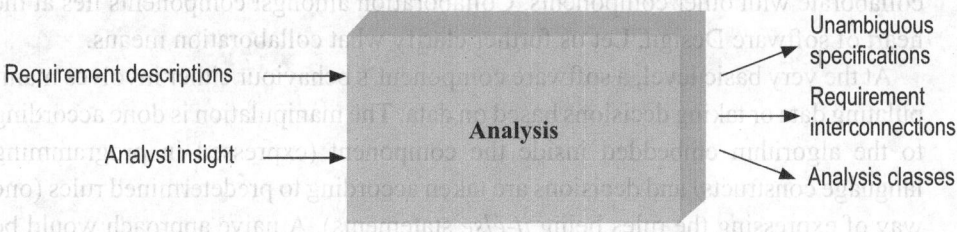

Figure 14.8   Analysis: inputs and outputs

14.5.3 Design

'Design' is a word of many meanings. For some, Design relates to beauty or aesthetics, for others it is about creating something new, while some others think of Design in terms of utility. The focus of our discussion here is the *Design workflow* of the software development life cycle.

Designing a software system is about making choices. We had the Requirements; we have distilled and disambiguated them during Analysis, now we need to decide how software components will fulfill the Requirements. As mentioned earlier, a software system is a set of interacting components, each of which needs to do its own tasks, as well as help other components with their tasks. The choices to be made during Design are ramifications of one core question: Who will do what in the system? In successful teams, success depends not only on how well each member does his or her own work, but also on the level of cooperation between the members. Similarly, in a software system, collaboration of components is essential. The act of assigning responsibilities to components and deciding how they should collaborate is largely subjective and there are no fit-all formulas. Thus, Design is deeply iterative. No matter whether iterative or sequential development approach is being followed, Design undergoes many refinements and counter-refinements. Design needs not be right the first time. Designers, who try to get it right the first time, usually get it wrong.

As Design takes off from Analysis, the focus shifts from the abstract to the concrete. During Analysis, our interest was more on understanding the full extent of what the users want from the system. With Design, we need to start deciding how the system is going to deliver what the user wants. This is involved with a Transition from the *static* to the *dynamic* view. For a software system to *run*, there has to be some movement. This movement is very different from the mechanical movement of car wheels or airplane wings. Movement in software system is the movement of information, as data and decisions. So, the dynamic view developed during Design needs to specify communication between components. The Analysis classes that were placeholders of functionality are now made more concrete by adding data members (attributes) and behaviours (methods). This equips them to collaborate with other components. Collaboration amongst components lies at the heart of software Design. Let us further clarify what collaboration means.

At the very basic level, a software component's behaviour either involves manipulating data or taking decisions based on data. The manipulation is done according to the algorithm embedded inside the component (expressed in programming language constructs) and decisions are taken according to predetermined rules (one way of expressing the rules being *if-else* statements). A naive approach would be to assign one component per Requirement or use case, expecting it to do all that is needed to deliver that specific functionality. But since all the Requirements relate to the system's common context, they are interrelated, as was discovered during Analysis. So it is very likely the data manipulated and the decisions taken by a particular component will affect other components as well. Thus each component delivering a piece of functionality by itself, will lead to needless duplication of code and effort. The other extreme is to have one component deliver *all* the Requirements—the classic *monolithic* system. Here the problem is primarily with enhancement and maintenance. Every time Requirements change, the same single component will need to be modified, and inadvertent errors are likely to creep into code that implements other (unchanged) Requirements. So we do not want to have one component for each Requirement, or a single component for all Requirements. We need to have a set of components somewhere in between, each of which fulfills its own responsibilities, and exchanges information with other components. A component fulfills its own responsibility by running the code inside it, and it exchanges information by calling methods on other components, or when other components call its method. This is how components collaborate. During Design we need to decide what each component will do, and which other components it needs to get help from.

Some of the important activities during the Design workflow [Jacobson et al. 1999] are discussed in the following subsections.

14.5.3.1 Look at Specific Technology

Till the Design workflow, the development life cycle is more or less technology agnostic. Although developers may have some ideas about which programming language, database, or application server may be used, these options should be deliberately not delved into. During Design however, specific technology has to be looked at and decided upon. Technology may not only mean a language or a database product, it may also be a question of paradigm (see Chapter 12). What we said in the last section about collaboration of components remains true irrespective of the technology. However, each technology has its own mechanisms to facilitate such collaboration. The earlier a specific technology is decided upon during Design, the easier it is for developers to leverage its benefits.

14.5.3.2 Decompose System Into Implementation Units

Design leads into Implementation. Thus Design workflow is the time when components have to be organized into *subsystems* or *packages* based on their collaboration structure and functional coherence. This will later aid delegation of work amongst development and Testing teams. Additionally, the segregation of systems into subsystems helps detect and clarify crucial interfaces within the systems, as well as with external systems. Interfaces can be thought to represent the contracts which parts of the system must honor amongst themselves, for the system to function smoothly.

14.5.3.3 Engage in High-Level Design

In Chapter 11, we have discussed how architecture represents a set of decisions regarding the structure and interfaces of a software system. Using the insights gained during Analysis, high-level Design builds on architectural guidelines to specify all components, and their interfaces with other components in terms of a specific technology. It creates something of a *shell* of the system, to be filled in during low level Design.

14.5.3.4 Engage in Low-Level Design

Low-level Design is the last logical step before Implementation. Ideally, low-level Design should specify every Design detail, down to the last data element and method signature. It is quite likely, and permissible, for low-level Design and Implementation to overlap to a significant extent. Even the most experienced designer will leave some aspects of the Design to be refined during Implementation. If such modifications maintain the architectural integrity of the system, they should certainly be allowed. The distinction between high-and low-level designs is essentially one of the levels of specificity and details.

14.5.3.5 Design in Summary

Figure 14.9 shows what goes into the Design workflow and what comes out of it. Although the most focused Design effort is invested during this time, there is an element of Design in every workflow. A good Design is one which serves the present purpose effectively, while allowing the system to evolve elegantly in the future. These two aims are some times contradictory; the designer has to use best judgement to resolve the tension.

At the end of the Design workflow, we have a *blueprint* for building the system. Just as a building's blueprint lists physical details such as the thickness of walls, width of doors and windows etc. and the configurations in which these elements are arranged, the software system's Design lists the internal structure of the components

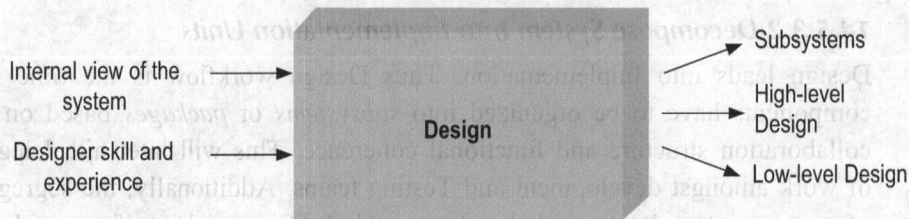

Figure 14.9 Design: inputs and outputs

and how they collaborate. Based on this blueprint, the Implementation workflow builds the system.

14.5.4 Implementation

The output of the Implementation workflow is by far the most tangible of what comes out of the entire development life cycle: working code. A large part of Implementation is programming, although Implementation is not only about programming. The following subsections outline the major focus areas of the Implementation workflow [Jacobson et al. 1999].

14.5.4.1 Implement Design Classes and Subsystems

As we have said earlier, Design creates the blueprint for building the system. If low-level Design is complete, we should have the skeleton of all the components in place—for object- oriented systems, data members-and methods for every class. For methods, the skeleton consists of the input parameters and the return value. During Implementation, the skeleton is added with muscle: For a method it is the steps of the algorithm that support the method's signature. As we implement, the Design that was so far only on paper comes to life in software components ready to collaborate with other components. The medium of Implementation is the programming language (see Chapter 13). The aim of Implementation is to convert the *intent* of Design into programs that can run and be tested. Testability is the most important criteria of Implementation.

14.5.4.2 Unit Test Components

As we have underscored earlier, unit Testing needs to be integral to the act of programming. Whether Implementation has lived up to Design, can only be proven through unit Testing (see Chapter 17). The Implementation workflow cannot ignore unit Testing under the assumption it will be addressed during the test workflow. Implementation's mandate is to produce software components that function and interact with other components as per the Design. However, individual components fulfilling their own tasks do not ensure the whole system will function as required. Here comes the need for integration planning.

14.5.4.3 Plan System Integrations

During Implementation, we not only need to build individual components, we also have to figure out ways of tying them together, so that the parts combine seamlessly into the whole. In Chapter 17, we have outlined several strategies for integration (top-down, bottom-up etc.). It is during Implementation that the integration plan has to be laid out. For iterative and incremental development, the integration strategy revolves around what is to be delivered in the next incremental release. If an iteration was started with the aim of delivering a specific piece of functionality, the integration plan has to include all the pieces of the system that will deliver that functionality.

14.5.4.4 Devise the Deployment Model

The objective of Implementation is to have running code. But code cannot run in vacuum. The components that are completed during Implementation must function on some hardware substratum. Components are compiled (see Chapter 13) into *executables* which run on physical machines. The process of putting software components on hardware platform to make them run is referred to as *deployment*. Deployment details need to be decided upon during the Implementation workflow. For systems with very high usage load, issues like load balancing (having more than one piece of hardware to run the same code) will also need to be addressed during this time. The deployment model consists of details such as mapping of software to hardware machines, how machines are best configured to run the code etc.

14.5.4.5 Implementation in Summary

As the Implementation workflow draws to a close, we have groups of components fulfilling particular lines of functionality. With unit Testing, we have sought to ensure each component behaves as designed. We have also planned system integrations and devised the deployment model. Figure 14.10 shows what goes into the Implementation workflow and what comes out of it.

Next, we head into the test workflow.

14.5.5 Test

Testing has been covered at length in Chapter 17. There is no other activity in software development whose importance is so chronically underestimated. This is precisely why inadequate Testing causes a large majority of software projects

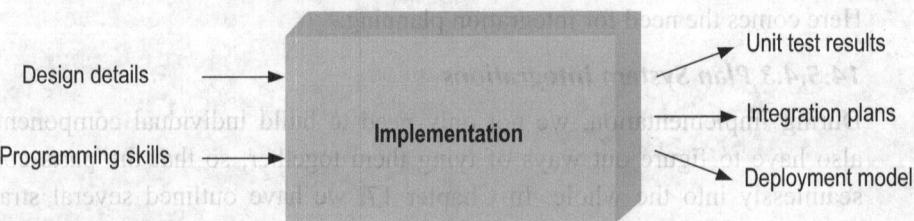

Figure 14.10 Implementation: inputs and outputs

to fail. And that is why we have devoted a large portion of Chapter 17 to Testing. We will not repeat what has already been discussed, other than reiterating that a software system can never be *completely* tested. Thus at best, the Testing workflow can aim to run a clever subset of all possible test cases, in the hope of detecting the most egregious of errors. Activities of the test workflow are discussed in the next subsections [Jacobson et al. 1999].

14.5.5.1 Create Test Cases

A test case is a Testing scenario corresponding to a particular line of functionality, often associated with a specific use case. In a test case we specify what the input to the system needs to be, what the expected output is, and what conditions or constraints the system must be functioning under. Instead of arbitrarily assigning test cases, it is helpful to map them out to use cases, so that important lines of functionality are not overlooked during Testing.

14.5.5.2 Execute Test Procedures

A test case needs to be run either by a human tester or an automated test program. A test procedure describes the steps which must be performed to run the test case. It has to be at a level of detail that enables someone not familiar with the Design and Implementation of the system to run the test cases effectively.

14.5.5.3 Analyse Test Results

Once the results of Testing are available, they have to be analysed. When an unexpected test result comes up, it can either be due to a bug in the system, or a Design feature that was not accounted for. However, when a test executes as expected, it does not guarantee the system is free from all errors. This dichotomy (discussed at length in Chapter 17)—a negative result implying something is wrong, but a positive result not necessarily signifying everything is right—lies at the heart of the Testing challenge. Very often Testing will result in revisiting the earlier workflows, to rectify or refine something. This is a natural dynamic of software development, irrespective of the development methodology.

14.5.5.4 Test in Summary

In the test workflow, we expect to check whether the system is functioning as designed. Due to its very nature, Testing is never complete. Test cases need to be chosen such that they help us test the most vulnerable parts of the system. In Figure 14.11, we show the input and output of the test workflow.

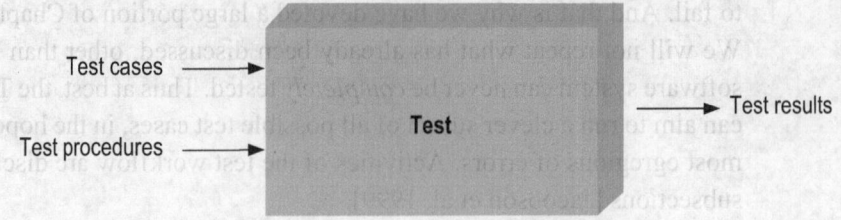

Figure 14.11   Test: inputs and outputs

14.6 PHASES

As described in the preceding sections, workflows delineate the steps in building a software system, from a mere idea to a working product. Software engineers need to ensure software development is on time and within budget. Towards that end, the phases—Inception, Elaboration, Construction, Transition—guide the division of work and the shifting of focus and priorities across the development life cycle. However, as shown in Figures 14.12 and 14.13, each phase has different time and

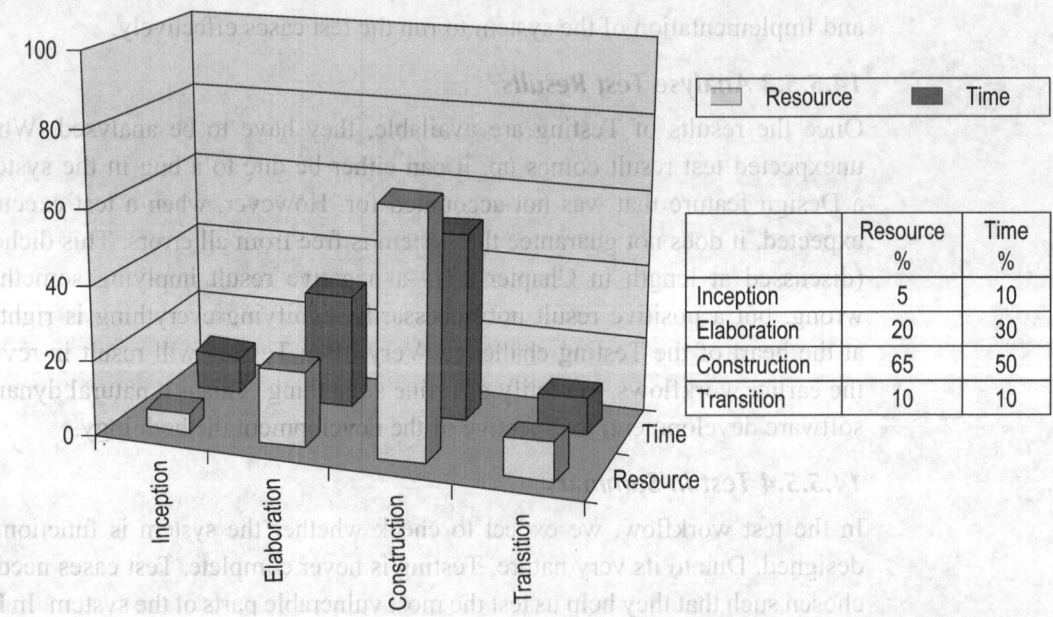

	Resource %	Time %
Inception	5	10
Elaboration	20	30
Construction	65	50
Transition	10	10

Figure 14.12   Time and resources needs (in appropriate units) across phases for typical projects (inspired by Jacobson et al. 1999)

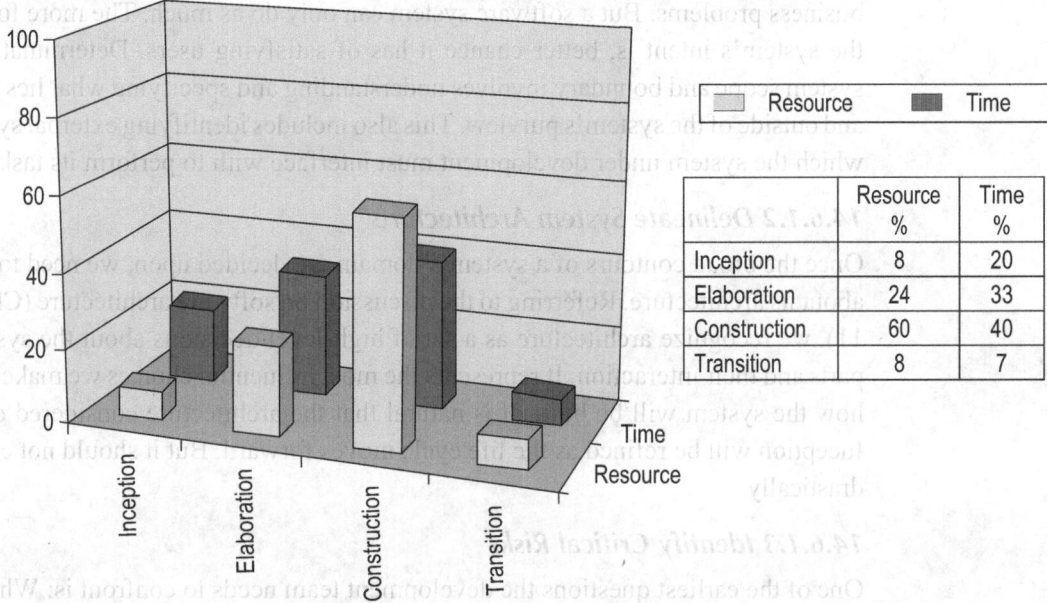

	Resource %	Time %
Inception	8	20
Elaboration	24	33
Construction	60	40
Transition	8	7

Figure 14.13   Time and resources needs (in appropriate units) across phases for difficult projects (inspired by Jacobson et al. 1999)

resource needs, which in turn depend on whether a project is *typical* or *difficult* [Jacobson et al. 1999]. (The percentages in the figure are broadly illustrative, exact time and resource needs of specific projects may differ.) As evident, for difficult projects, relatively more time and resource is spent during Inception and Elaboration. Difficult projects often have many unknown factors. Thus more attention needs to be focused early on in the life cycle.

Each phase ends with a milestone, which indicates the progress made so far in the development life cycle. We will now discuss the objectives and activities of each phase.

14.6.1 Inception

The aim of the Inception phase is to *establish feasibility*. A software development project succeeds only when the delivered system brings satisfaction to users. Whether the commissioned system can be built within the constraints of time, budget, and other factors is an essential element of that satisfaction, so it has to be decided upon early. Towards establishing feasibility, the steps to be taken during Inception [Jacobson et al. 1999] are described in the following subsections.

14.6.1.1 Determine System Scope and Boundary

When customers commission a new software system, they usually want the moon. It is not unusual for them to expect the system to solve a widely disparate range of business problems. But a software system can only do as much. The more focused the system's intent is, better chance it has of satisfying users. Determination of system scope and boundary involves understanding and specifying what lies inside and outside of the system's purview. This also includes identifying external systems which the system under development must interface with to perform its tasks.

14.6.1.2 Delineate System Architecture

Once the broad contours of a system's domain are decided upon, we need to think about its architecture. Referring to the discussion on software architecture (Chapter 11), we recognize architecture as a set of high-level decisions about the system's parts and their interaction. It represents the most influential choices we make about how the system will be built. It is natural that the architecture considered during Inception will be refined as the life cycle moves forward. But it should not change drastically.

14.6.1.3 Identify Critical Risks

One of the earliest questions the development team needs to confront is: What can possibly go wrong with the development life cycle? The drivers of what can go wrong are the *risks*. These are factors which can disrupt development effort to the extent that schedule and budget is affected. The earlier risks are seen, the easier they are handled. But all risks are not visible upfront; the more insidious ones appear only as we go deeper into the life cycle. During Inception, the focus is not on identifying all risks, but the most *critical* ones.

14.6.1.4 Build a Proof of Concept

The best way to demonstrate the extent of a system's scope, its architectural framework, as well as potential impact of critical risks is to build a proof of concept prototype. Such a prototype also serves another important purpose: eliciting Requirements. When users are able to see a semblance of the real system in front of their eyes it helps clarify much of what they want (and do not want) from the system. But as we discussed in Chapter 4, the prototype has to be seen as what it is: A quick and dirty scale model of the system covering only a subset of its functionality. The prototype should not be modified to be turned into the system.

14.6.1.5 Inception in Summary

The objective of the Inception phase is to establish feasibility for the system under development. The end of Inception is marked by the *life cycle objective* milestone—sometimes called the LCO. The LCO presents the insights gained from the Inception activities described above, and makes a case for either going ahead with the system or aborting development without further expenditure of time and resource.

14.6.2 Elaboration

As discussed, during Inception we are concerned with *whether* the system can be built. Assuming the question has been settled in favour of moving forward, attention next moves to *do-ability* in the Elaboration phase. The next subsections describe activities that impact how development proceeds.

14.6.2.1 Create Architectural Baseline

After identifying an architectural paradigm during Inception, it is now time to supplement the choice with more details. How are the most important system features sought to be implemented? How would parts of the system interact with one another to deliver the system's overall functionality? Has a similar architecture been used for a similar system elsewhere? Which parts of system will be the most liable to change? These are some of the questions the architectural baseline addresses. We are now trying to find out how well the chosen architecture will serve as a framework for the system's development.

14.6.2.2 Identify Significant Risks

As we dig deeper, we find more factors that can derail development. These are the *significant* risks, which in addition to the critical ones discovered during Inception can cause problems. At this time, we also need to come up with a risk mitigation plan, which lists each risk by its significance, how it can be addressed, and whether the risk has already been mitigated.

14.6.2.3 Specify Quality Attributes

As discussed in Chapter 17, software quality is difficult to define, but poor quality software is easy to identify. During Elaboration, it is important to sense which attributes users will perceive to be most indicative of the system's quality. Very often the non-functional Requirements of usability, reliability, performance, supportability etc. come to be regarded as key elements of user satisfaction. The quality parameters within which the system's functionality must be fulfilled are now specified.

14.6.2.4 Extend Coverage of Use cases

User needs come as Requirements. As the user's view of the system is slowly translated into the developer's view, Requirements are converted into use cases. We have described use cases in brief in our discussion of the Requirement workflow and they will be illustrated in the case study of the next chapter. During Elaboration, use case coverage is extended to roughly 80% of the Requirements. The remaining 20% awaits further clarification from users.

14.6.2.5 Prepare Schedule and Cost Estimate

During Elaboration, the scope of the system has been clarified and major risks identified. This information is used to make an estimate of how long the project is going to take and how much resource it is going to consume. The estimation techniques discussed in Chapter 6 are now used to arrive at an estimate, that is sent for the customer's approval. If the customer agrees, the project is formally under way.

14.6.2.6 Elaboration in Summary

The *life cycle architecture* milestone—often called LCA– marks the end of the Elaboration phase. Elaboration expands on the insights gained during Inception, to come up with a clear strategy of how development will proceed.

14.6.3 Construction

After Inception and Elaboration, the next phase is Construction. As the name suggests, the aim of Construction is to *build* the system. The primary activities of Construction [Jacobson et al. 1999] are described below.

14.6.3.1 Extend Use case Identification

As mentioned, during Elaboration use case coverage of the Requirements is around 80%. The remaining 20% is covered during Construction. Once the coverage is complete, the development team has a clear picture of the entire range of user requests to the system, as well the expected system responses. The complete set of use cases need to be described in detail, so that development can proceed based on them.

14.6.3.2 Commence or Complete Analysis, Design, Implementation, and Test

As mentioned, Construction subsumes a major part of the actual building of the system. When Construction starts, it is likely that more than half of the use cases are still to be analysed, and when it ends, about 90% of Testing still remains to be done [Jacobson et al. 1999]. Thus during this phase, the remaining Analysis is completed, Design and Implementation conducted from start to finish, and Testing commenced. A major portion of Construction is thus low-level Design and Implementation.

14.6.3.3 Maintain Integrity of Architecture

While the system is being built during this phase, many new issues surrounding Analysis, Design, and Implementation come up. Only after getting one's hands dirty with the system's inner workings does one realize the importance of these details. As these concerns come to the fore—often manifested in, bugs—the system's architecture has to be revisited and modified. One of the key concerns during Construction is to ensure that in spite of the numerous changes, the integrity of the architecture is maintained. When serious Implementation related difficulties come up during Construction, it is always better to revise the chosen architectural style, instead of trying to make ad hoc patches work.

14.6.3.4 Monitor and Mitigate Risks

The crucial and significant risks that were identified during Inception and Elaboration are continually monitored during Construction. Depending on how pragmatic one has been while identifying risks, some of the risks may not materialize. However, some of the risks will surely persist and they have to be addressed as per the mitigation plan drawn up during Elaboration. The risks which cannot be successfully managed by the end of Construction will either have to be insignificant ones, or they will be serious enough to undermine schedule and budget estimates.

14.6.3.5 Construction in Summary

The *initial operational capability* milestone—abbreviated as IOC—marks the end of the Construction phase. During Construction, a major part of building the system takes place. Activities of all workflows, other than test, are significantly completed in this phase.

14.6.4 Transition

During the Transition phase, developers hand over the system to the users. The major Transition activities [Jacobson et al. 1999] are described below.

14.6.4.1 Prepare the User for the System

Prior to the Web, software systems had to be *installed* at customer premises like an air-conditioner or a refrigerator. The users had to be trained with instructions on how and how not to use the system. Now, a large majority of software systems have a Web interface. This has removed the need for specific installation procedures at user locations. However, users still need to be prepared for using the system. It is not unusual for users of a newly released system to be dissatisfied with one aspect or another, simply because they have not been trained on the new system's capabilities. Preparing users to handle the software system is a key activity during the Transition phase.

14.6.4.2 Tune the Software for Actual Operating Conditions

While a software system is built, certain assumptions are made about its usage patterns. For example, the number of concurrent users the system has to handle or the number of transactions per minute etc. can strongly influence the way a system functions in its actual operating conditions. But unless the system goes into the full usage mode, there is no way to know the exact range of these parameters. After Implementation and Testing are complete, there often arises the need to tune the system for optimal performance. This tuning is mostly incremental, and it needs to be done in every iteration rather than during final release.

14.6.4.3 Correct Defects Found after User Feedback

Not all serious defects will be detected until real users get to use the system in large numbers. During the Transition phase, development team needs to correct defects based on user feedback. This is a key step in transitioning the software system from the developer domain to the user domain.

14.6.4.4 Transition in Summary

The end of the Transition phase is marked by the product release milestone. What happens next? A software system in use is always a work in progress. So Transition is not the end of the road. For life after Transition, see Chapter 18.

14.7 WORKFLOWS ACROSS PHASES

As is evident from the above discussions, the phases provide the context for the workflows to run. How much of a particular workflow is done in a given phase varies from one phase to another. Figures 14.14 to 14.17 show the distribution

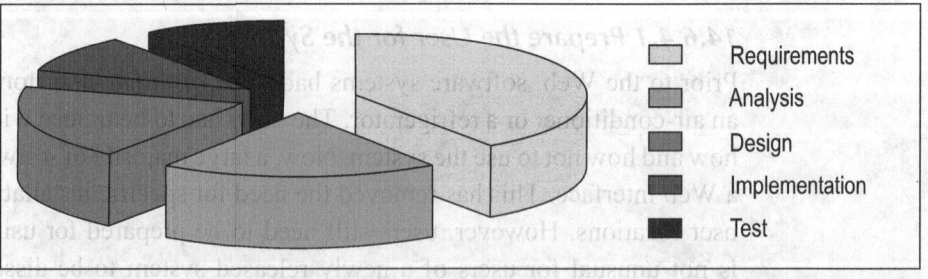

Figure 14.14 Distribution of Workflow Resources (in appropriate units) across Inception (adapted from Jacobson et al. 1999)

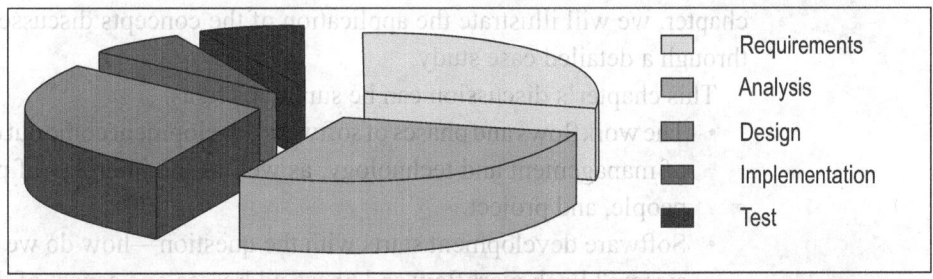

Figure 14.15   Distribution of Workflow Resources (in appropriate units) across Elaboration (adapted from Jacobson et al. 1999)

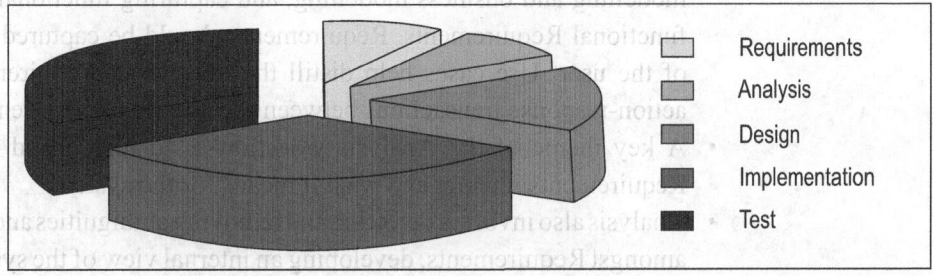

Figure 14.16   Distribution of Workflow Resources (in appropriate units) across Construction (adapted from Jacobson et al. 1999)

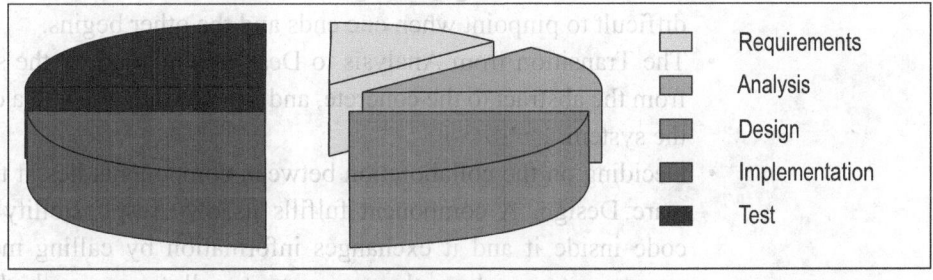

Figure 14.17   Distribution of Workflow Resources (in appropriate units) across Transition (adapted from Jacobson et al. 1999)

of workflow resources across the phases [Jacobson et al. 1999]. Once again, the figures are for illustrative purposes only, individual projects may have different distributions.

SUMMARY AND TAKE-AWAYS

In this chapter, we have discussed how a software system is developed across phases and workflows. The discussion has been mainly theoretical. In the next chapter, we will illustrate the application of the concepts discussed in this chapter through a detailed case study.

This chapter's discussion can be summarized as:

- The workflows and phases of software development come out of the interaction of management and technology, as well as the interplay of product, process, people, and project.
- Software development starts with the question—how do we build a software system? Each workflow and phase addresses one aspect of this question.
- During the Requirements workflow, the primary activities include listing candidate Requirements, understanding the system context through domain modelling and business modelling, and capturing functional as well as non-functional Requirements. Requirements should be captured in the language of the user. Use cases help distill the essence of Requirements as sets of action-response transactions between the user and the system.
- A key theme of the Analysis workflow is to understand how and where Requirements interact and what it means for the system.
- Analysis also involves detecting and removing ambiguities and inconsistencies amongst Requirements, developing an internal view of the system, and identifying the Analysis classes and their collaborations. Analysis classes are preliminary placeholders of functionality.
- Analysis and Design, though distinct, are closely related activities. It is difficult to pinpoint when one ends and the other begins.
- The Transition from Analysis to Design is marked by the shifting of focus from the abstract to the concrete, and from a static view to a dynamic view of the system.
- Deciding on the collaboration between components lies at the heart of software Design. A component fulfills its own responsibility by running the code inside it and it exchanges information by calling methods on other components, or when other components call its own methods.
- The Design workflow involves considering specific technologies, decomposing the system into Implementation units, and engaging in high-level and low-level designs.
- At the end of the Design workflow, we should have a blueprint for building the system.
- During Implementation, the Design of a software system is transformed into code. A large part of Implementation is programming. Implementation also involves unit Testing, planning system integrations, and devising the deployment model.

- Due to its very nature, Testing is never complete. The primary activities of the test workflow include creating test cases, running test procedures, and analysing test results.
- The Inception phase aims at establishing feasibility for the system under development. The main activities include determining system scope, delineating system architecture, identifying critical risks, and building a proof of concept.
- Elaboration comes up with a clear strategy on how development should proceed, by creating an architectural baseline, identifying significant risks, specifying quality attributes, extending use case coverage, and preparing schedule and cost estimates.
- During Construction, a major part of building the system takes place. The primary activities are extending use case identification, commencing or completing the workflows, maintaining integrity of architecture, and monitoring and mitigating risks.
- Transition moves the software system from the developer domain to the user domain. The major activities of Transition include preparing the user, tuning the system to the actual operating conditions, and correcting defects discovered by users.

WHERE TO LOOK FOR MORE

- OpenUP is a lean Unified Process that applies iterative and incremental approaches within a structured lifecycle.—http://epf.eclipse.org/wikis/openup/. This is specially suited to applying the Unified Process for small to medium projects.

EXERCISES

Review Questions

Review Questions test your understanding of the key concepts presented in this chapter.

1. The four P's of software development are
   (a) Process, purpose, project, people
   (b) People, project, procurement, process
   (c) Product, process, people, project
   (d) None of the above
2. The interplay of the four P's give rise to which of the following?
   (a) Phases and workflows

   (b) Inception and Elaboration
   (c) Analysis and Design
   (d) Implementation and Testing
3. The dimensionality of software development arises from the intersection of
   (a) business and performance
   (b) regulation and competition
   (c) legacy and modern systems
   (d) technology and management
4. One of the key concerns of software development relates to

(a) keeping Requirements unchanged

(b) delivering high performance software

(c) breaking down larger problems into smaller, more manageable problems

(d) all of the above

5. The broad objective of the Requirements workflow is to ensure

(a) development is aimed at the right system

(b) development is aimed at the most economic system

(c) development is aimed at the most reliable system

(d) development is aimed at the most easy-to-use system

6. A feature list is the same as the list of final Requirements.

(a) True

(b) False

(c) Depends on the feature list

(d) Depends on the list of final Requirements

7. The so-called planning values such as risk, priority etc assigned to each member of the feature list are decided by

(a) developers

(b) users

(c) consultation between developers and users

(d) none of the above

8. Business modelling and domain modelling are approaches to understand

(a) the system's context

(b) performance demands

(c) dependencies on other systems

(d) all of the above

9. Functional Requirements relate to

(a) what the system does for the user

(b) how the system's functionality is determined by business concerns

(c) performance, supportability etc

(d) system specific Requirements

10. Non-functional Requirements are

(a) functional Requirements that have been retired

(b) Requirements put forward by users but which can not be implemented due to system constraints

(c) implicit Requirements such as usability and reliability that can affect the user experience

(d) none of the above

11. The idea behind use cases is to capture

(a) the various types of users who will be using the system

(b) the various ways in which users can use the system

(c) user preferences for future system features

(d) all of the above

12. The output of the Requirements workflow is the input to

(a) iteration planning

(b) Design

(c) Implementation

(d) Analysis

13. During Analysis, the focus is on describing the system using the

(a) language of the user

(b) language of the customer

(c) language of the developer

(d) none of the above

14. Which of the following is likely to be used during Analysis

(a) Java

(b) C++

(c) Z

(d) all of the above

15. Analysis needs to determine

(a) how independent Requirements are

(b) how computer savvy the system's users are

(c) how interconnected Requirements are

(d) none of the above

16. Analysis classes are

   (a) final components that can be implemented using a programming language

   (b) initial placeholders of functionality

   (c) classification of Requirements

   (d) a special type of objects

17. Ambiguities and inconsistencies are likely to remain in Requirement descriptions when

   (a) they are written by those who have no idea what the final system will be

   (b) they are written by users

   (c) they are written by developers

   (d) they are written in natural language

18. One objective of Analysis is to develop

   (a) an external view of the system

   (b) an internal view of the system

   (c) external or internal view, depending on the system

   (d) a third party view of the system

19. The different types of Analysis classes are

   (a) boundary, internal, external

   (b) storage, persistence, entity

   (c) responsibility, collaboration, objects

   (d) boundary, entity, control

20. CRC cards help

   (a) specify a class' responsibilities

   (b) specify a class' collaborations

   (c) both (a) and (b)

   (d) none of the above

21. In the Design workflow, 'Design' relates to

   (a) beauty of the software system

   (b) utility of the software system

   (c) activities aimed at assigning responsibilities to components and specifying their collaboration

   (d) drawing UML diagrams

22. Which of the following statement is true?

   (a) Analysis is concrete, Design is abstract.

   (b) Design is iterative, Analysis is sequential.

   (c) Analysis and Design can be done in any order.

   (d) Analysis is abstract, Design is concrete.

23. A software component's behaviour involves which of the following?

   (a) Manipulating decisions

   (b) Accessing data

   (c) Taking decisions

   (d) Manipulating data and/or taking decisions based on data

24. During Design, we need to

   (a) start looking at technology

   (b) stop looking at technology

   (c) find more about new technology

   (d) none of the above

25. High-level Design and low-level Design differ in their

   (a) level of detail

   (b) choice of high-level programming language

   (c) choice of low-level machine language

   (d) none of the above

26. Decomposition of work into Implementation units is done during

   (a) Implementation

   (b) Analysis

   (c) Testing

   (d) none of the above

27. The most important artefact coming out of Implementation is

   (a) working code

   (b) deployment plan

   (c) UML diagrams

   (d) work units

28. Unit Testing is a part of which of the following workflows?
    (a) Analysis
    (b) Design
    (c) Implementation
    (d) Testing

29. During Implementation, system integration has to be planned for each
    (a) phase
    (b) workflow
    (c) iteration
    (d) increment

30. Deployment involves
    (a) assigning work to developers
    (b) assigning code components to hardware which they will run on
    (c) assigning functionality to iterations
    (d) all of the above

31. Test cases and use cases
    (a) may map one-to-one
    (b) are orthogonal
    (c) are independent of one another
    (d) none of the above

32. A test procedure
    (a) is likely to be run by a tester
    (b) outlines the specific details of running a test case
    (c) should be written such that it can be run without any knowledge of the internal workings of the system being tested
    (d) all of the above

33. Who is the best person to test a particular functionality of a software system?
    (a) One of the developers
    (b) Any stakeholder
    (c) One of the developers not associated with the development of that particular functionality

    (d) All of the above

34. Division of the software development life cycle in phases guides the
    (a) division of labor
    (b) shifting of focus
    (c) clarification of Requirements
    (d) (a) and (b)

35. A feasibility study for a project is likely to be done during
    (a) Elaboration
    (b) Transition
    (c) Inception
    (d) Construction

36. Which of the following artefacts are likely to come out of Inception?
    (a) Risk list
    (b) Proof of concept
    (c) Architectural overview of the system
    (d) All of the above

37. The life cycle architecture milestone marks the end of
    (a) Inception
    (b) Elaboration
    (c) Construction
    (d) Transition

38. The initial operation capability milestone marks the end of
    (a) Inception
    (b) Elaboration
    (c) Construction
    (d) Transition

39. The life cycle objective milestone marks the end of
    (a) Inception
    (b) Elaboration
    (c) Construction
    (d) Transition

40. Quality attributes of the system under development are specified during
    (a) Inception
    (b) Elaboration
    (c) Construction
    (d) Transition
41. Which of the following need to be actively done during Construction?
    (a) Risk identification
    (b) Risk monitoring
    (c) Risk mitigation
    (d) (b) and (c)
42. Transition involves
    (a) system Analysis
    (b) system Design
    (c) system tuning
    (d) none of the above
43. Educating users on how to use the system needs to be done during
    (a) Analysis
    (b) Elaboration
    (c) Testing
    (d) Transition

Reflective Questions

Reflective Questions require you to think more deeply about some of the ideas and come up with your own interpretations and answers.

1. While introducing the idea of phases and workflows, we used the metaphor of reading for pleasure versus studying for an examination. Every metaphor has certain assumptions. Can you identify the assumptions in this case and comment on their validity? Can you suggest another metaphor that is closer to the relationship between phases and workflows?

2. We have identified phases and workflows to represent two dimensions of software development. What do you think can be construed as a third dimension?

3. While discussing the Requirements workflow, we introduced use cases. Can you suggest a process for converting Requirement to use cases? How do you think we can associate non-functional Requirements with use-cases?

4. Analysis and Design should be done by the same set of people. Give three points for and against this statement.

5. 'Designing a software system is ultimately about choices.' Do you think this statement is justified in its current form, or should it be modified or qualified?

6. Do you think the way an electrical circuit is designed is different from the way a software system is designed? If yes, why do you think so? If not, how are they similar?

7. 'Iteration is an inherent aspect of designing a software system.' Is this statement true if the software system is being developed using the Waterfall methodology?

8. What do you think drives the quest for the 'right' number of software components to balance between a single monolithic component and one component for each Requirement?

9. In the software systems you have built so far, what has been the approximate percent of Implementation effort, in terms of the total development effort? Do you think the same percent is likely to hold true for large-scale software systems? If yes, why do you think so? If no, why not?

10. Figures 14.12 and 14.13 show the variation of time and resources across the phases for two categories of projects. List three specific factors you think are responsible for this variation.

11. Implementation and Construction essentially cover the same set of activities. Do you agree with this statement? If yes, give three reasons in support. If no, give three reasons in opposition.

12. Why do we need to wait till Elaboration to finalize time and cost estimates?

13. You are leading a team of developers in a software project. An object-oriented system is being developed, using Java. In the middle of the Construction phase, you find a major architectural flaw. You can rectify the flaw and rework the cascading changes, leading to major schedule slippage and cost increase. Or, you can put a patch (violating architectural integrity), continue development and plan to address the flaw in the next iteration. What would you do? What do you understand as architectural integrity of a software system?

REFERENCES

Booch, G., Rumbaugh, J., and Jacobson, I. (2005), *The Unified Modeling Language User Guide,* Second Edition, Addison-Wesley.

Grady, R.B. (1992), *Practical Software Metrics for Project Management and Process Improvement*, Prentice Hall.

Jacobson, I., Booch, G., and Rumbaugh, J. (1999), *The Unified Software Development Process*, Addison-Wesley.

Building a Software System: An Extended Case Study

> ## Learning Objectives
>
> In this chapter, we will demonstrate how a software system is built across the entire software development life cycle. Specifically, we will cover:
> - How the material of this chapter relates to other parts of the book
> - Overview of the example system, *Kuber Bank*
> - Understanding and Analysing Requirements; Designing, Implementing, and Testing the system along one line of functionality—overview of the workflows
> - Phase milestones
> - A discussion of the case study's limitations

15.1 MOTIVATION

Software engineering is no exact science. In some sense *no* engineering is exact science. But software engineering's inexactitude stems from many of software's peculiarities, which we have discussed at length earlier in this book. The aim of this book is to impart an understanding of the theory as well as practice of software engineering. Thus almost all of the ideas we have presented have been motivated and often illustrated by practical applications. This is an entirely application chapter; we introduce no new theory, but show how the concepts learnt so far can be applied to the actual building of a software system.

Applying theory to practice can be approached at many levels. At one level, it is important to appreciate how theory merges with practice; at another, the nitty-gritty of practicality assumes significance. For example, let us suppose we have been tasked with illuminating a room. One level of concern is to design an electrical circuit depending on the specific lighting Requirements of the room. Another level is to decide on the types of light bulbs, wires, switches, and other elements of the circuitry. Still another level is to build the bulbs, wires, switches etc. All these levels play their own parts in fulfilling the end-user Requirement (in this case, lighting the room). This chapter aims at giving a sectional view across corresponding levels for building a software system.

The example system we will build in this chapter is an online banking application for a bank, which we will call *Kuber Bank* (KB). The choice of the example system is driven by the ubiquity of online banking; many of us do not recollect when we last went to a 'physical' bank. The name of the bank is inspired by Indian mythology; the bank aspires to be the custodian of Kuber's riches! The protagonist of the case study's story will be Preeti whom we already know from earlier chapters. Other characters will be introduced as we go along. The use of 'story' in a previous sentence is deliberate.

Readers are encouraged to treat this chapter as a story. A good story, while still being contrived, tells us much about reality. It exposes insights, while hiding clutter. At the end of the chapter, we will see where and how our story has diluted reality. But by that time, you, the reader, would have grasped the key points.

15.2 EXAMPLE SYSTEM: AN OVERVIEW

As remarked earlier, we know what a typical online banking application offers. It allows the bank to record and modify transaction records, offers account holders the facility to check their balances, and communicate with the bank; and so forth. As will soon be apparent, something seemingly as straight forward can still present a number of subtleties.

Let us now delve into our example.

Preeti is leading a team of software engineers engaged in building the system. Kuber Bank (KB) has been in the banking business for over thirty years with branches all over India. About 15 years ago, the branches were computerized and they are currently connected through a nation-wide network. To keep up with the competition, KB needs to offer online banking facilities to its customers. Preeti's organization has won the contract to 'Webify' KB. An initial meeting was arranged between KB's senior management and Preeti's team to understand the scope of the projects.

– 'So, as a first step, it would be good for us to know the broad Requirements you have in mind', began Preeti.

– 'Well, we want to let the customers do online all they can do at a branch', Sanjeev Kumar, KB's senior vice-president, replied.

This snippet of conversation is very telling. It is very common, and makes things very difficult for the software engineer. Every customer wants the moon, and start with the feeling that software can solve all their problems. This is to a large extent due the *cognitive gap* we discussed earlier.

If we dissect what Sanjeev said '…we want to let customers do online all they can do at a branch', it does not make full sense. 'All' that customer's can do at a bank's physical branch cannot be replicated online—withdrawing and depositing cash is one of the many examples. What Sanjeev must have *meant* was that the online system needs to offer a rich subset of KB's services so that customers can do many of the things online they can only do at the branch now. But that is not what Sanjeev *said*. More often than not, customers will say things very different from what they mean, and it is the job of the software engineer to sense the essence of what actually the customer needs to get done.

So the essence of Sanjeev's statement boils down to: The software system *Kuber Bank online* (KBO) needs to provide functionalities that will allow KB's customers to fulfill a subset of their banking needs online. This is a very broad statement, something of a vision for KBO. But a system built on this statement alone will satisfy neither the Kuber Bank, nor its customers. Thus, Preeti's team needs to drill down into more specific Requirements.

15.3 REQUIREMENTS

Preeti, with the help of Sumit – the systems analyst – elicited the broad lines of functionality for KBO as presented in Table 15.1.

For definitions of the words/phrases (in **bold** when they occur for the first time in Table 15.1 and the discussion below), in the KBO context, please refer to Table 15.2: Glossary.

Evidently, Table 15.1 show very high-level Requirements. In fact, they can hardly be called *Requirements* in the software engineering sense of the term. These represent more the needs/wishes/wants of the **customer**. (Please see Table 15.2 for the distinction between 'user' and 'customer'.)

With reference to Figure 14.7, we may note that user needs/wishes/wants go into the Requirements workflow and come out as the business and domain model, and functional and non-functional Requirements. We will see how that transformation takes place. (Some authors distinguish between 'needs' and 'wants', but we will use the terms synonymously here.)

Table 15.1 KBO – High-Level Functionality

1. **Administrator** can:
   - **View transactions** across all **accounts**.
   - **Add** transactions to any account.
   - **Delete** transactions from any account.
   - View all **messages** sent by **users**.
   - **Reply** to messages sent by users.

2. User can:
   - View their **profile** information.
   - View list of their accounts with the **bank**.
   - View transactions for each of their accounts.
   - Send messages to the bank.
   - View **history** of messages they have sent to the bank and replies from the bank.

Preeti and Sumit need to determine the scope as a first step in turning the broad lines of functionality of Table 15.1 into specific Requirements. We have taken an online banking example as similar systems are widely used. However even such systems can be very complex when drilled down to details. Just something as basic as ensuring that when I log into my online bank account I get to see *my* details only, and not someone else's (or more disturbingly, someone else gets to see my details!), takes a lot of under-the-hood engineering. To do justice to the complexity of software development (see section 12.4), we will only take one line of KBO's functionality for further Elaboration in this chapter.

Let us take the high-level functionality of the administrator being able to add transactions details to any account as our candidate for attention. Referring once again to Figure 14.7, we note that business and domain model is one of outputs of the Requirements workflow. Often, a full-blown business and domain model is not necessary if reasonable understanding exists amongst stakeholders as to what business need the system addresses. However, even in the absence of a formal business model, what is necessary is a glossary.

In a vast majority of software systems, Requirements are specified in a natural language. (Please refer to section 13.6 for other specification approaches.) And natural languages, although very intuitive, can also be very ambiguous. While specifying Requirements in natural languages, a word can have a contextual meaning quite distinct from what it means in casual usage. Take for example, the word 'user' as mentioned in the high-level functionality of Table 15.1. In the KBO context, who exactly is a user? Is it someone from KB who will run the user-acceptance tests? Is it someone from Preeti's team who will use the system while it is being built? Is it the *end-user*, that is, KB's customer who will use KBO for online banking facility? Is the administrator also a user?

A glossary captures the precise meaning and scope of an idea or entity *in the context* of the system being developed. It serves as a single point of reference for domain knowledge and also a parking lot for concepts that have not yet been fully explored or understood. A glossary is very much a living document, new entries are added as the life cycle moves forward over iterations, and old entries are refined. For projects whose problem domains are not radically new to the stakeholders, a glossary is sufficient to capture the key concepts of the business and domain model. A glossary may also serve as a placeholder for unanswered questions. As development proceeds and those questions are clarified, the glossary becomes richer with the domain knowledge.

Let us now go deeper into our chosen line of functionality: Administrator can add transactions to any account. Taking this one sentence of seven words as the starting point and after countless meetings with KB officials, Sumit elaborated the Requirement into Table 15.3.

Table 15.2 Glossary

Word/phrase	Contextual meaning	Remarks/open issues
Account	The bank account held by a user with Kuber Bank (KB), identified by a unique account number.	Can one user have multiple accounts?
Add	Recording of information by administrator or users in the Kuber Bank Online (KBO) system.	
Administrator	A special type of user who has unique privileges in manipulating information relevant to KBO.	
Bank	Kuber Bank	
Delete	Removal of information by administrator or users in the Kuber Bank Online (KBO) system.	
History	Information regarding past transactions recorded in any media in Kuber Bank's archives.	
KB	Kuber Bank	
KBO	The Kuber Bank Online system, consisting of all software components and the infrastructure required for the functioning and support of such components.	
Master Ledger (ML)	An entity within Kuber Bank that contains complete information regarding users, accounts, transactions, and messages.	What is the format of ML? Is it a physical recording medium or an electronic database? How is it accessed?
Message	A unit of information containing a subject line, body, and unique identifier that can be sent by user to administrator and vice versa within KBO.	
Profile	A unit of information uniquely identifiable with a user that contains a set of relevant data such as name, user id, email address etc.	
Reply	A unit of information containing a subject line, body, and unique identifier that can be sent by user to administrator and vice versa within KBO in response to a message.	
System	Kuber Bank Online	
Transaction	A unit of information that completely encapsulates all the data required to uniquely represent a banking service offered by Kuber Bank to its account holders.	
User	An account holder of Kuber Bank who is allowed by Kuber Bank to use the Kuber Bank Online system.	Can the same individual be an user and the administrator?
View	A facility offered by Kuber Bank Online to users for visually inspecting a unit of information, without the ability to change it.	

Table 15.3 Description of Requirement KBO_Req_01

Requirement ID	KBO_Req_01
Requirement Description	Administrator shall access transaction information for a particular account on a periodic basis and record the same in the KBO system such that user can view them through the Web interface. The transaction information will be available to be accessed by the Administrator in the existing Master Ledger (ML) system. Details for a transaction need to be available at KBO for viewing by a user within 48 hours after the transaction has been initiated by the user.

The fleshing out has led us from one sentence to three sentences, from seven words to 69 words. How much more does Preeti's and Sumit know about the Requirement now? Well, quite a lot. They know that to fulfill this Requirement, the main actor, i.e. Administrator has to interact with at least one more entity outside KBO: the Master Ledger (ML). They know the Administrator has to fetch transaction information from ML and record it in KBO. They also know that the information for a particular transaction has to be available for a user to view at KBO within 48 hours of the transaction initiation.

However, there are also things the development team does not know at this time. How will the Administrator *access* ML and *record* transaction information in KBO? Are accessing and recording manual processes? Will KBO need to automate them? How will failed transactions—like dishonored checks—be handled? Need failed transaction details also be available for user to view at KBO within 48 hours?

Preeti and Sumit debated these questions for a while, spoke with KB functionaries, and came up with the expanded version of KBO_Req_01 (Table 15.4)

It is interesting to note that the expanded version of KBO_Req_01, has in fact *contracted* the system's scope. This is a key observation. During the Requirements workflow, it is very important not only to specify what the system will do, but also what it will *not* do. What the system will not do, very often gives key insights into how to build the system for what it will do. The phrase 'outside the scope of KBO's current release' has been used twice in Table 15.4, and it has helped clarify the development team's questions.

Still, at this point, the development team does not know as much as it would have liked to know. And this is very natural. Preeti, being a seasoned campaigner, did not fret too much over it. And she also did not commit the usual error of going back and forth endlessly with the customer, trying to thresh out every small detail. Stakeholders from KB are unlikely to have much understanding at this point, beyond what the Requirement says in words, which they are expected to have read and agreed to. In software engineering, being able to move forward with incomplete knowledge or missing information is vital. The Requirement workflow is not meant to answer all questions; it is the time to ask the right questions, elicit some answers, identify issues which can only be addressed later, and move forward.

Based on the development team's understanding so far, the functional and non-functional parts of the Requirement are shown in Table 15.5.

It needs to be underscored that in every non-trivial system, there will be non-functional Requirements. Sometimes they are spelt out explicitly, often they are not. In either case, the development team needs to detect and understand them fully, and do it as early as possible in the project. No matter how well and truly the functional Requirements are implemented; even one unfulfilled non-functional Requirement can cause significant user dissatisfaction. There is no widely accepted mechanism to associate non-functional Requirement with use cases (which address the functional part); [Datta 2005] gives one approach.

The functional part of KBO_Req_01 will now be represented as a use case. The key difference between a (functional) Requirement description and its corresponding use case is that the former is narrative while the latter is interactive. Along a particular line of functionality, a use case sees the system as a black-box, and specifies all that the user can do with the system in an action-reaction format. Sometimes pre-conditions and post-conditions—what must happen before the use case plays out, and what would be the state of the system after it is complete, are also specified. Table 15.6. gives the use case KBO_UC_01: Record Transaction, corresponding to KBO_Req_01.

The use case flow in Table 15.6 is relatively simple. Yet, it is hardly straightforward. Evidently, KBO_UC_01, representing one line of functionality among several in KBO, cannot run all by itself, but needs to interact with a number of other use cases. Also, this one use case

Table 15.4   Expanded Version of KBO_Req_01

Requirement ID	KBO_Req_01
Requirement Description	Administrator shall manually access the transaction information for a particular account on a periodic basis from the existing Master Ledger system. (Automation of this access mechanism is outside the scope of KBO's current release and may become a Requirement for a future release.) Administrator shall record the transaction information for a particular account accessed from the ML system, in the KBO system. Only successful transactions need to be available at KBO for viewing by users within 48 hours after they have been initiated by the user. Users will be notified of failed transactions through procedures outside the scope of KBO's current release.

Table 15.5   Functional and Non-functional Parts of KBO_Req_01

Functional part	Non-functional part
Recording transaction details in KBO, by administrator	• Ensuring transactions are recorded in a format that makes it reasonably easy to be understood by users. • Ensuring transactions are correctly recorded for the appropriate user accounts. • Ensuring successful transactions are recorded at KBO within 48 hours of their initiation. • Ensuring large volumes of transaction for a particular account within a particular period of time can be handled.

description brings to light entities external to KBO (the 'system') that influence the system's functioning, such as Master Ledger and Failed Transaction Queue.

So, we have the Requirement KBO_Req_01 and its corresponding use case KBO_UC_01: Record Transaction. We have already seen how this one relatively simple Requirement has deeper ramifications; in terms of interactions with other parts of the system as well as entities outside the system, and the need to conform

Table 15.6 Use Case KBO_UC_01: Record Transaction

ID	KBO_UC_01
Name	Record Transaction
Objective	To record transaction information for a particular account in KBO.
Pre-conditions	Administrator must be logged into KBO and must be able to access transaction information from Master Ledger.
Post-conditions	• Success 1. Transaction information is successfully recorded in KBO. • Failure 1. Transaction has failed. 2. Transaction has succeeded but transaction information cannot be recorded in KBO.
Actors	• Primary 1. Administrator • Secondary 1. Master Ledger 2. User Notifier
Trigger	Periodic reminder to Administrator
Normal flow	1. Administrator requests transaction information for a particular account for a particular period of time from Master Ledger via the Get Transaction Information use case. 2. Administrator accepts information supplied by Master Ledger. 3. Administrator checks status of each particular transaction, whether success or failure via the Check Transaction Status use case. Administrator sends failed transaction information to failed transactions queue. System processes failed transaction information via the Notify User use case. 4. Administrator records succeeded transactions in the system. System confirms successful recording of transaction.
Alternative flow	1. a Master Ledger is not able to supply transaction information; use case concludes with error notification. 3. a. No successful transactions to record; use case concludes with error notification. 4. a. Successful transactions can not be recorded in KBO; use case concludes with error notification.
Interacts with	Get Transaction Information, Check Transaction Status, Record Transaction, Login, Notify User use cases.
Open issues	1. How quickly must Master Ledger respond to request for transaction information? 2. How will the system identify Administrator from other users?

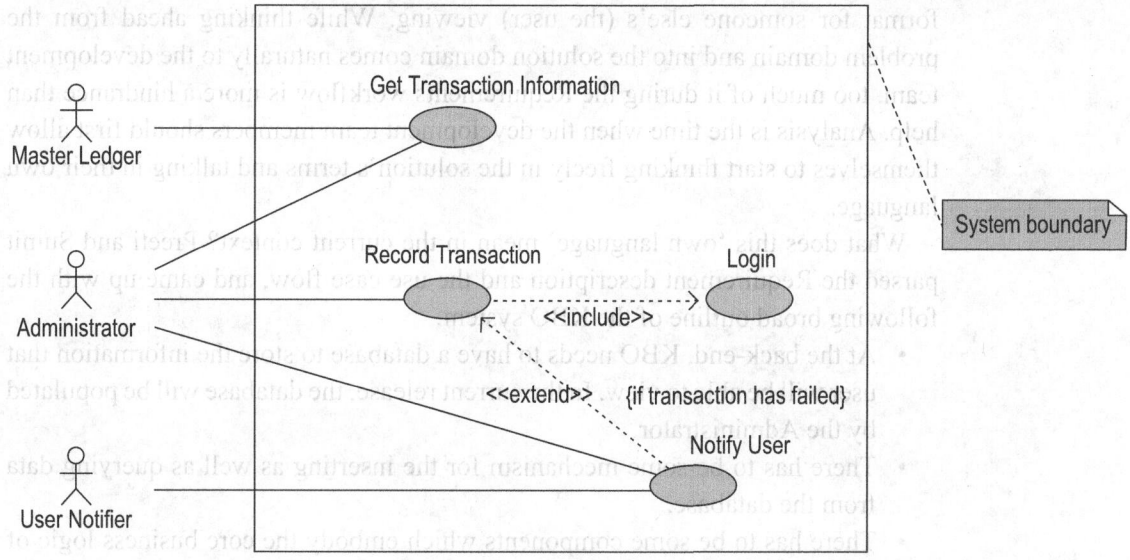

Figure 15.1   Use case diagram of KBO_UC_01

to non-functional parameters. Figure 15.1 is the use case diagram for KBO-UC-01. Two use cases are connected by the 'include' relationship if the behaviour of the inclusion use case is important for the base use case. On the other hand, the 'extend' relationship is used to model an optional or conditional part of system behaviour.

It is time now for Preeti and her team to get into Analysis.

15.4 ANALYSIS

With reference to section 14.5.2, we note that the main activities during Analysis are specifying Requirements in the developer's language, identifying Requirement interconnections, detecting ambiguities and inconsistencies in Requirements, developing an internal view of the system, and based on all of the above, identifying Analysis classes and their collaborations. As depicted in Figure 14.8, viewing Analysis as a black-box, what goes in are Requirement descriptions (including use cases) and analyst insight; what comes out are specifications, Requirement interconnections, and Analysis classes.

When developers elicit Requirements from users, at the back of their minds the process of translating domain concepts into software constructs have already begun. For the development team, KBO_UC_01: Record Transaction boils down to someone (the administrator) updating a database with some information (transaction details), and that information being made available in a predefined format for someone else's (the user) viewing. While thinking ahead from the problem domain and into the solution domain comes naturally to the development team, too much of it during the Requirements workflow is more a hindrance than help. Analysis is the time when the development team members should first allow themselves to start thinking freely in the solution's terms and talking in their own language.

What does this 'own language' mean in the current context? Preeti and Sumit parsed the Requirement description and the use case flow, and came up with the following broad outline of the KBO system:

- At the back-end, KBO needs to have a database to store the information that user will be able to view. In the current release, the database will be populated by the Administrator.
- There has to be some mechanism for the inserting as well as querying data from the database.
- There has to be some components which embody the core business logic of KB—like calculating the interest amount and deposit tenure etc. (This aspect is not directly relevant to KBO_UC_01: Record Transaction, but typically, Analysis has to spread across the entire range of use cases.)

Notice how there has already been a subtle but clear shift in the jargon; from talking about transactions and accounts, we are now speaking in terms of databases and components. The specification of Requirements in the developer's language has already begun. The user's language and the developer's language are not two disjoint sets; they are part of a continuous spectrum, at one end of which lies user expectations from the system, often not even clearly articulated in words, while at the other end lies the system in its most concrete form, code.

A key aspect of Analysis is also about identifying how Requirements interconnect. Each Requirement represents a particular line of functionality distinct from other Requirements, yet the Requirements *together* make up the whole system. As each Requirement is developed, the aim is to converge into one cohesive system, rather than a set of independent modules callously strung together. Understanding connections between Requirements early is the key to fulfilling this aim.

In this example, we are focusing on only one Requirement. Yet, its interconnections with other Requirements can be discerned, even if at a superficial level. Given its scope, KBO_Req_01 also connects with the Requirements of administrators being able to view transactions and deleting transactions across all user accounts, and users being able to view transactions for all their accounts using the same user id and password.

The next step is to detect ambiguities and inconsistencies in Requirements. Preeti and Sumit have already done round one of it, between the first and the second versions of the Requirement description (Tables 15.3 and 15.4). In the initial description, one of the issues not specified was how data was to be fetched from ML; this was an ambiguity which could have caused problems later. The expanded description clarifies what exactly is within the scope of KBO's current release. The Requirement also says that successful transactions will be available for viewing by users only 48 hours after they were initiated. But KBO_Req_01 does not say what should happen if a user tries to view a transaction less than 48 hours after it was initiated. This is an example of typical inconsistencies that need to be caught and resolved during the Analysis phase. In this case, Preeti clarified with a KB manager that KBO needs to show *nothing* related to a particular transaction until 48 hours after its initiation.

With ambiguities and inconsistencies weeded out, the internal view of the system can now be developed. That view has already begun crystallizing with the shift from the domain language—'transactions' and 'accounts' – into the development language—'databases' and 'components'. The internal view is what developers see when they peer into the black-box that will always be 'the system' for users and all other external entities. For an object-oriented system, the internal view will consist of objects interacting via messages to fulfill the system's functionality. The task now is to form an initial conception about the nature of these objects and the messages they need to send. With reference to earlier discussion, Preeti and Sumit identified the *significant* nouns (or noun phrases) and verbs (or verb forms) from the description of KBO_Req_01. These are highlighted in **bold** below, the first time they occur:

Administrator shall manually **access** the **transaction** information for a particular **account** on a periodic basis from the existing **Master Ledger system**. (Automation of this access mechanism is outside the scope of KBO's current release and may become a Requirement for a future release.) Administrator shall **record** the transaction information for a particular account accessed from the ML system, in the **KBO system**. Only succesful transactions need to be available at KBO for **viewing** within 48 hours after they have been initiated by the user. **Users** will be **notified** of failed transactions through procedures outside the scope of KBO's current release. The nouns/noun phrases identified are:

- administrator
- transaction
- account
- Master Ledger system
- KBO system
- users

The verb/verb forms identified are:

- access
- record

- viewing
- notified

Now, what do we do with these nouns/noun phrases and verbs/verb forms?

As discussed in section 14.5.2.5, these serve as heuristics for identifying components and their behaviours. As a thumb rule, a noun/noun phrase is a candidate for being a class or a data member, whereas a verb/verb form is a method call. Whenever there is a doubt about whether a noun/noun phrase will be a class or a data member, it is best to make it a class and refine later if necessary. Iteratively refining from coarse to fine grains is a common thread running through all of software development, and Analysis is no different. As with every other development activity, it should not be imperative for Analysis is to be right the first time.

Table 15.7 gives a summary of the tentative Analysis classes and their methods, based on the identification of the nouns/noun phrases and verbs/verb forms earlier.

In Table 15.7, note how Preeti and Sumit used words such as 'may' and 'probably'. This indicates the fluid nature of their thinking at this point, preliminary placeholders are being identified, and nothing is concrete yet. Many of the Analysis classes or their methods will not remain as they are now during Design and Implementation. And that is how it should be.

In section 14.5.2.5, we have also discussed CRC cards as helpful artefacts in capturing Analysis insights. Figure 15.2 gives a sample CRC card for one of the Analysis classes.

With the Analysis artefacts at hand, the development team is ready to get into the Design workflow. While performing Analysis, a little bit if Design has already been done; the development team will go deeper in that direction now.

At the beginning of the Design workflow, Preeti and Sumit welcomed Vikram into the team. Vikram has several years of experience in developing and deploying Web based systems in Java.

15.5 DESIGN

As we have said several times in this book, 'Design' is an overloaded word, and it means many things to many people in many contexts. The current context of 'Design' should be clear, we are entering the Design workflow in the development of KBO. As discussed in section 14.5.3, this workflow is concerned with looking at specific technologies, decomposing the system into Implementation units, and then engaging in high-level Design followed by low-level Design. Let us join Preeti and her team as they navigate these activities.

The technology of software development comes in layers. At the low-level Design and Implementation layer it is about the choice of programming language,

Table 15.7 Summary of noun and verb Analysis

SL	Noun phrase or verb form	Class or method	Remarks
1.	administrator	class – control or entity	An actor of the use case KBO_UC_001; a component driving the functionality of Requirement KBO_Req_01; a component encapsulating the information of uniquely identifying the administrator.
2.	transaction	class – entity	A component encapsulating the information of a transaction.
3.	account	class – entity	A component encapsulating the information of an account.
4.	Master Ledger system (ML)	–	System external to KBO
5.	KBO system	–	The system itself
6.	user	class – control or entity	An actor of the use case KBO_UC01; a component participating in the functionality of KBO_Req_01; component encapsulating the information uniquely identifying a user.
7.	access	method – access() of some boundary class interacting with ML	May not be relevant now as administrator is manually accessing ML in the current scope.
8.	record	method – record() of some boundary class, that interfaces with KB's database	Probably a method on a data accessor class that serves as a wrapper for the database calls.
9.	viewing	method – view() on some boundary class, that allows users to see their transaction records	Probably viewer will be a class that interacts with the user interface components.
10.	notified	method – notify() on some control class that manages communications with the user	May not be relevant now as user notification is outside KBO's current scope.

Class: Administrator	
Responsibility	**Collaboration**
Access Master Ledger	Master Ledger, user, transaction, account
Communicate with Transaction Status Checker	Transaction Status Checker
Record transaction	KBO
Facilitate user notification for failed transaction	User Notifier

Figure 15.2 CRC card for the Analysis class administrator

data structures, and algorithms. At the high-level Design layer, it is more about selecting an architectural paradigm. Let us first see which architectural paradigm is the best fit for KBO. In the context of Analysis and Design, Exhibit 15.1 reflects on plans and planning.

An enduring credo of engineering is not to re-invent the wheel, but to put the wheel to ever inventive uses. Software engineering—though a very young engineering—has already accumulated a growing collection of reusable constructs. With reference to the discussion of Chapter 11, architectural patterns are such reusable ideas.

From what we have seen of KBO so far, several architectural observations can be made. KBO needs to have a Web interface as the front-end. The system also involves recording and fetching data from a back-end database. And the logic of how the data must be accessed needs to be captured somewhere in between the front-end and the back-end. So we implicitly recognize three *tiers* in the system. The model-view-controller (MVC) architectural pattern discussed in Chapter 11 closely matches this view of the system. The model is the back-end database and

---

Exhibit 15.1   Plans and Planning: Onions and Flavour

In my early days as a software engineer, I expected every workflow to be complete when it ended, and the next workflow to build on what the previous one had produced, discarding or distorting nothing. Thus, my disappointment was severe with Analysis and Design, because much of what Analysis churns out invariably fall by the way side during Design. Did that mean Analysis, as I knew it, was always improper or incomplete? Or were there more adequate ways of analysing I knew nothing of? No book gave me these answers and I remained frustrated. With experience, my confusion and the reality begun to reconcile.

There is an anecdote about General Dwight Eisenhower, who led the Allied forces to victory in the Second World War and subsequently became the President of the United States of America. Eisenhower was once asked whether all the detailed pre-battle plans come in handy when the actual fighting starts. He answered that the plans do not help, but the planning does. Weinberg in his book on writing [Weinberg 2005] recounts the recipe for a kind of flavoured butter: Onions are to be fried in the butter and then the butter served with the fried onions thrown out. He then draws a parallel to the act of writing; words, phrases, and sentences are written and then rejected or refined during editing; the finished piece still retains the flavour of the thinking that went into writing all that was subsequently removed or changed.

Whether it is planning for war, flavouring butter, or writing, the key message is to be able to throw away some of the intermediate products—plans, fried onions, inapt phrases—but retain their essence. This is also a central theme of Analysis. Many of the artefacts of Analysis will morph or be dropped during Design, but the thought that led to them will guide subsequent workflows.

components that facilitate data access, the view is the front-end user interface and the controller lies in the middle, brokering the interaction between the model and the view.

With the architectural pattern in place, the system can now be decomposed into smaller subsystems. Not only do the subsystems represent logical parts that make up the system; they also facilitate delegation of work amongst the development team's members. To represent the subsystems, a block-diagram is usually drawn. A block-diagram is hardly anything more than what it says it is, a diagram consisting of blocks connected by lines. The blocks represent chunks of the system whose internal entities are logically related to one another, and the lines indicate interactions between the blocks.

Figure 15.3 shows the high-level architectural diagram of KBO and maps the tiers to the MVC architecture pattern. (See Table 15.9 for the explanation of the smaller numbered rectangles inside the larger rectangles denoting the subsystems).

At the Implementation level, Java is the programming language chosen for KBO.

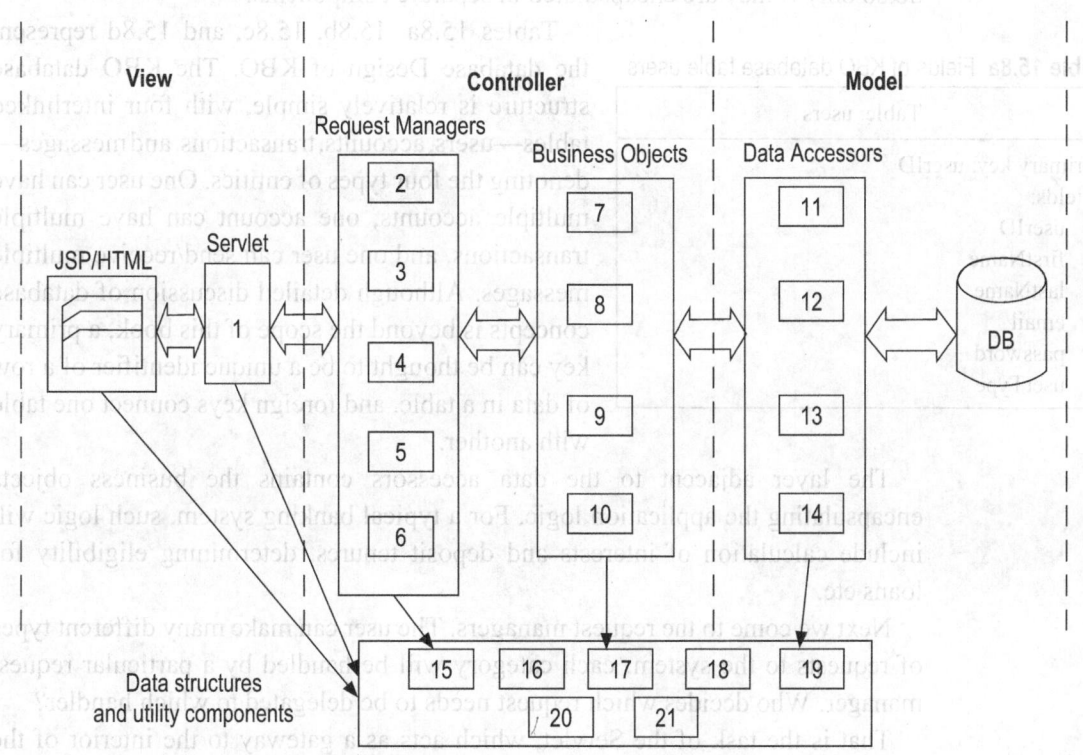

Figure 15.3   High-level architecture of KBO

Figure 15.3 also represents the significant architectural and high-level Design choices already made. Proceeding from right to left, let us briefly discuss each subsystem.

First we have the database, and then the set of data accessors. A software program usually talks to a database using some special purpose language such as structured query language (SQL). A programming language such as Java provides an application programming interface (API) for database communication to act as wrappers around SQL commands. The database access logic is different form the core business logic of an application. For example, a part of the core business logic of KBO will involve deciding what interest rate to offer to a particular account holder, which may depend on factors such as the amount of money in the account, how long the account has been with the bank etc. But the database access logic will be concerned with issues such as optimizing database calls such that queries on large volume transaction details can be quickly completed. It is a recommended Design practice to segregate business logic and data access logic in different components. This segregation recognizes that business logic and data access logic are likely to change independent of one another. The two can have the freedom to do so only if they are encapsulated in separate components.

Tables 15.8a, 15.8b, 15.8c, and 15.8d represent the database Design of KBO. The KBO database structure is relatively simple, with four interlinked tables—users, accounts, transactions, and messages—denoting the four types of entities. One user can have multiple accounts, one account can have multiple transactions, and one user can send/receive multiple messages. Although detailed discussion of database concepts is beyond the scope of this book, a primary key can be thought to be a unique identifier of a row of data in a table, and foreign keys connect one table with another.

Table 15.8a Fields of KBO database table users

Table: users
Primary key: userID Fields: • userID • firstName • lastName • email • password • userType

The layer adjacent to the data accessors contains the business objects encapsulating the application logic. For a typical banking system, such logic will include calculation of interests and deposit tenures, determining eligibility for loans etc.

Next we come to the request managers. The user can make many different types of requests to the system; each category will be handled by a particular request manager. Who decides which request needs to be delegated to which handler?

That is the task of the Servlet, which acts as a gateway to the interior of the system. It intercepts incoming user requests and hands them over to the appropriate request manager for further processing.

Table 15.8b   Fields of KBO database table accounts

Table: accounts
Primary key: accNumber
Foreign key: userID
Fields:
• accNumber
• userID
• initialBalance
• openingDate

Table 15.8c   Fields of KBO database table transactions

Table: transactions
Primary key: txnId
Foreign key: accNumber
Fields:
• txnId
• accNumber
• txnDate
• amount
• transactionType
• transactionDesc

Table 15.8d   Fields of KBO database table messages

Table: messages
Primary key: msgID
Foreign key: userID
Fields:
• msgID
• userID
• subject
• query
• response
• isAnswered

Finally, Java Server Pages and static HTML pages form the user interface. Figure 15.3 also shows how the three tiers of the architectural view superimpose on the blocks and how it is mapped to the model, view, and controller of the MVC pattern.

At this point, Preeti's team can be said to have completed high-level Design. Starting with the identification of an architectural pattern, they have decomposed the system into subsystems and their interactions. Now comes low-level Design.

While identifying Analysis classes, we found 'transaction' to be a candidate for an entity class encapsulating the information of a transaction. Let us now drill down into the core functionality of recording transaction information in KBO and understand what all needs to be done, and then who should be entrusted with doing it.

Here is how KBO_UC_001 flows in terms of the system components:

The human administrator has accessed a particular transaction's information from the ML system. He now approaches KBO to record the transaction information. The administrator's request is intercepted by the BankFacadeServlet.java, who hands it over to AdminTransactionsRequestManager.Java based on some parameter that identifies the nature of the request. The AdminTransactionsRequestManager.java next hands over control to TransactionController.java, which in turn uses services of TransactionDataAccessor.java, TransactionDataBean.java and BankResultBean.java to complete the task of recording transaction information in KBO.

Table 15.9 gives the entire set of Java classes and Java Server Pages used in KBO.

The Analysis class 'transaction' comes closest to TransactionDataBean.java; and the functionality around recording transaction information has spawned all the

Table 15.9 Set of code components for KBO (Refer to Figure 15.3 for high-level architecture of KBO involving these components.)

Category of component	Functionality	Members
Servlet	Intercepts user requests and delegates them to the appropriate request manager.	1. BankFacadeServlet.java
Request manager	Manages user requests by invoking the appropriate business object.	2. LoginRequestManager.java 3. AdminTransactionsRequestManager.java 4. AdminMessagesRequestManager.java 5. UserTransactionsRequestManager.java 6. UserMessagesRequestManager.java
Business object	Encapsulates business logic of the application.	7. UserController.java 8. AccountController.java 9. TransactionController.java 10. MessageController.java
Data accessor	Provides wrappers to database calls.	11. UserDataAccessor.java 12. AccountDataAccessor.java 13. TransactionDataAccessor.java 14. MessageDataAccessor.java
Data structure	Holds data during transfer between other components.	15. UserDataBean.java 16. AccountDataBean.java 17. TransactionDataBean.java 18. MessageDataBean.java 19. BankResultBean.java
Utility component	Provides miscellaneous services.	20. Log.java – A logger class which has the functionality of turning output printing on/off based on a flag. 21. CONSTANTS.java – An interface having the "hard coded" constants (such as database connection information etc.) as its fields.
User interface	Enables information to presented and received from users.	• login.jsp • adminMenu.jsp • adminTransactions.jsp • adminMessages.jsp • customerMenu.jsp • customerTransactions.jsp • customerMessages.jsp • confirmation.jsp • kboError.jsp

Exhibit 15.2   Why Are All UML Artefacts Not Used?

In our discussion of UML in section 13.5.2 we mentioned several artefacts. While building KBO, we find just few of them are being used used. Why?

This is a question fresh software engineers often ask. The answer simply is that we do not need to. Just as we do not need to use every word in the English language whenever we write in English, there is no obligation to use every UML artefact in every system we build. UML, with all its expressive power, is ultimately just a vehicle for communicating and documenting the constructs of software development. Only those parts of UML that are relevant for the system being built need to be used. Everything else is a distraction.

other classes mentioned above. Evidently, low-level Design has transformed quite a bit from the classes and methods identified during Analysis. The objective of Analysis is to create a framework for answering the critical question of *who does what?* Design builds upon the tentative responses to that question by asking, *how?* The answer to that 'how', comes at varying levels of detail. At the top-most level it is specified in the architectural diagram of Figure 15.3; at the lowermost level it is the code of Table 15.10.

Between these levels lie different dimensions of describing the system, two of which are expressed in Figure 15.4 and Figures 15.5a, 15.5b, 15.5c, 15.5d, 15.5e. (refer to section 13.5.2 for a discussion of UML). Exhibit 15.2 highlights why we use a subset of UML artefacts in this example.

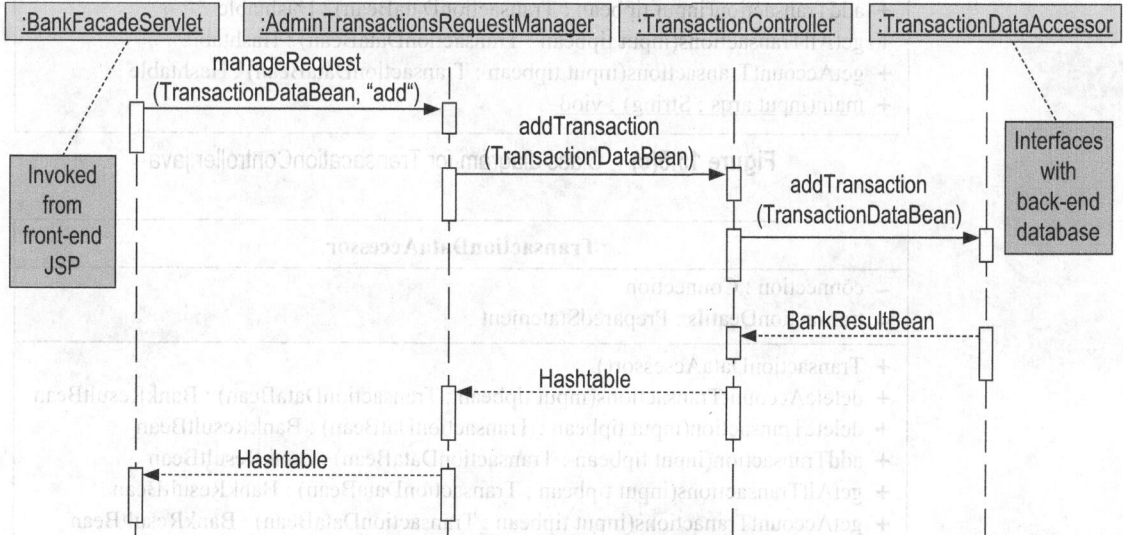

Figure 15.4   Sequence diagram for KBO_UC_001

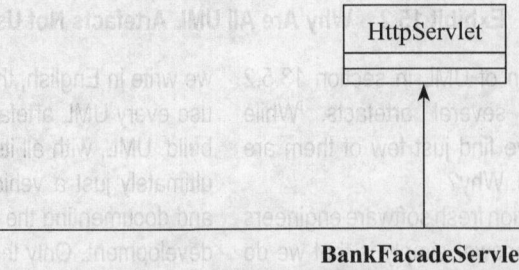

HttpServlet

BankFacadeServlet
doGet(input request : HttpServletRequest, input response : HttpservletResponse) : void

Figure 15.5(a) Class diagram for BankFacadeServlet.java

AdminTransactionsRequestManager
+ AdminTransactionsRequestManager()
+ manageRequest(inout tipbean : TransactionDataBean, in task: String) : Hashtable
+ <u>main(input arqs : String) : void</u>

Figure 15.5(b) Class diagram for AdminTransactionRequestManager.java

TransactionsController
+ TransactionsController()
+ deleteAccountTransactions(input tipbean : TransactionDataBean) : Hashtable
+ deleteTransaction(input tipbean : TransactionDataBean) : Hashtable
+ addTransaction(input tipbean : TransactionDataBean) : Hashtable
+ getAllTransactions(input tipbean : TransactionDataBean) : Hashtable
+ getAccountTransactions(input tipbean : TransactionDataBean) : Hashtable
+ <u>main(input arqs : String) : viod</u>

Figure 15.5(c) Class diagram for TransacationController.java

TransactionDataAccessor
− connection : Connection
− transactionDeatils : PreparedStatement
+ TransactionDataAcsessor()
+ deteleAccountTransactions(input tipbean : TransactionDataBean) : BankResultBean
+ deleteTransaction(input tipbean : TransactionDatBean) : BankResultBean
+ addTransaction(input tipbean : TransactionDataBean) : BankResultBean
+ getAllTransactions(input tipbean : TransactionDataBean) : BankResultBean
+ getAccountTranactions(input tipbean : TransactionDataBean) : BankResultBean

Figure 15.5(d) Class diagram for TransactionDataAccessor.java

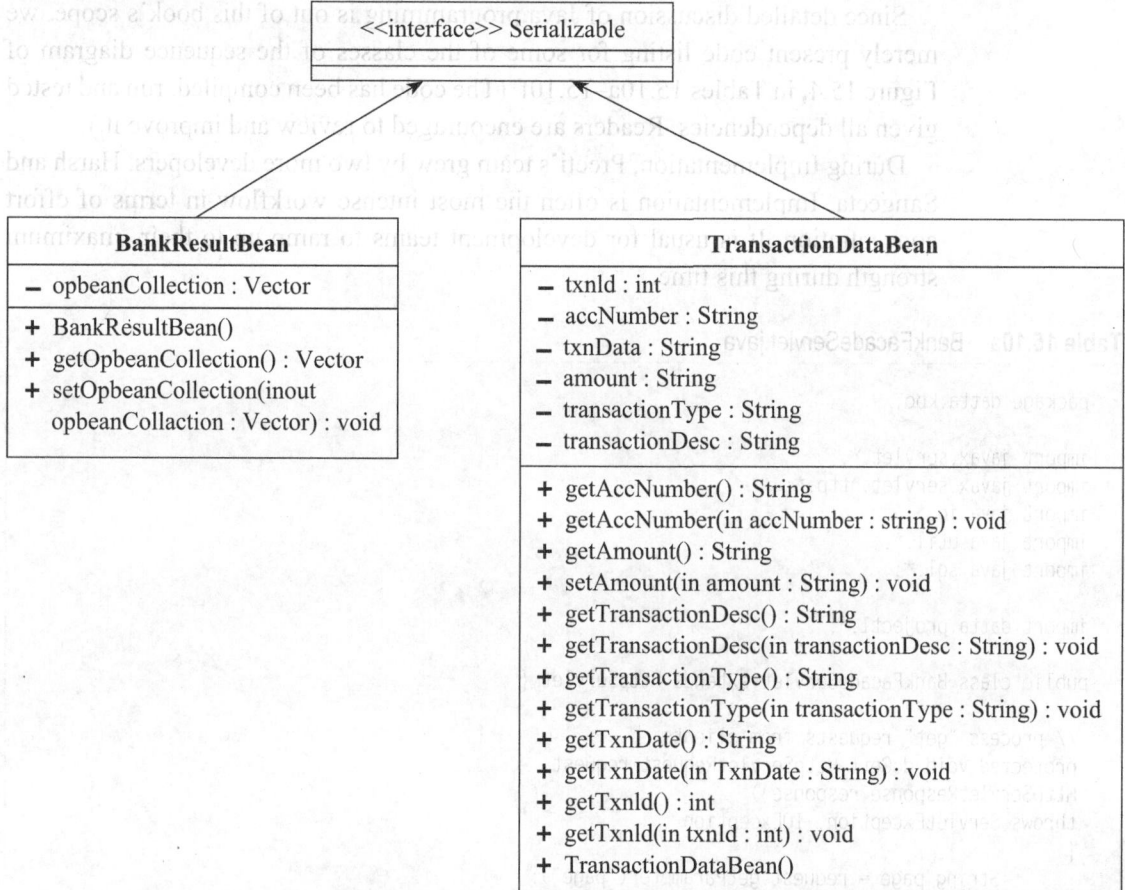

Figure 15.5(e)   Class diagram for BankResultBean.java and TrasactionDataBean.java

15.6 IMPLEMENTATION

Once we have Design artefacts at the level of detail shown in Figures 15.4 and 15.5(a), 15.5(b), 15.5(c), 15.5(d), and 15.5(e); Implementation seems relatively straightforward; translate the Design into code. However, to anyone with even a little programming experience, it should be clear that Implementation is far more involved. With reference to the class diagrams of Figures 15.5, it is apparent that only the method signatures are specified, that is, what the input parameters are and what the method returns. The body of the method, which are the set of steps that compute the return value given the input parameters, needs to be filled in during Implementation. This 'filling in' is the actual 'writing' of code, and it involves a variety of considerations: choice of data structures and algorithms, coding styles, performance issues, unit Testing etc.

Since detailed discussion of Java programming is out of this book's scope, we merely present code listing for some of the classes of the sequence diagram of Figure 15.4, in Tables 15.10a–15.10f. (The code has been compiled, run and tested given all dependencies. Readers are encouraged to review and improve it.)

During Implementation, Preeti's team grew by two more developers, Harsh and Sangeeta. Implementation is often the most intense workflow in terms of effort concentration. It is usual for development teams to ramp up to their maximum strength during this time.

Table 15.10a   BankFacadeServlet.java

```java
package datta.kbo;

import javax.servlet.*;
import javax.servlet.http.*;
import java.io.*;
import java.util.*;
import java.sql.*;

import datta.project1.*;

public class BankFacadeServlet extends HttpServlet {

// process "get" requests from clients
protected void doGet( HttpServletRequest request,
HttpServletResponse response )
throws ServletException, IOException
{
        String page = request.getParameter("page");

        if(page.equalsIgnoreCase("login"))
                {
                        RequestDispatcher dispatcher = request.getRequestDispatcher ("/jsp/
login.jsp");
                        dispatcher.forward(request, response);
                } // end if(page.equalsIgnoreCase("login"))
        if(page.equalsIgnoreCase("verify"))
                {
                        LoginRequestManager lrm = new LoginRequestManager();
                        UserDataBean ipbean = new UserDataBean();
                        ipbean.setSsn((request.getParameter("ssn")).trim());
                        ipbean.setPassword((request.getParameter("password")).trim());

                        Hashtable ht = new Hashtable();
                        ht = lrm.manageRequest(ipbean);
                        //Log.sop("ht", ht);
                        String userType = "noUserType";
                        userType = (String)(ht.get("userType"));
                        //Log.sop("userType", userType);
```

```
                    if(userType.equalsIgnoreCase("admin"))
                    {
                            RequestDispatcher dispatcher = request.getRequestDispatcher ("/
jsp/adminMenu.jsp");
                            dispatcher.forward(request, response);
                    }
                else if (userType.equalsIgnoreCase("customer"))
                        {
                            UserDataBean userDataBean = new UserDataBean();
                            userDataBean = (UserDataBean)(ht.get("userDataBean"));
                            //Log.sop("First Name", opbean.getFirstName());
                            request.setAttribute("userDataBean", userDataBean);

                            BankResultBean bankResultBean = new BankResultBean();
                            bankResultBean = (BankResultBean)(ht.
get("bankResultBean"));
                            request.setAttribute("bankResultBean", bankResultBean);

                            RequestDispatcher dispatcher = request.
getRequestDispatcher ("/jsp/customerMenu.jsp");
                            dispatcher.forward(request, response);

                        }
                        else
                        {
                                request.setAttribute("pageMessage", "The system
could not verify your login credentials succesfully. Please try again.");
                                RequestDispatcher dispatcher = request.
getRequestDispatcher ("/jsp/kboError.jsp");
                                dispatcher.forward(request, response);

                        }
            } // end if(page.equalsIgnoreCase("verify"))
            if(page.equalsIgnoreCase("viewCustomerTransactions"))
            {
                            TransactionDataBean tipbean = new TransactionDataBean();
                            UserTransactionsRequestManager utrm = new
UserTransactionsRequestManager();
                            Hashtable ht = new Hashtable();
                            BankResultBean bankResultBean = new BankResultBean();

                            tipbean.setAccNumber((request.getParameter("accNumber")).trim());
                            ht = utrm.manageRequest(tipbean);
                            //Log.sop("ht from UserTransactionsRequestManager", ht);
                            bankResultBean = (BankResultBean)(ht.get("bankResultBean"));

                            request.setAttribute("bankResultBean", bankResultBean);
                            RequestDispatcher dispatcher = request.getRequestDispatcher ("/
jsp/customerTransactions.jsp");

                            dispatcher.forward(request, response);
```

```
                } //end if(page.equalsIgnoreCase("viewCustomerTransactions"))

            if (page.equalsIgnoreCase("customerMessages"))
            {
                    String task = request.getParameter("task");
                    if (task.equalsIgnoreCase("addMessage"))
                        {

                                Hashtable ht1 = new Hashtable();
                                UserMessagesRequestManager umrm = new
UserMessagesRequestManager();

                                MessageDataBean mipbean = new MessageDataBean();
                                mipbean.setSsn(( request.getParameter("ssn") ).trim());
                                mipbean.setSubject(( request.getParameter("subject")
).trim());
                                mipbean.setQuery(( request.getParameter("query")
).trim());
                                mipbean.setResponse(( request.getParameter("response")
).trim());
                                mipbean.setIsAnswered(( request.getParameter("isAnswered"
) ).trim());

                                ht1 = umrm.manageRequest(mipbean, "addMessage");

                                request.setAttribute("pageMessage", "Your message has
been recorded. KBO will revert to you as soon as possible.");
                                RequestDispatcher dispatcher = request.
getRequestDispatcher ("/jsp/confirmation.jsp");
                                dispatcher.forward(request, response);
                        }
                    if (task.equalsIgnoreCase("getAllMessagesForACustomer"))
                        {
                                Hashtable ht2 = new Hashtable();
                                UserMessagesRequestManager umrm = new
UserMessagesRequestManager();

                                MessageDataBean mipbean = new MessageDataBean();
                                mipbean.setSsn(( request.getParameter("ssn") ).trim());

                                ht2 = umrm.manageRequest(mipbean,
"getAllMessagesForACustomer");

                                BankResultBean bankResultBean = new BankResultBean();
                                bankResultBean = (BankResultBean)(ht2.
get("bankResultBean"));

                                request.setAttribute("bankResultBean", bankResultBean);
                                RequestDispatcher dispatcher = request.
getRequestDispatcher ("/jsp/customerMessages.jsp");
                                dispatcher.forward(request, response);
                        }
```

```
                } // if (page.equalsIgnoreCase("message"))
             if(page.equalsIgnoreCase("adminTransactions"))
     {
                 AdminTransactionsRequestManager atrm = new
AdminTransactionsRequestManager();
                 String task = request.getParameter("task");

                 if (task.equalsIgnoreCase("view"))
                 {
                         TransactionDataBean tipbean1 = new TransactionDataBean();
                         Hashtable ht1 = new Hashtable();

                         ht1 = atrm.managerRequest(tipbean1,"view");

                         BankResultBean bankResultBeanTxn = new BankResultBean();
             bankResultBeanTxn = (BankResultBean)(ht1.get("bankResultBeanTxn"));
                         request.setAttribute("bankResultBeanTxn", bankResultBeanTxn);

                         BankResultBean bankResultBeanAcc = new BankResultBean();
             bankResultBeanAcc = (BankResultBean)(ht1.get("bankResultBeanAcc"));
                         request.setAttribute("bankResultBeanAcc", bankResultBeanAcc);

                         //request.setAttribute("pageMessage", "You can view, add or delete
transactions");

                         RequestDispatcher dispatcher = request.getRequestDispatcher ("/
jsp/adminTransactions.jsp");
                         dispatcher.forward(request, response);

                 }
                 if (task.equalsIgnoreCase("delete"))
                 {
                         TransactionDataBean tipbean2 = new TransactionDataBean();
                         tipbean2.setTxnId( (new Integer( ( request.getParameter("txnID")
).trim() ) ).intValue() );

                         Hashtable temp1 = new Hashtable();
                         temp1 = atrm.managerRequest(tipbean2,"delete");

                         Hashtable ht2 = new Hashtable();
                         ht2 = atrm.managerRequest(tipbean2,"view");

                         BankResultBean bankResultBeanTxn = new BankResultBean();
             bankResultBeanTxn = (BankResultBean)(ht2.get("bankResultBeanTxn"));
                         request.setAttribute("bankResultBeanTxn", bankResultBeanTxn);

                         BankResultBean bankResultBeanAcc = new BankResultBean();
             bankResultBeanAcc = (BankResultBean)(ht2.get("bankResultBeanAcc"));
                         request.setAttribute("bankResultBeanAcc", bankResultBeanAcc);

                         request.setAttribute("pageMessage", "The transaction has been
deleted");
```

```java
                            RequestDispatcher dispatcher = request.getRequestDispatcher ("/
jsp/adminTransactions.jsp");
                    dispatcher.forward(request, response);

            }
            if (task.equalsIgnoreCase("add"))
            {
                    TransactionDataBean tipbean3 = new TransactionDataBean();

                    tipbean3.setAccNumber(( request.getParameter("accNumber") ).trim()
);

                    tipbean3.setTxnDate(( request.getParameter("txnDate") ).trim());
                    tipbean3.setAmount(( request.getParameter("amount") ).trim());
                    tipbean3.setTransactionType(( request.getParameter("transactionTyp
e") ).trim());
                    tipbean3.setTransactionDesc(( request.getParameter("transactionDes
c") ).trim());

                    Hashtable temp2 = new Hashtable ();
                    temp2 = atrm.managerRequest(tipbean3,"add");

                    Hashtable ht3 = new Hashtable();
                    ht3 = atrm.managerRequest(tipbean3,"view");

                    BankResultBean bankResultBeanTxn = new BankResultBean();
            bankResultBeanTxn = (BankResultBean)(ht3.get("bankResultBeanTxn"));
                    request.setAttribute("bankResultBeanTxn", bankResultBeanTxn);

                    BankResultBean bankResultBeanAcc = new BankResultBean();
            bankResultBeanAcc = (BankResultBean)(ht3.get("bankResultBeanAcc"));
                    request.setAttribute("bankResultBeanAcc", bankResultBeanAcc);

                    request.setAttribute("pageMessage", "The transaction has been
added");
                    RequestDispatcher dispatcher = request.getRequestDispatcher ("/
jsp/adminTransactions.jsp");
                    dispatcher.forward(request, response);

            }
        } //end if(page.equalsIgnoreCase("adminTransactions"))

        if(page.equalsIgnoreCase("adminMessages"))
    {
            AdminMessagesRequestManager amrm = new AdminMessagesRequestManager();
            String task = request.getParameter("task");

            if (task.equalsIgnoreCase("getAllMessages"))
            {
                    Hashtable ht1 = new Hashtable ();
```

```
                    MessageDataBean mipbean1 = new MessageDataBean();

                    ht1 = amrm.manageRequest(mipbean1, "getAllMessages");

                    BankResultBean bankResultBean1 = new BankResultBean();
                    bankResultBean1 = (BankResultBean)(ht1.get("bankResultBean"));

                    request.setAttribute("bankResultBean", bankResultBean1);

                    RequestDispatcher dispatcher = request.getRequestDispatcher ("/
jsp/adminMessages.jsp");
                    dispatcher.forward(request, response);
                }
                if (task.equalsIgnoreCase("addReplyToMessage"))
                {
                    Hashtable temp = new Hashtable ();
                    MessageDataBean mipbean2 = new MessageDataBean();


                    mipbean2.setMsgId( (new Integer( ( request.getParameter("msgId")
).trim() ) ).intValue() );
                    mipbean2.setResponse( (request.getParameter("response") ).trim());
                    mipbean2.setIsAnswered( (request.getParameter("isAnswered")
).trim());

                    temp = amrm.manageRequest(mipbean2, "addReplyToMessage");

                    Hashtable ht1 = new Hashtable ();
                    MessageDataBean mipbean1 = new MessageDataBean();

                    ht1 = amrm.manageRequest(mipbean1, "getAllMessages");

                    BankResultBean bankResultBean1 = new BankResultBean();
                    bankResultBean1 = (BankResultBean)(ht1.get("bankResultBean"));

                    request.setAttribute("bankResultBean", bankResultBean1);

                    request.setAttribute("pageMessage", "Bank's response to the
message has been recorded.");
                    RequestDispatcher dispatcher = request.getRequestDispatcher ("/
jsp/adminMessages.jsp");
                    dispatcher.forward(request, response);

                }

        } // end if(page.equalsIgnoreCase("adminMessages"))



 }
}
```

Table 15.10b AdminTransactionsRequestManager.java

```java
package datta.kbo;

import java.util.*;

public class AdminTransactionsRequestManager {

    /**
     *
     */
    public AdminTransactionsRequestManager() {
        super();
        // TODO Auto-generated constructor stub
    }
    public Hashtable manageRequest(TransactionDataBean tipbean, String task)
    {
        Hashtable ht = new Hashtable();

        if(task.equalsIgnoreCase("delete"))
        {
            TransactionController tc = new TransactionController();
            ht = tc.deleteTransaction(tipbean);

        }
        if(task.equalsIgnoreCase("add"))
        {
            TransactionController tc = new TransactionController();
            ht = tc.addTransaction(tipbean);

        }
        if(task.equalsIgnoreCase("view"))
        {
            Hashtable transactionData = new Hashtable();
            TransactionController tc = new TransactionController();
            transactionData = tc.getAllTransactions(tipbean);

            Hashtable accountData = new Hashtable();
            AccountController ac = new AccountController();
            AccountDataBean aipbean = new AccountDataBean();
            accountData = ac.getAllAccounts(aipbean);

            ht.put("bankResultBeanTxn", transactionData.get("bankResultBean"));
            ht.put("bankResultBeanAcc", accountData.get("bankResultBean"));


        }
        return ht;
    }
```

```
public static void main(String[] args) {

        AdminTransactionsRequestManager atrm = new AdminTransactionsRequestManager();
        TransactionDataBean tipbean 3 = new TransactionDatabean();
        Hashtable ht3 = new Hashtable();
        ht3 = atrm.managerRequest(tipbean3,"view");


        Vector v1 = new Vector();
        BankResultBean bankResultBeanTxn = new BankResultBean();
        bankResultBeanTxn = (BankResultBean)(ht3.get("bankResultBeanTxn"));

        v1 = bankResultBeanTxn.getOpbeanCollection();

        for(int i=0; i<v1.size(); i++)
        {
                TransactionDataBean topbean = new TransactionDataBean();
                topbean = (TransactionDataBean)(v1.elementAt(i));

                Log.sop("txnID", new Integer(topbean.getTxnId()));
                Log.sop("accNumber", topbean.getAccNumber());
                Log.sop("txnDate", topbean.getTxnDate());
                Log.sop("amount", topbean.getAmount());
                Log.sop("transactionType",topbean.getTransactionType());
                Log.sop("transactionDesc", topbean.getTransactionDesc());


        }
        Vector v2 = new Vector();
        BankResultBean bankResultBeanAcc = new BankResultBean();
        bankResultBeanAcc = (BankResultBean)(ht3.get("bankResultBeanAcc"));

        v2 = bankResultBeanAcc.getOpbeanCollection();

        for(int i=0; i<v2.size(); i++)
        {
                AccountDataBean aopbean = new AccountDataBean();
                aopbean = (AccountDataBean)(v2.elementAt(i));

                Log.sop("accNumber",aopbean.getAccNumber());
                Log.sop("ssn",aopbean.getSsn());
                Log.sop("initialBalance",aopbean.getInitialBalance());
                Log.sop("openingDate",aopbean.getOpeningDate() );

        }

    }
}
```

Table 15.10c TransactionController.java

```
package datta.kbo;

import java.util.Hashtable;

public class TransactionController {

        /**
         *
         */
        public TransactionController() {
                super();
                // TODO Auto-generated constructor stub
        }
        public Hashtable deleteAccountTransactions(TransactionDataBean tipbean)
        {
                Hashtable ht = new Hashtable ();
                BankResultBean rbean = new BankResultBean();

                ht.put("bankResultBean",rbean);
                return ht;
        }
        public Hashtable deleteTransaction(TransactionDataBean tipbean)
        {
                Hashtable ht = new Hashtable ();
                TransactionDataAccessor tda = new TransactionDataAccessor();
                BankResultBean rbean = new BankResultBean();

                rbean = tda.deleteTransaction(tipbean);

                ht.put("bankResultBean",rbean);
                return ht;

        }
        public Hashtable addTransaction(TransactionDataBean tipbean)
        {
                Hashtable ht = new Hashtable ();
                TransactionDataAccessor tda = new TransactionDataAccessor();
                BankResultBean rbean = new BankResultBean();

                rbean = tda.addTransaction(tipbean);

                ht.put("bankResultBean",rbean);
                return ht;

        }
        public Hashtable getAllTransactions(TransactionDataBean tipbean)
        {
                Hashtable ht = new Hashtable ();
                TransactionDataAccessor tda = new TransactionDataAccessor();
                BankResultBean rbean = new BankResultBean();
```

```
                rbean = tda.getAllTransactions(tipbean);

                ht.put("bankResultBean",rbean);

                return ht;

        }
        public Hashtable getAccountTransactions(TransactionDataBean tipbean)
        {
                Hashtable ht = new Hashtable ();
                TransactionDataAccessor tda = new TransactionDataAccessor();
                BankResultBean rbean = new BankResultBean();

                rbean = tda.getAccountTransactions(tipbean);
                ht.put("bankResultBean",rbean);
                return ht;



        }
        public static void main(String[] args) {

        }
}
```

Table 15.10d   TransactionDataAccessor.java

```
package datta.kbo;

import java.sql.*;
import java.util.*;

/**
 * @author Administrator
 *
 * TODO To change the template for this generated type comment go to
 * Window - Preferences - Java - Code Style - Code Templates
 */
public class TransactionDataAccessor {
        private Connection connection;
        private PreparedStatement transactionDeatils;

        /**
         *
         */

        public TransactionDataAccessor() {
                super();

                }
```

```
        public BankResultBean deleteAccountTransactions(TransactionDataBean tipbean)
        {
                BankResultBean rbean = new BankResultBean();

                return rbean;

        }
        public BankResultBean deleteTransaction(TransactionDataBean tipbean)
        {
                Connection connection;
                PreparedStatement accountTransaction;
                BankResultBean rbean = new BankResultBean();

                try
                {
                        Class.forName( CONSTANTS.dbDriver );
                 connection = DriverManager.getConnection(CONSTANTS.dbConnection) ;
                 accountTransaction = connection.prepareStatement("DELETE FROM transactions WHERE
txnId = ?");

                        accountTransaction.setInt(1,tipbean.getTxnId());

                        accountTransaction.executeUpdate();

                        accountTransaction.close();
                        connection.close();
                }
                catch (Exception e)
                {
                        e.printStackTrace();
                }
                return rbean;

        }
        public BankResultBean addTransaction(TransactionDataBean tipbean)
        {
                Connection connection;
                PreparedStatement accountTransaction;
                BankResultBean rbean = new BankResultBean();

                try
                {
                        Class.forName( CONSTANTS.dbDriver );
                 connection = DriverManager.getConnection(CONSTANTS.dbConnection) ;
                 accountTransaction = connection.prepareStatement("INSERT INTO transactions (accNum
ber,txnDate,amount,transactionType,transactionDesc) VALUES (?,?,?,?,?)");

                        accountTransaction.setString(1,tipbean.getAccNumber());
                        accountTransaction.setString(2,tipbean.getTxnDate());
                        accountTransaction.setString(3,tipbean.getAmount());
                        accountTransaction.setString(4,tipbean.getTransactionType());
```

```
                accountTransaction.setString(5,tipbean.getTransactionDesc());

                accountTransaction.executeUpdate();

                accountTransaction.close();
                        connection.close();
        }
        catch (Exception e)
        {
                        e.printStackTrace();
        }
        return rbean;


}
public BankResultBean getAllTransactions(TransactionDataBean tipbean)
{
        Connection connection;
        PreparedStatement allTransactions;
        BankResultBean rbean = new BankResultBean();

        try
        {
                Class.forName( CONSTANTS.dbDriver );
        connection = DriverManager.getConnection(CONSTANTS.dbConnection) ;
        allTransactions = connection.prepareStatement("SELECT * FROM transactions");


        ResultSet results = allTransactions.executeQuery();

        Vector opbeanCollection = new Vector();

        while ( results.next() )
                {
                        TransactionDataBean topbean = new TransactionDataBean();
                        topbean.setTxnId(results.getInt(1));
                        topbean.setAccNumber(results.getString(2));
                        topbean.setTxnDate(results.getString(3));
                        topbean.setAmount(results.getString(4));
                        topbean.setTransactionType(results.getString(5));
                        topbean.setTransactionDesc(results.getString(6));

                        opbeanCollection.add(topbean);
                }

                rbean.setOpbeanCollection(opbeanCollection);

                allTransactions.close();
                connection.close();
        }
        catch (Exception e)
        {
```

```
                                    e.printStackTrace();
                }
                return rbean;

        }
        public BankResultBean getAccountTransactions(TransactionDataBean tipbean)
        {
                Connection connection;
                PreparedStatement accountTransactions;
                BankResultBean rbean = new BankResultBean();

                try
                {
                        Class.forName( CONSTANTS.dbDriver );
                 connection = DriverManager.getConnection(CONSTANTS.dbConnection) ;
                 accountTransactions = connection.prepareStatement("SELECT * FROM transactions
WHERE accNumber = ?");
                        accountTransactions.setString(1, tipbean.getAccNumber().trim());

                ResultSet results = accountTransactions.executeQuery();

                Vector opbeanCollection = new Vector();

                while ( results.next() )
                        {
                                TransactionDataBean topbean = new TransactionDataBean();
                                topbean.setTxnId(results.getInt(1));
                                topbean.setAccNumber(results.getString(2));
                                topbean.setTxnDate(results.getString(3));
                                topbean.setAmount(results.getString(4));
                                topbean.setTransactionType(results.getString(5));
                                topbean.setTransactionDesc(results.getString(6));

                                opbeanCollection.add(topbean);
                        }

                rbean.setOpbeanCollection(opbeanCollection);

                accountTransactions.close();
                connection.close();
        }
        catch (Exception e)
        {
                e.printStackTrace();
        }
        return rbean;
    }

}
```

Table 15.10e BankResultBean.java

```
package datta.kbo;

import java.io.*;
import java.util.*;

public class BankResultBean implements Serializable {

    /**
     *
     */
    public BankResultBean() {
        super();
        // TODO Auto-generated constructor stub
    }
    private Vector opbeanCollection;
    /**
     * @return Returns the opbeanCollection.
     */
    public Vector getOpbeanCollection() {
        return opbeanCollection;
    }
    /**
     * @param opbeanCollection The opbeanCollection to set.
     */
    public void setOpbeanCollection(Vector opbeanCollection) {
        this.opbeanCollection = opbeanCollection;
    }
}
```

Table 15.10f TransactionDataBean.java

```
package datta.kbo;

import java.io.*;

public class TransactionDataBean implements Serializable {

    /**
     * @return Returns the accNumber.
     */
    public String getAccNumber() {
        return accNumber;
    }
    /**
     * @param accNumber The accNumber to set.
     */
    public void setAccNumber(String accNumber) {
        this.accNumber = accNumber;
    }
    /**
```

```
       * @return Returns the amount.
       */
      public String getAmount() {
             return amount;
      }
      /**
       * @param amount The amount to set.
       */
      public void setAmount(String amount) {
             this.amount = amount;
      }
      /**
       * @return Returns the transactionDesc.
       */
      public String getTransactionDesc() {
             return transactionDesc;
      }
      /**
       * @param transactionDesc The transactionDesc to set.
       */
      public void setTransactionDesc(String transactionDesc) {
             this.transactionDesc = transactionDesc;
      }
      /**
       * @return Returns the transactionType.
       */
      public String getTransactionType() {
             return transactionType;
      }
      /**
       * @param transactionType The transactionType to set.
       */
      public void setTransactionType(String transactionType) {
             this.transactionType = transactionType;
      }
      /**
       * @return Returns the txnDate.
       */
      public String getTxnDate() {
             return txnDate;
      }
      /**
       * @param txnDate The txnDate to set.
       */
      public void setTxnDate(String txnDate) {
             this.txnDate = txnDate;
      }
      /**
       * @return Returns the txnId.
       */
      public int getTxnId() {
```

```
            return txnId;
    }
    /**
     * @param txnId The txnId to set.
     */
    public void setTxnId(int txnId) {
            this.txnId = txnId;
    }
    /**
     *
     */
    private int txnId;
    private String accNumber;
    private String txnDate;
    private String amount;
    private String transactionType;
    private String transactionDesc;
    public TransactionDataBean() {
            super();
            // TODO Auto-generated constructor stub
    }

}
```

15.7 TESTING

In Chapter 17, we discussed many of the challenges associated with Testing and section 14.7.5 outlined the test workflow. As discussed, test cases for user Testing are driven by use cases. In Table 15.11 the test case corresponding to KBO_UC_01 is presented.

Table 15.11   Test Case KBO_TC_01: Record Transaction

ID	KBO_TC_01	
Name	Record Transaction	
Objective	To test whether Administrator is able to record transaction details for a particular account, if Administrator has all relevant information.	
Steps	**Expected Outcome**	**Actual Outcome**
1. Enter user id and password in the login page.	If user id and password combination is correct, allow access into the system.	
2. Identify specific user account for which transaction has to be recorded.	Present selected user account to Administrator and placeholder for recording transaction information.	
3. Record transaction information.	Present confirmation screen for successful recording.	

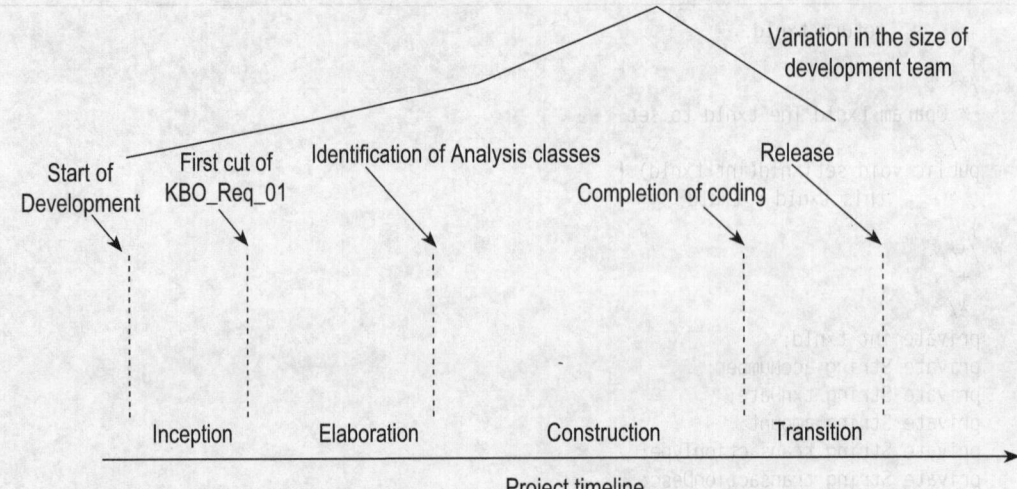

Figure 15.6 Phase boundaries and variation of project team size for first iteration of KBO

15.8 PHASE MILESTONES

So far we have seen how the development team traversed the software development life cycle across the workflows while developing KBO. With reference to Chapter 14, the software development life cycle is a matrix of workflows and phases. So, what about the phases?

The system we have considered is really a 'toy' example—the next section discusses many of its limitations. Moreover, on demands of brevity and simplicity, we have just taken one slice of functionality of that example for detailed discussion. The phases of Inception, Elaboration, Construction, and Transition can be realistically identified only at a certain scale of development. For our example, a demarcation of the phase boundaries will be more illustrative than exact. Still, for the sake of completeness, Figure 15.6 shows how the phases map into our development scenario alone with variation of the development team size. For KBO, only the Requirements workflow can be said to have had more than one iteration—developing the initial and expanded versions of KBO_Req_01. As discussed in section 14.8, for a real project however, every phase can have several iterations across the workflows within it.

15.9 LIMITATIONS OF CASE STUDY

As already indicated, the KBO example is really a much watered down version of a real-life software system. No textbook system can match a real one in terms of scope, subtlety, and complexity. One learns about the real world only by stepping into the real world; textbooks can only partially prepare us. In this section we will highlight some limitations of KBO when compared to a similar system in the real world.

In any financial system, security is a vital concern. The real world KBO needs to be designed to ensure that security needs are met. When we log on to any online banking application, it is quite likely many other users are also logged in at the same time. How is it ensured that we only see information that is relevant to us? Supporting the needs of concurrent users—tens, hundreds, or thousands at a time—is a key demand on any online application. Moreover, in the Record Transaction use case we considered, there was no provision for the common contingency of network error. What happens if the network connectivity fails when the administrator is in the process of recording these details? Are the details recorded partially or rolled back completely? This involves the very important issue of *transactional integrity* for financial systems, which was ignore in KBO. (In a bank transaction, if an account is being debited by Rs X to credit another account by the same Rs X, the debit and corresponding credit must either *both* happen, or *neither* should happen. This is a very simple example of transactional integrity.) Financial systems have high demands of performance and other non-functional Requirements. Some of them are identified in Table 15.5, but no specific Design choices in KBO were made to ensure these Requirements were met.

The above are only few of the characteristics of a real-life online banking application we have chosen to ignore in KBO to keep life simple. But real life is not simple, and as argued in section 12.4, addressing the complexity of the real world is one of the basic challenges of software development. In this chapter's case study, that complexity has been willfully diluted to better demonstrate the basic activities that go into building a software system from scratch. The objective has been to acquaint readers with the nuances of the development process in a simple scenario.

SUMMARY AND TAKE-AWAYS

In this chapter no new concept has been introduced. Some of the key ideas from earlier chapters have been illustrated by building part of an example software system—Kuber Bank Online (KBO). We have followed the development team from Requirements to Testing of a slice of KBO's functionality. Development artefacts such as Requirement description, use case, class diagram, sequence diagram, code, test cases etc have been presented. Allowing for the very modest scale of the example, we have also identified phases of development life cycle. Finally, several of KBO's limitations, when compared to a corresponding real-life system, were highlighted.

WHERE TO LOOK FOR MORE

- [Larman 1997] presents a detailed and rigorous case study of building a software system using UML constructs.

EXERCISES

Reflective Questions

Reflective Questions require you to think more deeply about some of the ideas and come up with your own interpretations and answers.

1. During Analysis we identified transaction as an entity class. Did this observation remain valid during Design? If not, why?

2. For a system of KBO's scope, identification of the workflows in easier than segregation of the phases. Do you agree with this statement? Justify your answer.

3. Pick out one limitation of KBO vis-à-vis a corresponding system in the real world. How would you modify KBO's Design to address this limitation?

REFERENCES

Datta, S. (2005), 'Integrating the furps+ model with use cases – a metrics driven approach', In *ISSRE 2005: Supplementary Proceedings of the 16th IEEE International Symposium on Software Reliability Engineering*, pp. 4.51–4.52, IEEE Computer Society.

Datta, S. (2007), *Metrics-Driven Enterprise Software Development: Effectively Meeting Evolving Business Needs*, J. Ross Publishing.

Larman, C. (1997), *Applying UML and Patterns*, Prentice Hall.

Weinberg, G.M. (2005), *Weinberg on Writing: The Fieldstone Method*, Dorset House Publishing Company, Inc.

Tricks of the Trade

Learning Objectives

In this chapter we present some tricks of the trade—a set of ideas not confined to a specific area of software development, but of significant importance in making better software. Specifically, we will discuss:

- The spectrum of refactor, reuse, refine
- How refactoring helps improve the structure of existing code
- The importance of reuse in software development
- How software development is influenced by continual refinement
- Basic ideas of structured Analysis, modular Design, transform and transaction mapping and real-time software Design

16.1 MOTIVATION

The last few chapters sought to impart an understanding of software architecture, paradigms and languages of software development, and how software is developed across phases and workflows. In this chapter we present a set of ideas which underpin or extend much of what we have said so far and will say in the following chapters, yet do not closely fit into the narrative of any other chapter of the book. In the absence of a better name, this set of ideas is being called the tricks of the trade.

Every trade has a bag of tricks—growing over time—that come from the ever enriching mix of practitioner skill, experience, and intuition. This is especially important in software engineering, given the relative immaturity of our discipline. With time, some tricks become obsolete due to changes in technology or methodology, while others evolve into accepted principles of the discipline.

'Trick' has more than one shade of meaning. It carries a sense of ingenuity and utility; a quick and clever way to get something done. It is also at times taken to mean duplicity or disingenuousness (as in the phrase '... tricked into believing'). The title of this chapter uses 'trick' in the former sense!

16.2 REFACTOR, REUSE, REFINE

In this chapter, we will first highlight the artifices—refactor, reuse, and refine—from software engineering's growing collection of tricks. Let us first see how refactor, reuse, and refine are positioned amongst the workflows and phases, in the spectrum of concrete to abstract.

What do we mean by concrete? In this context, concrete is something that gives clear guidelines of what is to be done in a particular situation of software development, to solve a problem, or to improve the solution. The key point about 'concrete' is that it relates to some advice that can be readily implemented in practice. As we shall see, refactoring lies towards the concrete end of the spectrum. Abstract, on the other hand, embodies a general way of thinking underlying the activities of software development. It helps us discern patterns and trends, and endows a deeper perception of the problem at hand. The idea of refinement is more towards the abstract end of the spectrum. Reuse comes in between concrete and abstract. In Figure 16.1, refactor, reuse, and refine are positioned along the concrete-abstract axis, as well as the intersecting dimensions of the workflows and phases. As we will discuss in the following sections, refactor is most relevant for artefacts such as code, which is produced largely in the later workflows and phases. Reuse is most apt during the later workflows and the early phases; and refine is

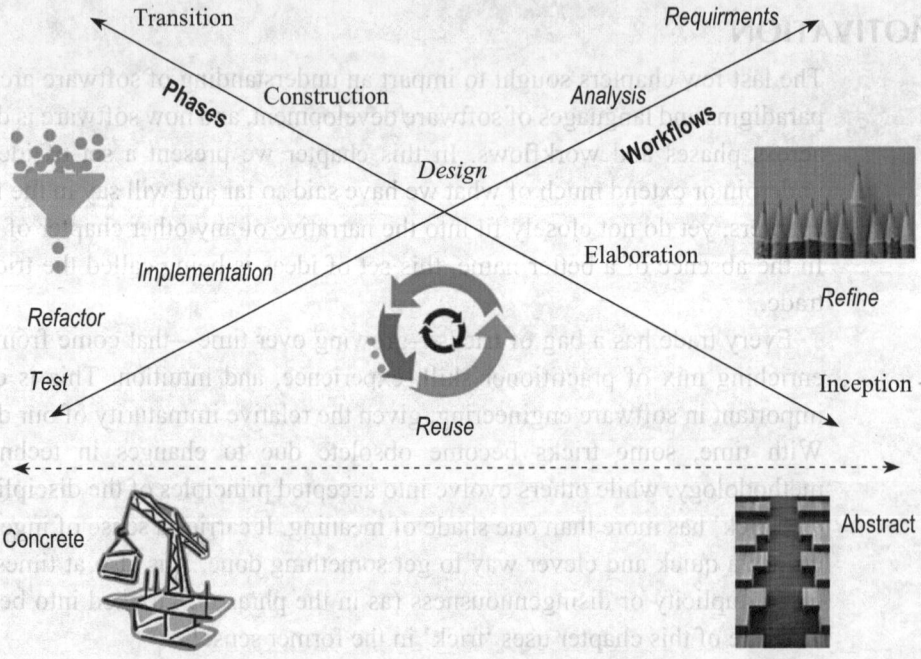

Figure 16.1 The positioning of refactor-reuse-refine

most relevant in the early workflows and phases. This is not to say refactor, reuse, or refine do not have any value at places other than those shown in Figure 16.1; but generally their biggest benefit comes at these junctures.

We will now explore each of these individually.

16.3 REFACTOR

Any system that is in use will be subjected to change (see Chapter 18). If we live in a house for long, we make it more habitable by adding a verandah here, breaking down a partition wall there, turning a terrace into a room at still another place. Similarly, a software system undergoes gradual accretion of functionality above and beyond the original intent of its Design. The aim of refactoring is to ensure that even as these functional changes happen, the structural soundness of the system is preserved.

Fowler, one of the pioneers of refactoring, says, 'Refactoring is a disciplined technique for restructuring an existing body of code, altering its internal structure without changing its external behaviour. Its heart is a series of small behaviour preserving transformations. Each transformation (called a 'refactoring') does little, but a sequence of transformations can produce a significant restructuring' (www. refactoring.com).

It is worth reiterating that refactoring *does not* seek to alter behaviour, but seeks to modify the structure of the code to offset some of the inevitable degeneration due to changing functionality. Thus discipline is of paramount importance in refactoring. It is easy to fall into the trap of adding bits and pieces of new functionality while making changes to code. But refactoring prohibits adding new functionality. The focus needs to be on restructuring the code, while keeping functionality constant. It is also important to refactor incrementally; the entire body of code cannot be restructured in one fell swoop; trying to do that would inevitably lead to commensurately large failure. Not only should refactoring not introduce new functionality, it also should not break the existing functionality. Refactoring is essential for keeping large and complex software systems functioning for large durations of time.

According to Booch, '... Without adding energy by means of refactoring, a complex software-intensive system will indeed become increasingly irregular and thus increasingly chaotic over time. Sometimes, when that system's entropy becomes too great, you simply have to blow it up and start over' [Booch 2008]. This is an interesting observation, as it brings into software the energy-entropy metaphor that is relevant to so many other systems. The natural tendency of systems is to descend into disorder—a state of increasing entropy—and only via a continuous injection of energy through some ordering process can the disorder

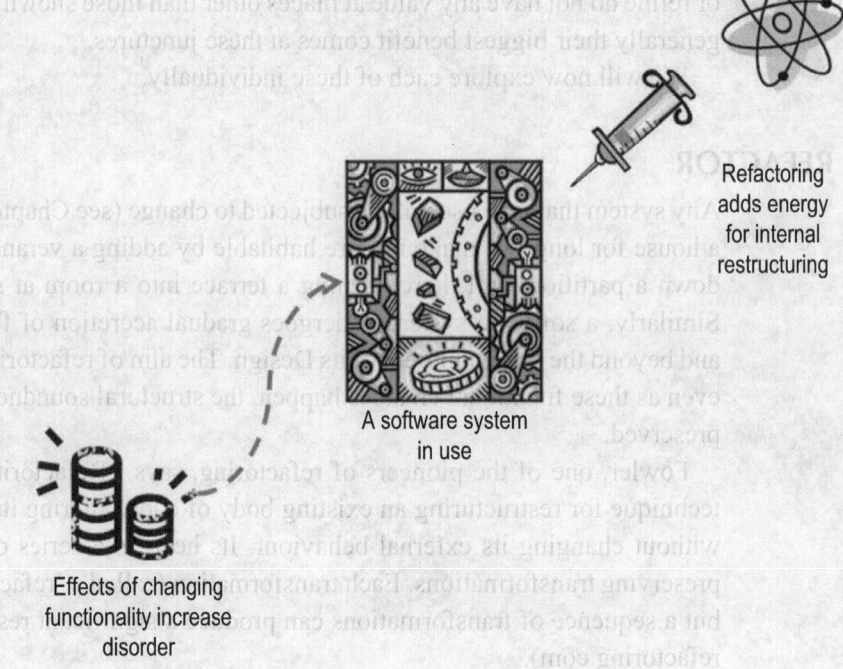

Refactoring
adds energy
for internal
restructuring

A software system
in use

Effects of changing
functionality increase
disorder

Figure 16.2 The context of refactoring

be kept in check. The idea of software entropy is introduced in Chapter 18. With reference to Figure 16.2, refactoring can be seen as an organized way of managing software entropy.

The reason why refactoring is near the concrete end of Figure 16.1 is because there are clear refactoring guidelines for variety of code conditions. The definitive guide for refactoring at—www.refactoring.com—has a number of such guidelines, two of which are presented in Figure 16.3 and Figure 16.4.

As Figure 16.3 and Figure 16.4 indicate, refactoring may not necessarily be something esoteric. While remaining an essential influence on a system's long term health, refactoring frequently starts with the recognition of degenerating structure of a piece of code, and then applying common Design and coding best practices towards restructuring.

16.4 REUSE

Even 10 years ago it was commonly heard that software engineers will soon—in the next couple of years, that is—no longer labour to build software from scratch. Software will be built by combining off-the-shelf components, LEGO-like. One still hears this prophesy, though less frequently. 'Next couple of years' has become a decade, and software engineers still labour to build software from scratch. The proponents of the prophecy point(ed) out that in every other engineering complex structures are built by combining standard off-the-shelf components. If a house is being built, no one worries about building each window frame, electrical wiring, or bathroom fixture from scratch. Appropriate ones are chosen from a variety of readily available ones, and put together as per the design. The essential skill lies in architecting and designing the house according to its inhabitant's needs, and not slogging to fabricate each component from its basic constituents. By analogy, shouldn't the focus of software engineers be on the essence of architecting and designing, while the building of software becomes the assembly of off-the-shelf components?

As an argument, this is sound. But soundness of argument often does not translate to ease of service. Software reuse—which makes the case that every time a new software system is built it should reuse software that is already built and tested—has

A complicated if-else statement:

```
if (price.moreThan(LOWEST_PRICE) && price.lessThan(HIGHEST_PRICE)) {
    orderAmount = quantity * price;
}
else {
    orderAmount = quantity * price - specialDiscount;
}
```

Method extracted from the condition; the if part and the else part:

```
if (withinRange(price)) {
    orderAmount = standardOrder(quantity);
}
else {
    orderAmount = specialOrder(quantity);
}
```

Figure 16.3 Decompose conditional (adapted from http://www.refactoring.com/catalog/decompose Conditional.html)

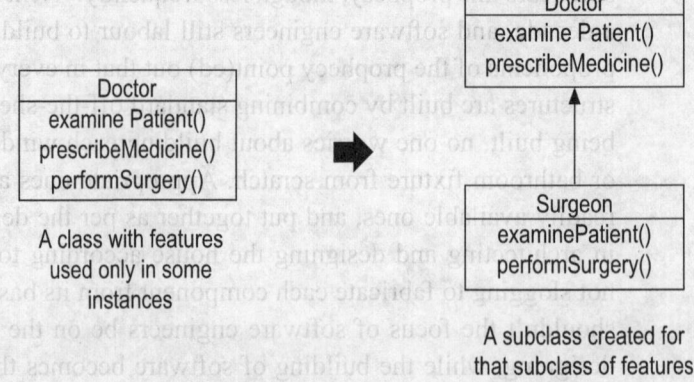

Figure 16.4   Extract subclass (adapted from http://www.refactoring.com/
catalog/extractSubclass.html)

long been a holy grail of software engineering. Every new development paradigm comes with bold new promises of increasing reuse by orders of magnitude. Object orientation is no exception; however, its promises of enabling heightened reuse have remained largely unfulfilled.

Reuse is necessary for software engineering to be closer to engineering than ad-hoc hacking. At the same time, reuse is not easy. Whenever the word is mentioned, the implicit assumption is often the reuse of code. Reuse of code is commonly seen to be picking whole or parts of existing programs and implanting them in the required context, with the expectation that everything will work fine, thus saving us all the hassle of writing that code out ourselves. The problem is that contexts are very specific; or rather a body of code is seldom general enough to be apt for implant anywhere as it is.

The place where the analogy between reuse in software vis-à-vis other branches of engineering breaks down relates to the identification of atomic components of the build structure. Code may be the final truth of software; but what are the atomic components of code—like the way a brick, beam, or bracket is for structural engineering? If the atomic components are the statements of a programming language, we are reusing them all the while. A good IDE (integrated development environment) ensures that we do not need to type out each line of the program. With the use of libraries offered by programming languages, we are also reusing data structures and algorithms. But the so-called holy grail of reuse is something more broad; it is reusing parts of the solution of one business problem in the context of another business problem. This kind of reuse needs looking beyond code.

Actually, even though prophecies of reuse to the extent of obviating any fresh development have not yet come true, reuse in software development does already happen at several levels. The architectural patterns discussed in Chapter 11, as well as the Design patterns (see Exhibit 16.1) are nothing but modules of insight reused across a variety of development scenarios.

Even for reuse at the component and code level, there has been considerable progress in the recent past. The Web as a development medium (Chapter 19), as well as the open source paradigm (Chapter 22) have contributed significantly to code reuse. Today, if one needs code for something like say *quicksort*, it is usually just a Web search away. A bigger challenge is reusing components in the solution context of a business problem, but strides have also been made in that direction. (See 'Where to Look for More' at the end of this chapter.)

So in summary, we are far from the point where software engineers can develop a system exclusively by reusing earlier constructs and building nothing new. There is no guarantee we will reach that point in the foreseeable future. But significant progress has been made through the use of patterns in reusing architecture, Design,

---

Exhibit 16.1   Design Patterns

With reference to the discussion of architectural patterns in Chapter 11, we observe that *patterns* offer a significant scope of the reuse of ideas and insights in software development. A 'pattern' in the software context is similar to our common understanding of the term. In geometry, a pattern can be said to be an arrangement of lines and shapes which is uniquely identifiable in a variety of backgrounds. We talk of patterns of events or people's behaviour when we recognize a schema repeating across situations.

The recognition of pattern of software development was inspired to a large extent by patterns in the architecture of cities, buildings, and other physical spaces, as explored by Alexander in *The Timeless Way of Building* [Alexander 1979] and other books. The book by Gamma, Helm, Johnson, and Vlissides,— the authors collectively called the Gang of Four or GoF—*Design Patterns: Elements of Reusable Object-Oriented Software* [Gamma et al. 1995]—is one of the earliest and most detailed introductions to software Design patterns. Newer patterns of software Design have also been proposed since then.

Very simply, a pattern consists of three parts: a context, a problem, and a solution. A pattern by itself need not necessarily be something abstruse; in fact, we use patterns all the time just for going about our daily lives. For example, if it is raining (context), we will get wet when we go out of the house (problem), so we need to carry an umbrella (solution). This is very much a pattern. But for a pattern to be widely relevant, the context needs to be sufficiently general, the problem sufficiently pressing, and the solution sufficiently effective. Usually, for these conditions to be met the pattern has to come out of experience, and after significant amounts of trial. A pattern basically tells a practitioner: I have seen this situation before, I know what your problem is, and this is the solution that has worked in the past. Thus a pattern offers the scope of reusing wisdom from the past.

A well-known Design pattern is the *singleton*. Sometimes it is necessary to have exactly one instance of an object. For example, an object for logging, database access, coordination of tasks etc can have such a need. But object-oriented programming allows unlimited number of objects to be created from a class. The singleton pattern ensures that only *one* instance of an object can be created from its corresponding class template. An Implementation of the singleton pattern for Java is given in Figure 16.5.

Design patterns have been grouped into several categories depending on their scope: *creational patterns, structural patterns*, and *behavioural patterns*. The singleton pattern is a creational pattern.

The power and popularity of Design patterns have extended the scope of patterns in software development. Analysis patterns, that serve as conceptual models for understanding and expressing situations encountered during Analysis, have also been proposed [Fowler 1996].

Interestingly, along with the study and practice of patterns, there has also been significant exploration of *antipatterns* [Brown et al. 2001], [Laplante and Neill 2005]. 'Antipattern' is used in more than one sense. It is sometimes taken to mean a pattern that is applied out of context, thus forcing a solution into a problem that the pattern does not address in the first place. 'Antipattern' can also indicate a common pitfall that is widespread, and needs to be recognized and avoided. Like patterns, antipatterns also facilitate the reuse of experience and insights.

and even Analysis insights. The penetration of the Web and the popularity of open source software has also contributed largely to the reuse of code.

```
public class ExampleSingleton {

private static final ExampleSingleton SINGLE_INSTANCE =
new ExampleSingleton ();

// Private constructor prevents ensures ExampleSingleton
be instantiated from
// within ExampleSingleton only
private ExampleSingleton () {}
public static ExampleSingleton getInstance() {
return SINGLE_INSTANCE;
}
}
```

Figure 16.5 Example of Singleton.java

16.5 REFINE

Refine is by far the most abstract of the tricks of the trade we are discussing. Rather than a quick and clever way of getting something done, it is more of an underlying theme of software development. Refine encompasses refinement of physical artefacts, as well refinement of developers' perceptions across the software development life cycle.

Almost all software projects start with a barely defined problem, and the expectation that software will perfectly solve the problem. Refinement is necessary due to this inadequate problem definition as well as unreasonable expectations from the solution.

If software developers insist on answers to every outstanding question before committing to the next step, the next step will never start, and the project will stop. Successful refinement involves raising the pertinent questions, eliciting answers from stakeholders to the best of their knowledge, and dissecting the answers for the next set of questions. This process repeats over and over again as development proceeds.

Other than the most trivial context, there is be nothing called a *finished* software —any software that is in use is in constant flux. Refinement is the developer's response to this fact of the software's life. It is the continual maturity of the skills, intuition, and facility driving software development, in the light of new information and experience.

One of the primary reasons for the success of iterative and incremental development methodologies lie in their recognition of the importance of refinement. An iteration of development builds on the existent understanding of the problem, and the incremental release—where real users get a chance to test and give feedback —acts as a platform for refining the solution.

Successful software engineers master refinement to the point that it is no longer a tool they have to hunt for when the need arises, but integrated into their thinking. They need not pause to refine, it is happening all the time.

In the next section, we discuss a particular refinement technique.

16.6 STRUCTURED ANALYSIS AND DATA DICTIONARY

The technique of structured Analysis gained wide currency in the 1980s and it is still useful in some situations. The key theme of structured Analysis is to start at a high level of abstraction and iteratively refine to levels of finer granularity. In structured Analysis, a system is viewed in terms of the data coming into a system, being manipulated by the system, and flowing out of the system. Thus the system's function is essentially seen to be a set of processes that work on the data.

Towards addressing a system's functional Requirements, the output of structured Analysis is a set of diagrams, data definitions, and process descriptions.

A *context diagram* describes all the external entities interacting with a system; a *data flow diagram* is a depiction of the flow of information through the system's processes; and a *structure chart* is a low level illustration of system configuration, showing the hierarchy of module arrangements. *Structured Design* guides the development of these modules in their hierarchical arrangement.

A *data dictionary* is a key component of structured Analysis. It can be thought of as a glossary of a system's database, containing a list of database files, the records in each file, and the description of each field in each record. It should be underscored that data dictionary is *not* the repository of the system's data (the database contains such information), it is the placeholder for the *metadata* about the system's data. The data dictionary is not accessible to a system's users to prevent its accidental corruption. It is, however, essential for the database management system to refer to the data dictionary to be able to access the system's database.

16.7 MODULAR DESIGN

As discussed in Chapter 12, decomposition is a key strategy for addressing software's inherent complexity. Modular Design draws on the decomposition strategy to break down a complex system into smaller parts—modules—that can be developed and deployed independently in multiple contexts. Although primarily a systems engineering approach, modular Design has wide relevance in software development.

As we have mentioned earlier, a characteristic of a complex system is that it is greater than the sum of its parts. This gives rise to a basic challenge of modular Design—how best to subdivide a system into modules? Almost all the machinery we use in our day to day lives now is highly modularized. Random access memory (RAM) "sticks" can be used across computers, cars have tires that can be replaced without interfering with its engine, window grilles and bathroom fittings are bought off the shelf and integrated into a building. Modular Design seeks to leverage the benefits of standardization along with the facilities of customization.

It needs to be pointed out that the examples of modular Design given above are from conventional engineering disciplines—electronics, automobile, and structural—where standardization as well as the parameters of customization are well established. But this is not the case in software engineering yet. Modularity for software is deeply related to the concerns of reuse (see Section 16.4). We are far from reaching levels of modularity commonplace in other fields due the inherent nature of software (see Chapter 3) although it has been in focus for a long time [Parnas 1972].

For modular Design to be effective, we need to ensure each module has simple interface and successfully hides internal information.

16.8 TRANSFORM AND TRANSACTION MAPPING

As discussed earlier a data flow diagram (DFD) specifies the flow of information through a software system. Data flow is of two types—transform flow and transaction flow.

Data enters a software system in some external format through an input device such as keyboard or graphical user interface. For the software system to process the incoming data, it must be converted into some internal format. Incoming flow denotes the information entering the system in some external form to be converted into an appropriate internal form. To be converted to an internal form, incoming data passes through a transform centre inside the software system. Subsequently, it moves out of the system through outgoing paths as the outgoing flow. Transform centre represents part of the data flow diagram between the incoming and the outgoing flows, where the transformation of data between internal and external forms occurs; it is the seat of the transform flow.

Another view of the information flow through a system recognizes a single data item called a transaction, which can trigger other data flows along one among multiple paths. Which path will be followed in a particular situation is decided by the outcome of evaluating the transaction. The transaction centre is the hub from which the multiple paths originate; it is the seat of the transaction flow.

Transform and transaction mapping are steps in the Design process facilitating the mapping of a DFD with transform or transaction flow characteristics respectively, to be mapped into a predefined program structure. In both cases, transform or transaction centres are identified and the corresponding program structure is iteratively refined.

16.9 REAL-TIME SOFTWARE DESIGN

Real-time systems monitor and control the environment they operate in. They are invariably associated with hardware devices such as sensors—to collect data from the system's environment—and actuators—to influence the system's environment in some way. The essence of real-time systems is time criticality; they *must* respond within predefined time limits. Formally, "a real-time system is a software system where the correct functioning of the system depends on the results produced by the system and the time at which these results are produced" [Sommerville 2004].

When a real-time system receives a stimulus it must respond within a specific time; stimuli can either be periodic or aperiodic. Thus there needs to be provision for fast switching between stimulus handlers to be able to respond to diverse timing demands of different stimuli/responses. Accordingly, real-time systems are designed as cooperating processes under the control of a real-time executive [Sommerville 2004].

The conceptual elements of a real-time system are: sensor control processes, data processor, and actuator control processes. Design of real-time systems involves the Design of both hardware and software components associated with the system. Unlike a usual (non real-time) system, Design decisions have to be primarily based on non-functional Requirements of the system.

The Design process of a real-time system consists of the following steps: identifying the stimuli and required responses; identifying the timing constraint for each stimulus-response pair; aggregating the stimulus-response processing into concurrent processes—each process may address a particular class of stimulus-response; designing algorithms for each class of stimulus-response to deliver within the timing constraints; designing scheduling systems to ensure on-time Inception of the processes, and integrating all of the above through a real-time operating system [Sommerville 2004].

16.9.1 Real-Time Executive

As discussed above, a real-time executive controls the cooperating process of a real-time system. A preemptive programming approach forms the basis of real-time programming. The idea of preemption centres around the ability of the processor to pause execution of some parts of the code and start executing other parts of the code, based on some consideration. Preemption occurs when an interrupt arrives— a signal that some higher priority task requires attention. Our daily lives are full of interrupts; every other instance, a phone call or an email message is preempting our activity flow and forcing us to switch between contexts. As real-time systems closely mimic our lives in the real world, it is not surprising they operate in a preemptive mode.

Real-time executives offer sophisticated preemption mechanisms. Coordination between different tasks are managed through separate stacks or by reserving adequate space for each task on the same stack. *Tiny Exec* is a simple real-time executive written in C illustrating the basic principles of a real-time executive's functioning and is suitable for many low-complexity applications on standalone or MS-DOS based platforms (refer to http://www.ddj.com/cpp/184402613 for more details). Instead of running multiple programs, Tiny Exec runs multiple routines inside a single program. There are two types of routines: task routines and timer routines. Due to its inherent simplicity, Tiny Exec can be easily ported to other languages and platforms.

SUMMARY AND TAKE-AWAYS

In this chapter, we have discussed a set of disparate but relevant topics influencing software development. The discussion can be summarized as:

- Refactor is more concrete, refine is more abstract, and reuse lies in between.
- The aim of refactoring is to restructure code without changing functionality over a series of incremental operations, each referred to as a refactoring.
- Although reuse is commonly seen to be the reuse of code, it can happen at several other levels, as exemplified by the reuse of ideas and insights through architectural and Design patterns. We are still far from the point where a software system can be wholly built by reusing existing components. But reuse has increased in the recent years with the advent of the Web as a development medium and popularity of open source software.
- Design patterns offer an elegant and useful way to reuse insights. There are three major categories of Design patterns: creational, behavioural, and structural.
- Refinement involves raising the pertinent questions, eliciting answers from stakeholders to the best of their knowledge, and dissecting the answers for the next set of questions. This process repeats over and over again in the development life cycle. Refinement is a key skill for developers in the flux and ambiguity of real world software development.
- Structured Analysis, modular Design, transform and transaction mapping and real-time software Design are techniques useful in diverse software development scenarios.

WHERE TO LOOK FOR MORE

- www.refactoring.com serves as a definitive source for refactoring.
- www.koders.com has open source code for a wide range of programming problems.
- www.topcoder.com utilizes reuse of functional components for developing system, thus facilitating reuse of parts of the solution to a particular business problem.

EXERCISES

Review Questions

Review Questions test your understanding of the key concepts presented in this chapter.

1. Refactoring is
   (a) changing functionality of existing code
   (b) changing structure of existing code
   (c) changing performance of existing code
   (d) changing look and feel of existing code

2. The Extract subclass example of refactoring (Figure 16.4)
   (a) creates a subclass for a class' features used in specific situations
   (b) extracts a class from a set of subclasses
   (c) can be either of (a) or (b) depending on the situation
   (d) none of the above
3. Refactoring is important for
   (a) satisfying user needs
   (b) short-term performance improvement
   (c) long-term health of the system
   (d) meeting deadlines
4. Reuse is only concerned with code.
   (a) True
   (b) False
   (c) Depends on the particular system
   (d) Depends on the type of reuse
5. The most useful type of reuse is considered to be
   (a) reuse of code
   (b) reuse of Requirements
   (c) reuse of parts of solution of a business problem
   (d) all of the above
6. Which of the following is not a class of patterns?
   (a) Creational   (c) Conditional
   (b) Structural   (d) Behavioural

Reflective Questions

Reflective Questions require you to think more deeply about some of the ideas and come up with your own interpretations and answers.

1. What do you think is the most important characteristic of the concrete end of the refactor–reuse–refine spectrum? Is there a corresponding characteristic of the abstract end?

2. The energy-entropy metaphor has been invoked while discussing refactoring. In physical systems, adding energy usually has some external manifestations, like increase in temperature with the addition of heat. What do you think is the external manifestation of adding energy to a software system by refactoring? How can you measure it? Do you think the analogy between a software system and a physical system is being stretched too far? If so, where does the analogy start breaking down?

3. In the Decompose conditional example of refactoring (Figure 16.3), the number of lines of code is less in the refactored code. Do you think refactoring will always lead to a reduction in the number of lines of code? If not, why?

4. 'Reuse is necessary for software engineering to be closer to engineering than ad-hoc hacking.' Considering the prevalent level of reuse in software engineering today, are we closer to engineering or ad-hoc hacking?

5. It has been observed that the success of iterative and incremental development models comes to a large extent from the scope of refinement they offer. Is there scope of refinement if a project is following the Waterfall model of development? Justify your answer.

6. The most effective Design pattern is the one that is most general. Do you agree with this statement? Justify your answer.

7. Do you think the singleton pattern can also become an antipattern? Give an example. Why is it an antipattern in your example?

8. Explore www.topcoder.com. What kind of reuse—in terms of code, Design etc—do you think the 'components' are facilitating?

Programming Examples

Programming Examples require you to analyse or write a program or a program segment to understand a specific problem.

1. Can you think of a concrete situation when the singleton pattern can be helpful? Using the example singleton code given in this chapter, write a Java program using the singleton pattern.

2. You are given the task of sorting ten integers in ascending order. Write a program in Java to accomplish the task. Note how long it took you plan, write, test, debug, and run the program successfully. Now find some open source code from the Web which does something similar and modify it to solve the given problem. Note how long it took you plan, write, test, debug, and run the program successfully when you reused some parts of the code. Compare the time for the two cases. Which took longer? Why?

3. With reference to Figure 16.4, how would you extract a subclass Physician from the class Doctor?

4. With reference to Figure 16.3, write and run two Java programs, one using the original piece of code, and the other the refactored code. Do you think the refactored code is better than the original one? If you do not find it any better, why do you think that is so?

REFERENCES

Alexander, C. (1979). *The Timeless Way of Building*, Oxford University Press.

Booch, G. (2008), 'Measuring architectural complexity, *IEEE Softw.*, 25(4):14–15.

Brown, W.J., Malveau, R.C., McCormick, H.W.S., and Mowbray, T.J. (2001), *AntiPatterns: Refactoring Software, Architectures, and Projects in Crisis*, John Wiley and Sons.

Fowler, M. (1996), *Analysis Patterns: Reusable Object Models*, Addison-Wesley.

Gamma, E., Helm, R., Johnson, R., and Vlissides, J. (1995), *Design Patterns: Elements of Reusable Object-Oriented Software*, Addison-Wesley.

Laplante, P.A. and Neill, C.J. (2005), *Antipatterns: Identification, Refactoring, and Management*, Auerbach Publications.

Parnas, D.L. (1972), 'On the criteria to be used in decomposing systems into modules,' *Commun. ACM*, 15(12):1053–1058.

Sommerville, I. (2004), *Software Engineering*, Prentice Hall, 7th ed.

PART IV

TESTING, MAINTAINING, AND MODIFYING SOFTWARE SYSTEMS

Software Testing, Reliability, and Quality

17.1 MOTIVATION

The *light bulb effect* can point to the level of utility and penetration of a particular technology. When we turn on the switch for a light bulb, we expect light, and almost always, we do get light. This is such an ubiquitous occurrence in our lives now, there is hardly any scope to marvel at the great engineering feats of generating electricity, transmitting it over long distances, making it available at precise voltage and frequency for a light bulb to glow when its switch is turned on. The light bulb effect is something every great technology strives for—the extent of ease and familiarity for the end-user that makes the underlying engineering transparent. Electricity has long reached this level, commercial air travel too; and Internet and software systems are on their paths to the light bulb effect. But the path to this zenith of end-user utility is fraught with many trials and many errors.

Testing is an important aspect of this journey for software systems, reliability a way to measure how much we have come, and quality a way to ensure once we have crossed some milestones; we should never again lose our way.

Software testing, software reliability, and software quality are related, but distinct groups of ideas. They intermingle in unexpected ways; thus in this book we will treat them separately but in one chapter. It is naive to make sweeping comments about how Testing, reliability and quality connect with one another. But with a little bit of stretch, it may be said that Testing is one of the important ways to ensure reliability, which in turn helps establish consistent criteria of functioning for a software system, which is the primary concern of quality. As we go deeper into this chapter, we will appreciate these connections better.

17.2 SOME TESTING TERMS

As early as 1969, Dijkstra remarked, 'Testing can show the presence, but never the absence of errors in software' [Peters and Pedrycz 1999]. This statement brings out the essence of software testing in a very insightful way. It challenges the common notion about Testing being a way to certify that a system is free from defects.

Before we go into what software testing *is* and *is not*, it will be helpful to clarify some of the terms commonly associated with software testing:

- A *fault* is introduced as the result of a mistake [Schach 2005], usually human. Human mistakes resulting in faults are mostly associated with Design or Implementation, but as Figure 17.1 shows, such mistakes may originate at a number of levels—perception, understanding, Implementation and execution. Perceptual mistakes are the least concrete while those at the execution level, the most. Consequently, an execution fault is likely to have ready symptoms, which may be traced back to a perceptual mismatch between the user or the developer over a particular Requirement.
- According to the IEEE Standard 610.12 1990, a *failure* is 'The inability of a system or component to perform its required functions within specified performance Requirements.'
- An *error* is taken to be the amount by which a result is incorrect [Schach 2005].
- *Defect* is an all-encompassing term which covers faults, failures and errors.
- The word most closely associated with software testing is 'bug'. As explained in Exhibit 17.1, the word 'bug' has a checkered history. Anything going wrong with a software system (indeed, any system—a 'stomach bug' is taken to mean the cause of digestive disorder) is commonly referred to have been caused by a bug. In this chapter, we will use 'bug' to mean any *unintentional* feature that leads to an *unacceptable* behaviour of the software system. Unintentional means something that did not come from deliberate Design. Unacceptable is taken to be anything the end-users are not willing to accept.

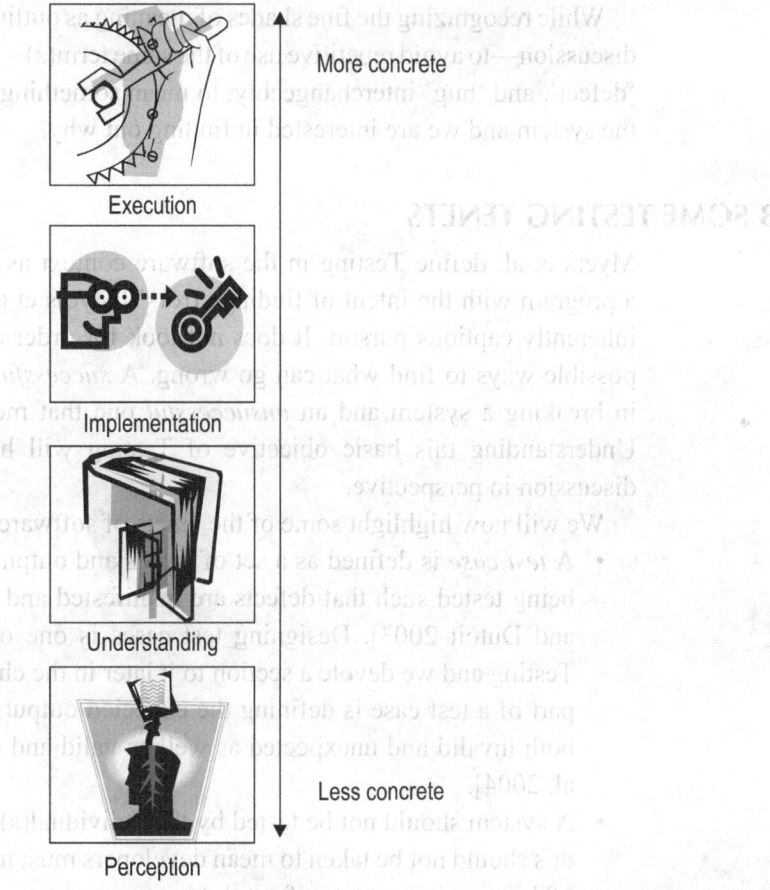

Figure 17.1 Origin of mistakes

More concrete

Execution

Implementation

Understanding

Less concrete

Perception

Exhibit 17.1 Origin of Bugs

The original use of 'bug' to denote the malfunctioning of a computer system is attributed to Rear Admiral Grace Murray Hopper (1906-1992). Hopper was a pioneering computer scientist, widely regarded as the creator of COBOL and reputed to have written the first compiler for a programming language. Legend goes that on September 9, 1945, a moth flew and lodged itself between the plates of a mechanical relay in the Mark II computer Hopper and her teams were working with at Harvard University. This was indeed an instance of a *real*, tangible bug in the system! Hopper extracted the moth and taped it into her log book with the comment, 'First actual case of bug being found' [Schach 2005].

Although this is most likely the first use of the word 'bug' in a software context, there is evidence the word was in use as an engineering jargon as early as the 19th century. Thomas Alva Edison had used it in 1878 to indicate 'faults and difficulties' [Schach 2005].

While recognizing the fine shades of meaning as outlined above, in the following discussion—to avoid repetitive use of the same term(s)—we will use 'fault', 'error', 'defect', and 'bug' interchangeably, to mean something that has gone wrong with the system and we are interested in finding out why.

17.3 SOME TESTING TENETS

Myers et al. define Testing in the software context as 'the process of executing a program with the intent of finding errors' [Myers et al. 2004]. So Testing is an inherently captious pursuit. It does not look for order and well-being; it tries all possible ways to find what can go wrong. A *successful* test is one that succeeds in breaking a system and an *unsuccessful* one that merely reports good health. Understanding this basic objective of Testing will help us put the following discussion in perspective.

We will now highlight some of the tenets of software testing:

- A *test case* is defined as a set of inputs and outputs that exercises the entity being tested such that defects are manifested and can be detected [Bruegge and Dutoit 2003]. Designing test cases is one of the central activities of Testing and we devote a section to it later in the chapter. The most important part of a test case is defining the expected output. The test case must cover both invalid and unexpected as well as valid and expected inputs [Myers et al. 2004].

- A system should not be tested by the individual(s) who develop it. However this should not be taken to mean developers must not *unit test* their code. One of the most important—if not *the* most important—elements of programmers' personal discipline is to make sure that the code that is being released is unit-tested. As we discuss in Exhibit 17.2, it is generally recommended that unit Testing code be written along with or even earlier than the code that is to be tested. Integration Testing or user acceptance Testing of a system should not be done by its developers, as developers are instinctively easy on parts of the system they know are most prone to breaking.

- 'The software is done. We are just trying to get it to work.' Bruegge et al. report this statement being made in an executive program review report [Bruegge and Dutoit 2003]. This illuminates a grave and widespread misconception: Testing can be done as an afterthought. Such a view arises from the tacit assumption—again, very wrong—that development will be so sound that no bugs will ever arise. Software engineering landscape is littered with many failed projects run under these beliefs. Testing has to be planned and budgeted (in terms of time, resources, cost and ideas) upfront. Not only will it not serve its purpose as an afterthought, it will end up being badly done and the system rejected by the end-users.

- Testing needs to verify *both* these criteria: Whether a system *does not* do what it is not supposed to do, as well as whether the system *does* something it is not supposed to do [Myers et al. 2004].
- A very interesting and counter-intuitive phenomenon related to software testing is summarized by the statement: 'The probability of the existence of more errors in a section of a program is proportional to the number of errors already found in that section' [Myers et al. 2004]. This seems odd, as we are accustomed to seeing Testing as a bug-removal process. This can be said to be the *vacuum cleaning* view—the more vacuum cleaning you do, you expect to find less dust. But an *earthquake* view is perhaps more apt for software testing: Places which have experienced more quakes in the past can expect to get more tremors in future.

17.4 TWO TESTING PHILOSOPHIES

17.4.1 Black-Box Testing

Black-box Testing addresses situations where the inner workings of the system being tested is not visible to the tester—as if the system lies wholly within a hypothetical box with *opaque* walls. The tester is only able to supply specific excitations and observe corresponding responses. The policy of black-box Testing is thus to apply a set of inputs to the system, and determine the presence or absence of bugs, based entirely on seeing the outputs. There is no way to know how the system *internally* functioned (or malfunctioned) to produce the output for a given input. As black-box Testing is driven entirely by the data one feeds into the system as input and then analysing the output, it is also referred to as *data-driven* or *input-output* Testing. Black-box Testing is a common Testing strategy for any system which we cannot see the inside of, or even if we are able to peep inside, we cannot make much sense of what we see. When a car makes an odd noise while running, our first instinct is to find out whether the noise is more at higher speeds or lower, or which gear the car is most noisy at. This is black-box Testing, the input data being the speed and the gear level, and the output being the extent of noise.

Evidently, as input data is the only leverage testers have in exercising the system, the variety of input has to be as wide as possible for black-box Testing. *Exhaustive* input Testing is thus a key criteria of black-box Testing—we not only have to try out all valid inputs, but also all possible inputs [Myers et al. 2004]. But exhausting all possible inputs is practically impossible; it should be clear from a simple example. Say, we have a program which supposedly tests whether a triangle is right-angled or not, given the length of three sides. The logic of the program is based on the Pythagorean Theorem we learnt in middle school geometry. But to exhaustively test the program as a black-box, we need to supply the potentially infinite combination of the lengths of the three sides, even the invalid but possible inputs of the length of a side being negative (going strictly by the criterion of exhaustive inputs, we have to cover the cases where the user *does not* know triangles can not have sides of negative lengths, or even if he or she knew, is liable to make a typographical error of entering a negative valued triangle side). From this example, it should be clear that exhaustive input Testing is impossible for anything other than very trivial programs. For Testing a reasonably complicated program, it cannot even be considered. The difficulty is even more compounded for systems whose present behaviour is influenced by what happened in the past. Take for example, an ATM machine; you will be able to withdraw Rs 1,000 from it only if your account balance at that time is greater than (or maybe equal to) Rs 1,000 (assuming credits or overdrafts are not allowed). Calculating the balance needs to take into account all past transactions since the account was opened. So to Design an exhaustive input test, we will not only have to feed the system all valid and possible combinations of inputs but also take into account the sequence of inputs [Myers et al. 2004]. The system should behave differently if you deposited Rs 500, Rs 200, Rs 400 and then tried to withdraw Rs 1,000, from the way it should behave if you deposited Rs 500, Rs 200, tried to withdraw Rs 1000, and then deposited Rs 400.

So, in summary, black-box Testing is an effective Testing strategy—the only way to test when we do not know how a system internally works. But exhaustive input Testing, which is a key criterion of black-box Testing, can never be practically fulfilled.

17.4.2 White-Box Testing

If we have black-box Testing, then by the principle of duality, it is expected we will have the *complementary* notion of white-box Testing. Engineering is full of dualities. We indeed have white-box Testing—the Testing philosophy drawing on the assumption that we can see and analyse the inner workings of a system, as if it was contained in a hypothetical *transparent* box. Taken literally, the terms black-box and white-box are somewhat misleading, you can very well have a box painted white that does not let you see inside. 'Opaque-box' and 'transparent-box' are surely closer to the essence of these Testing philosophies, respectively.

White-box Testing is also called *logic-driven Testing* or *exhaustive path Testing* [Myers et al. 2004]. Let us examine the implications of these phrases. 'Logic-driven' is based on our understanding about the structure of a computer program being governed by its *logic*—the sequence of program statements being executed to deliver its function. White-box Testing assumes we can see inside the 'box' that represents the system, we are free to scrutinize the system's logic and devise ways to best test it. Exhaustive path Testing, however, takes us into deeper waters. Evidently, exhaustive path means all unique paths of the control flow through a program need to be covered. Let us consider the following simple program segment:

```
...
for (int i=1; i≤5; i++)
{
if (some condition)
{do something}
else
{do something different}
}
...
```

Figure 17.2 shows the control-flow graph of this program statement. The number of unique paths in this program segment can be found by counting the number of unique ways one can go from *a* to *b* in the graph. The answer is 2. But note that the *if-else* branching is inside a *for* loop which iterates five times. So the number of unique paths effectively becomes, $2^5+2^4+2^3+2^2+2^1=62$. Exhaustive path Testing mandates the program must be made to run through *all* of these 62 unique paths.

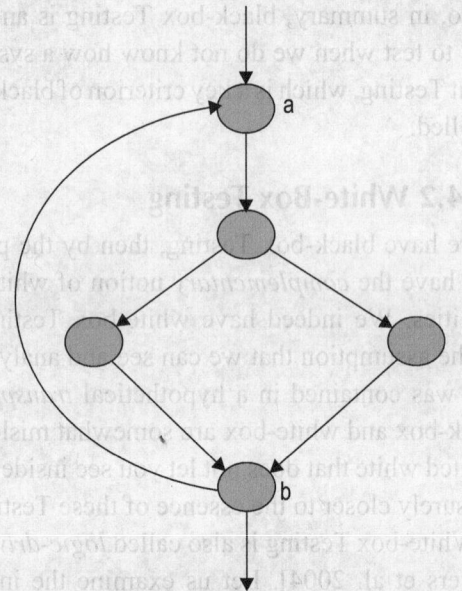

Figure 17.2 Control flow graph of program segment discussed in white-box Testing

Tedious, but not impossible, it appears. Let us increase the loop counter to 10. Calculating similarly, the number of paths is now 2046. If the loop counter is 20, the number of paths becomes more than 1,000,000. Assuming, (very optimistically) we can write a test case, run and verify it for each path in two minutes, it would take 347 days (of ceaseless toil) for us to exhaustively test all paths in the program segment for 20 passes of the *for* loop. Appreciating the fact that this is a near trivial example—real systems have thousands of paths and millions of iterations in the least—one understands the practical difficulties—almost impossibility—of *full* exhaustive path Testing.

For a moment, let us assume we have Aladin's genie at our service and instead of securing some other boon, we ask the genie to perform full exhaustive-path Testing for a system. Will this guarantee a complete test such that every error in the program will be detected? As Myers et al. point out, even full exhaustive path Testing *cannot* check whether the program has adhered to its specifications (that is, it tries to implement what it was meant to do), or if there are any missing paths, or whether there are data-sensitivity errors (for example, considering the *actual* difference between two variable values when the *absolute* value had to be taken into account) [Myers et al. 2004].

So, in summary, white-box Testing assumes we have more knowledge about the system's innards, but it still presents practical difficulties for doing in full, and even when done in full cannot ensure detection of all errors.

Now that we have an understanding of black-box and white-box Testing, we will discuss some other types of Testing, based on these general Testing philosophies.

17.5 DIFFERENT TYPES OF TESTING

Before going into the details of different types of Testing, it is important to point out that the classification of Testing types is largely perceptional; different authors have suggested different taxonomies. In this book, we will not worry too much about the hierarchies of the types of Testing. Instead we will highlight the importance of each type and where it fits into the development life cycle.

Some authors lay great emphasis on the distinction between *component Testing* and *system Testing*. While this may have been of importance earlier, with the predominantly *distributed* nature of software systems being developed now-a-days, it is often difficult to discern the boundary between component and system; what one sees as a system may be another (larger) system's component. In the following discussion we will side-step these issues and focus more on how different types of Testing address different Testing needs.

17.5.1 Unit Testing

Unit Testing is the Testing activity closest to actual programming. A *unit* may be a component, or even a procedure or a method that very few people—often one individual—are in charge of developing. As we remarked earlier, no disciplined programmer should allow his or her code to be used in a larger context without having unit tested it first. Errors which can be corrected easily in a unit test may become notoriously difficult to find and fix when the faulty unit has been integrated into a larger system. Among the many benefits of unit Testing, the primary are: Unit Testing makes the overall complexity of Testing activities more manageable by focusing on small pieces of code, which one—or few—individuals are closely familiar with; it leads to the precise detection of the locality of faults; and unit Testing allows for simultaneous Testing of different parts of the system, with each developer Testing the code he or she has developed [Bruegge and Dutoit 2003].

Prior to integration, units are stand alone, so their Testing sometimes need *drivers* and *stubs*. A driver is a program that allows for test data to be passed on to it, and using that data, it exercises the unit being tested and captures the output for evaluation. Stubs are dummy programs serving as proxies for components which are interfacing with the unit of code being tested; they supply whatever information is needed to test the latter [Pressman 2000]. Figure 17.3 shows a typical unit Testing scenario—the functionality of calculating the number of days between two dates is to be tested. The driver supplies test data and the stub performs the check if a leap year has to be accounted for. When the component is integrated into the actual

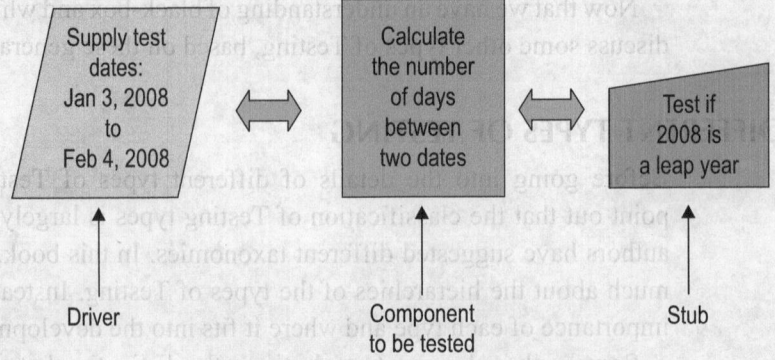

Figure 17.3 A typical unit Testing scenario

system, the driver and the stub's functionality will come from other interacting components.

Unit Testing strategies are deeply embedded within programming habits. To give a simple example, division by zero is not defined in arithmetic and all programming languages throw up nasty run-time errors when something is divided by zero. This is a very insidious error which can be prevented only by confirming the divisor is not zero *before* a division is carried out. A simple *if-else* statement can prevent the run time error. This is something even seasoned programmers often overlook, but a unit test can easily catch. Another very important aspect of unit Testing is to be able to quickly locate the error in a program and have a fair idea of what is causing it, from the execution trace log. Sophisticated exception handling mechanisms like that offered in Java can be harnessed to reduce the time and effort for finding and fixing unit Testing bugs. Not unexpectedly, setting up an efficient exception handling mechanism seldom works as an afterthought, it has to be planned and put in place before coding starts.

17.5.2 Integration Testing

From Testing units of code individually, we now move to the other end of the spectrum; when all the units have been integrated and all of them have to fit together like pieces of a jigsaw puzzle for the system to work as a whole. It is time now for integration Testing.

In a perfect world, if every unit of code was unit tested thoroughly, one could safely expect them to work together perfectly, when integrated. But the point about engineering is to make the best of an imperfect world (where there is nothing like a frictionless plane or a zero resistance wire). So in a real—that is, imperfect—world we have units of code which behaved so well on their own, end up causing all sorts of unexpected problems when asked to work together with other units of code.

Integration Testing is all about resolving these difficulties and ensuring the system works *as a whole*, as expected.

Integration Testing goes beyond mere Testing—*integration* is a vital aspect of this types of Testing. It is during integration Testing that the system is actually built from its building blocks by combining units of code, modules, components, and subsystems. While they are coming together, errors are most likely to occur at the interfaces, and integration Testing seeks to detect these errors. There are several approaches to integration Testing—incremental, non-incremental and hybrid—which we discuss next.

17.5.2.1 Top-down Integration Testing

With reference to Figure 17.4, top-down integration Testing is an incremental approach. It starts from the top level component(s) and moves down by integrating components in each subsequent level and Testing their behaviour. Top-down Testing helps test the most important decision and control points in the system (assuming they lie near the top of the component hierarchy, by Design) early. However, the start of top-down Testing may need to be delayed till the lower-level components have been developed and tested, since at every layer of the hierarchy, components depend on those in a lower layer.

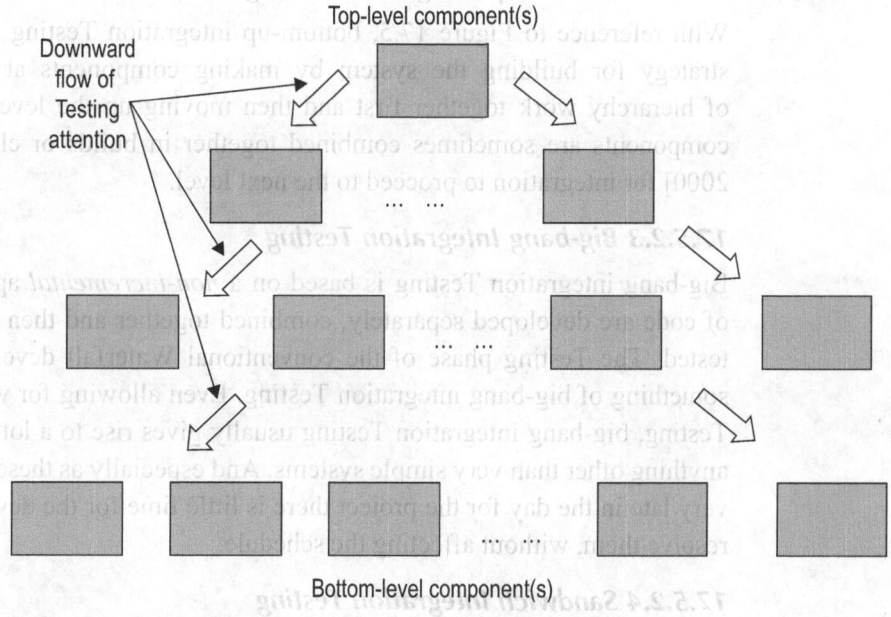

Figure 17.4 Top-down integration Testing

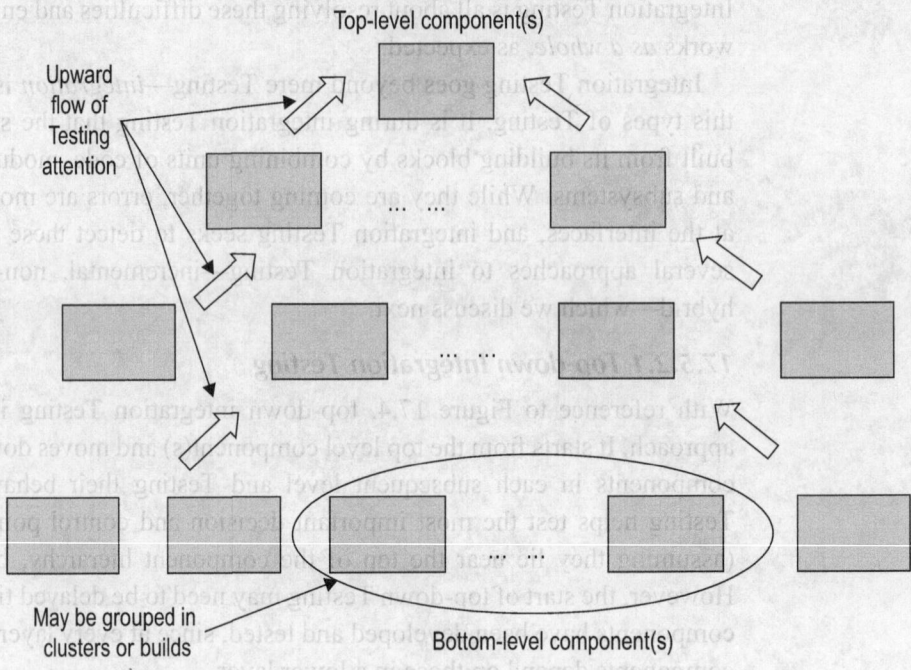

Top-level component(s)

Upward flow of Testing attention

...   ...

...   ...

...   .

May be grouped in clusters or builds

Bottom-level component(s)

Figure 17.5   Bottom-up integration Testing

17.5.2.2 Bottom-up Integration Testing

With reference to Figure 17.5, bottom-up integration Testing in an incremental strategy for building the system by making components at the lowest level of hierarchy work together first and then moving up the levels. The low level components are sometimes combined together in builds or clusters [Pressman, 2000] for integration to proceed to the next level.

17.5.2.3 Big-bang Integration Testing

Big-bang integration Testing is based on a *non-incremental* approach. All units of code are developed separately, combined together and then the whole system tested. The Testing phase of the conventional Waterfall development model is something of big-bang integration Testing. Even allowing for very thorough unit Testing, big-bang integration Testing usually gives rise to a lot of difficulties for anything other than very simple systems. And especially as these difficulties occur very late in the day for the project there is little time for the development team to resolve them, without affecting the schedule.

17.5.2.4 Sandwich Integration Testing

A form of Testing sometimes referred to as sandwich integration Testing [Peters and Pedrycz 1999] seeks to combine the benefits of top-down and bottom-up

Testing. A *target* layer of components in the hierarchy is first identified. The layers of components above the target layer are integrated and tested top-down, whereas bottom-up integration Testing applied to those below the target layer. This hybrid technique is often used where pure top-down or bottom-up Testing has practical difficulties like scheduling etc.

17.5.3 Regression Testing

Whenever a bug is detected in a software system, it needs to be fixed. The objective of regression *Testing* is to make sure that the process of repairing a defect has not introduced new defects. Every change in a software system can have potentially unwanted side effects and regression Testing tries to detect those side effects before the system is released to users. Regression Testing is an important but often neglected activity. This is unfortunate. Fixing errors—introducing new errors —fixing them—introducing yet new ones ... can quickly become a vicious cycle spinning out of control. Since it is often impractical to test every part of the system after each error has been fixed, regression Testing is often done by running a set of test cases that exercise the critical part of a system in addition to the part containing the error that was fixed. Regression Testing is a highly laborious job; to do regression Testing well, it is best to use automation. We outline automated test procedures in a later section.

17.5.4 Performance Testing

In many cases, merely ensuring that the software system delivers the required functionality is not enough. Let us take a simple example: We are Testing an authentication system that accepts user identifier and password and decides whether or not the user should be allowed into a website. Is it enough for the system to just authenticate correctly, that is, allow legitimate users and throw out illegitimate ones? We would be tempted to say yes. But what if authentication takes ten minutes? How would a user feel to be in the limbo for that long not knowing whether he or she will be allowed in? Would such a system—even as it does its work correctly—satisfy users? Most likely, no. Performance Testing's objective is to ensure whatever the system has to do has to get done within acceptable performance parameters. This is a vital aspect of a system's functioning. It was underlined in the previously mentioned IEEE Standard's definition of a failure being 'The inability of a system or component to perform its required functions within specified performance Requirements.' Performance Testing is exclusively concerned with the run-time performance of a system and it has to be carried out at several levels. Choice of a particular algorithm or data structure can have deep performance ramifications and these can only be detected during unit Testing.

But in many cases, major performance bottlenecks are manifested only during integration Testing, due to the compounding of minor performance issues at the component level. It is thus important that acceptable performance criteria are agreed upon between developers and users as part of non-functional Requirements at the beginning of the development cycle. Performance Testing can only succeed if acceptable performance levels are known during development.

17.5.5 Stress Testing

After unit Testing, integration Testing and performance Testing have established that the system is functioning acceptably within it usual operating circumstances, behaviour of the system under unusually arduous conditions may need to be tested. Some exampled conditions may be a million user hits on a website in an hour when the usual expectation is few thousand hits per hour, or thousands of users trying to access their account information simultaneously from a banking application, when usually hundreds of concurrent users are expected. Stress Testing addresses these situations. Its objective is often to find out the extent of maximum usage the system can handle before breaking down. This information is vital for load balancing and planning resources to handle unexpected spikes in usage. On January 26, 2001, western India was rocked by a major earthquake. Many Indians living across the world tried to access Internet news channels to get the latest news. This sudden and heavy increase of traffic had overwhelmed the websites of at least one major Indian newspaper, which had to temporarily shut down. Stress Testing helps foresee and plan for such unusual loads on the system. The kinds of Testing input required for stress Testing—millions of hits or thousands of simultaneous users—can hardly be manually generated. So stress Testing is almost always carried out by automated tools.

17.5.6 User-Acceptance Testing

All the types of Testing discussed so far are typically conducted within the development organization. For a software system to be finally released to the world, it must pass user-acceptance Testing.

User-acceptance Testing is done by a subset of *real* users; people who will finally use the system to derive value from its functionality. If the body of users is very large and diverse—as typical in a Web-based systems—user-acceptance Testing (often abbreviated as UAT) is carried out by a representative group of users. In iterative and incremental development, real users get to test the system at every increment and give their feedback so that next iteration can proceed. No matter what development paradigm is being followed, a system must have passed through UAT before it is released to the world.

UAT usually cannot and should not in any way be manipulated by developers. The most sporting way to go about UAT is to tell users: Here is the system, play with it, and break it if you can!

The terms *alpha Testing* and *beta Testing* are often used in the context of user-acceptance Testing. Alpha Testing is usually performed at the developer site by the customer, whereas beta Testing is performed at the customer site by the end-user of the software [Pressman 2000]. Obviously, for a Web-based system, considerations of developer site and customer site become irrelevant. It is now customary for software systems to have a beta launch before the final release.

17.5.7 Validation Testing

Closely related to some of the types of Testing discussed in the previous sections but often addressed separately, is validation Testing. Validation is concerned with the question, is the *right* system being built? Validation Testing is done at the end of the project, to establish whether a system fulfills its Requirements in the context of user needs and organizational goals. It is a high level activity which measures the delivered system against clearly defined completion criteria.

The other types of Testing described earlier primarily focus on the software system being developed. However, validation Testing is involved with ensuring that the software system—in addition to doing what it is supposed to do—also integrates successfully with the environment in which it has to function. It may seem trivial to belabour this point—after all, if a system fulfills its Requirements, should it not automatically function in its environment? But for large systems, this is hardly a trivial point. The complex interfacing between such a system and its environment often makes validation Testing a critical activity.

Closely related to validation, but different in its objective is verification. While the former relates to whether the right system is being built, the latter concerns with whether the system is being built *right*. Verification Testing is a low level activity, congruent with the other types of Testing discussed earlier. Unlike validation Testing, it is done within the software development life cycle and seeks to demonstrate consistency, completeness, and correctness of the system being built.

17.6 INSPECTIONS, WALKTHROUGHS, AND REVIEWS

Inspections, walkthroughs and reviews are mechanisms for Testing a program without actually running it. Although these are useful under certain situations—especially when the code is *self-documenting*, that is written in a way and supplemented by comments such that a human being can understand its intent and structure by reading it—they are no substitute for Testing. Inspections and walkthroughs involve groups of people reading the code, often line by line and reporting any errors and inconsistencies they may detect. Myers et al. give a detailed error checklist under the heads of *data reference errors*, *data-declaration errors*, *computation errors*, *comparison errors*, *control-flow errors*, *interface errors* and *input-output errors* [Myers et al. 2004]. Peer review of code, that is a piece of code being read by someone in the development team other than those who developed it, and being rated on the criteria of quality, maintainability, extensibility, usability and clarity, is also sometimes helpful as an evaluation mechanism [Myers et al. 2004].

17.7 DESIGNING TEST CASES

From what we have seen so far about software testing, it should be clear by now how difficult—practically impossible—exhaustive Testing is. Since we cannot reasonably test every possible scenario, the aim of Testing thus comes down to effectively *economizing*: How can we choose a representative set of test cases that best covers what needs to be tested with the least time and effort? This is the critical task of test case Design. In this section, we will outline some approaches to test case Design and illustrate them in the Case Study section.

Approaches for designing test cases are aligned to the two Testing philosophies discussed earlier: black-box and white-box Testing. Myers et al. classify the test case Design methods as [Myers et al. 2004]:

- Black-box: Equivalence partitioning, boundary-value Analysis, cause-effect graphing, error guessing.
- White-box: Statement coverage, decision coverage, condition coverage, decision-condition coverage, multiple-condition coverage.

A helpful strategy is to first devise test cases using the black-box methods, and then supplement them by the white-box methods [Myers et al. 2004]. We will first discuss the white-box Testing methods. which are collectively called *logic-coverage* Testing.

Statement coverage Testing seeks to execute every statement in a program at least once. This is adequate for non-decisional statements like manipulating the value of a variable. But for a decision statement, statement coverage Testing will not ensure all of the multiple branches are covered. So we need *decision or branch coverage* Testing, which makes sure all possible branches in the code are tested. Decisions in code are based on conditions. *Condition coverage* Testing ensures that all possible outcome of each condition towards reaching each decision is exercised. Stronger forms of condition coverage Testing are *decision-condition coverage* and *multiple-condition coverage* Testing. The former seeks to ensure each condition towards a decision takes on all possible outcomes at least once *and* each decision is taken all the different ways it can go at least once, *and* each point of entry is used at least once. The latter's objective is to have 'all possible combinations of condition outcomes in each decision, and all points of entry, ... invoked at least once' [Myers et al. 2004].

Moving now on to the black-box approaches, *equivalence partitioning* is a way of segregating the domain of input values (to a system) into *equivalent classes* such that it can be reasonably assumed that one value from each equivalent class fully represents all the values in that class and Testing with it as input would suffice for individually Testing with all the values from that class. *Boundary value Analysis* looks more closely at the edges of the equivalence classes, as these are most prone to causing unexpected errors. Neither equivalent partitioning nor boundary value Analysis can test the effects of various combinations of inputs. *Cause-effect graphing* addresses this deficiency. 'A cause-effect graph is a formal language into which a natural language specification is translated' [Myers et al. 2004]. It can be seen as an electronic logic circuit containing *gates* like OR, AND, NOT etc. In addition to helping the Design of test cases, cause-effect graphs also assist in detecting inconsistencies in specifications. Cause-effect graphing is an involved and powerful procedure that can address combinatorial complexities of inputs and outputs to some extent. Myers et al. call *error guessing* the use of experience and intuition for 'sniffing out errors' without the help of any of the specific methods described earlier. Software errors, when analysed in large numbers and across diverse systems, throw up cause and effect patterns; error guessing uses such insights to smell out what could have gone wrong.

Now we illustrate how some of these methods can be applied in practice.

CASE STUDY

Our protagonist Preeti is now into Testing. The program segment which Preeti is interested in is a part of an algorithm, as represented by the flow chart in Figure 17.6.

For covering every statement in the program it is sufficient to traverse the path *ABD*, which can be done by setting $x = 4$, $y = 1$ at *A*. But evidently this does not cover the branches *C*, *E*. To include those decisions in the test path, we must test with the additional values of $y = 3$. But this does not ensure all the four conditions in the code $x \geq 0$, $y < 10$, $x == 4$, $z > 5$ are being tested. To test both ways of each condition, we need the following range of values:

- At A: $x \geq 0$, $x < 0$, $y < 10$, $y \geq 10$
- At C: $x == 4$, $x \neq 4$, $z > 5$, $z \leq 5$

By setting the values, $x = 2$, $y = 2$ we can traverse the path *ABD* and by setting the values $x = 4$, $y = 3$ we can traverse the path *ACE*. So we have showed in sequence statement coverage, decision/branch coverage, and condition coverage

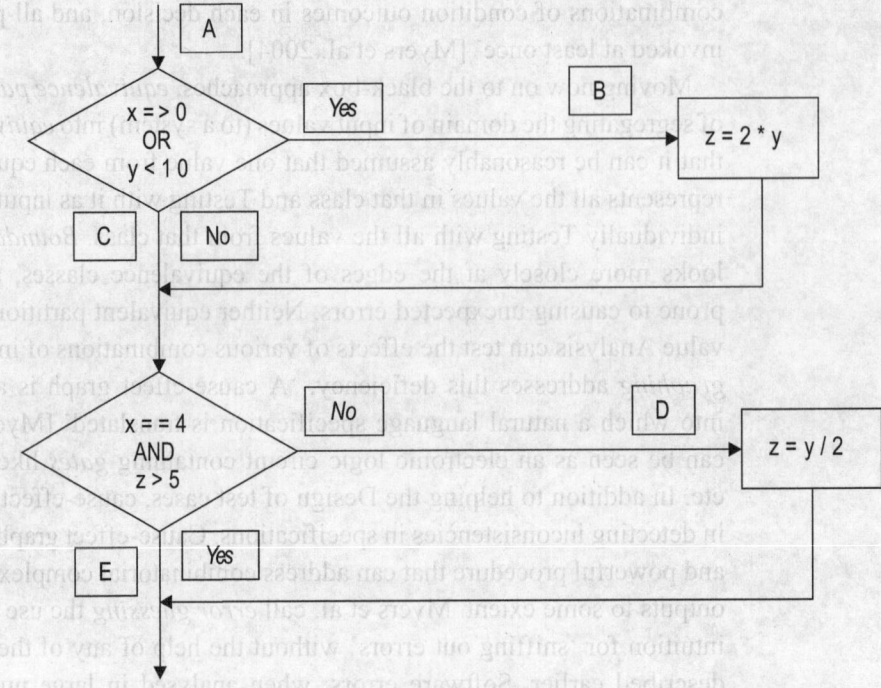

Figure 17.6   Flow chart for the program segment discussed in the case study

of the code segment. Refer to the exercises at the end of this chapter for thinking about more involved test case(s).

What are the equivalent classes for some input variable, say p, to this program segment, given the specification: p can range from 11 to 53, both inclusive? The valid equivalent class would be $\{11 \leq p \leq 53\}$ and the invalid equivalent classes are $\{-\infty \leq p < 11\}$ and $\{53 < p \leq \infty\}$. For boundary value Analysis, Preeti has to consider the *type* of the variable p. If it has been declared as an *integer*, some boundary values of interest may be 10, 12, 52, 54, etc. If p has been declared as a *float* or *double*, Preeti has to consider values like 10.01, 52.99, etc.

17.8 DEBUGGING TECHNIQUES

So far in this chapter, we have been concerned with detecting errors. But detecting errors is just half the battle. The other half—and arguably the more important half—is what needs to be done to resolve the errors. Debugging is all about what it says it is: de-bugging, that is systemic efforts to remove bugs from a system.

Debugging consists of two closely related but different concerns, finding the exact location of a bug and then fixing it. It would be hard to find a software engineer who enjoys debugging a program irrespective of whether it was written by him or her or someone else. But debugging must be done (as no programmer is perfect, except perhaps Linus Torvalds, who makes a specific claims on these lines [Torvalds and Diamond 2001]) and the best way to getting it done quickly is to do it well.

Some of the common approaches to debugging are: *debugging by brute force*, *debugging by induction*, *debugging by deduction*, and *debugging by backtracking* [Myers et al. 2004]. We will now briefly go over each of these methods.

17.8.1 Debugging by Brute Force

This is by far the most common debugging method and undoubtedly the most crude. Usual ways of going about this is to read through trace logs of the program execution, to find location of errors. A more sophisticated way is to use automated debuggers, which now comes built into almost all integrated development environments (IDE). The debugger lets you set *breakpoints* in the code and program execution is temporarily halted when the breakpoint is reached, so that values of specific variables can be checked. It is enticing to try out a little bit of brute force debugging in the hope of quickly finding and fixing a bug. Brute force debugging is most useful for small systems or when done by very experienced programmers, who know exactly where to look for bugs. We next discuss techniques which try to think through the debugging problem to a larger extent.

17.8.2 Debugging by Induction

The process of induction involves starting with an observation for specific instance(s) and extending it to a conclusion for a more general case. Debugging by induction goes from observing the symptoms of an error, organizing them so that some patterns can be found, making a hypothesis about the likely cause of the error and proving the hypothesis. On the schema of a detective story, debugging by induction starts with finding clues from the circumstances of the case, framing a theory from the clues on who committed crime, and then catching the culprit [Myers et al. 2004]. In the Worked-out Examples section, we give an example of debugging by induction.

17.8.3 Debugging by Deduction

In contrast to debugging by induction, debugging by deduction starts with identifying the probable causes for an error as hypotheses. The test data is used to eliminate some of the hypotheses while refining others to reach the most plausible one, which then needs to be proved conclusively. Evidently, success of this method strongly depends on the finesse of deciding on the list of probable causes at the beginning, and then systematically revising the list. For example, if we are trying to debug a text processing program which sometimes gives incorrect results when counting the number of characters of an input string, some of the initial hypotheses can be: The program does not know what to do with blank space, the program does not know how to handle special characters, the program prints out the results from an intermediate step in the count as the final output etc. Each of these hypotheses now needs to be refined or rejected by running more test cases on the program and finally proving the most likely hypothesis by visual examination of the code or further tests. Going back to the mystery story schema, debugging by deduction is starting with a list of suspects and pruning the list based on alibi etc, until you are left with the most probable culprit [Myers et al. 2004].

17.8.4 Debugging by Backtracking

Debugging by backtracking works well for programs of manageable scope. One starts where the error was first manifested and backtracks from that point upstream into the code, until the origin of the error is reached [Myers et al. 2004]. This is something similar to back-calculating to find out an error in arithmetic calculations. On the mystery story schema, this is analogous to what detectives call the reconstruction of the crime; starting with the outcome of the crime, going back step by step hoping to end up where it all started.

As great detectives either apply a combination of techniques or one technique in preference over another, based on the nature of the mystery, the choice of debugging technique also depends on specifics of the problem.

17.9 TEST AUTOMATION

Automating some aspects of Testing can be helpful, and for very large systems, essential. Sommerville has highlighted the following components of an automated Testing framework: *Test manager*—in charge of the overall coordination of the Testing activity, *Test data generator*—in charge of producing the data that is needed as Testing input, *Oracle*—in charge of predicting the output of a particular test, based mainly on historical data, *File comparator*—in charge of detecting differences between the outputs of tests carried out at different points of time (especially important for results of regression tests before and after a bug is fixed) *Report generator*—in charge of reporting test results, *Dynamic analyser*—in charge of injecting additional Testing code into a program, *Simulator*—in charge of simulating the environment and operational circumstances for the system [Sommerville 2004].

There are a number of automated Testing tools, such as *Quick Test Professional*—used mainly for regression Testing and Testing a system's functionality,

Exhibit 17.3 Correctness Proofs

In the software context, a *correctness proof* is a mathematical way of proving whether a program functions correctly. 'Functioning correctly' is open to many interpretations, but the 'correctness' these proofs seek to verify is defined strictly as adhering to the program specifications. Correctness proofs seem to promise an abstract way to know if the program has been written according to specifications, without getting caught in the quagmires of Testing. There are many strong feelings, both for and against using correctness proofs in large-scale industrial software.

One way of correctness proving is to place *assertions* before and after each statement in a piece of code, which are claims on certain mathematical properties holding at those points in the code. Assertions are checked with the help of *input and output specifications*. In addition to assertions, another aspect of the correctness proving process is the designation of *loop invariants* —which are mathematical expressions that must be true at specific points in a loop irrespective of the number of times the loop has been executed [Schach 2005]. Detailed discussion of correctness proving is beyond the scope of this book. A difficulty with correctness proof comes from determining input and output specifications that properly capture the correctness stakeholders are interested in. In spite of the limitations of applying correctness proving for large and complex systems, it holds promise for building error free software. Correctness proving is sometimes considered part of the formal methods discussed in Chapter 13.

LoadRunner—used mainly for performance and stress Testing, *Winrunner*—used mainly for Testing user interfaces. The website http://www.opensourcetesting.org/ has a list of open-source tools for automated software testing.

Before discussing the basics of software reliability, it is interesting to note idea of correctness proving in Exhibit 17.3.

17.10 BASIC IDEAS OF SOFTWARE RELIABILITY

In the preceding sections on software testing, we dealt with how to find and fix bugs. Now we go up one level, from detecting individual errors, to understanding the dynamics of groups of errors. *Software reliability* deals with such concerns. *Reliability* for software is defined as 'the probability of execution without failure for some specified interval of time' [Musa 1998]. As probability enters into the very definition of software reliability, it is evident we are dealing with collections of errors rather than specific ones. As software systems handle more and more critical tasks, the need for higher levels of reliability has made *software reliability engineering* an important field of study. Detailed discussion of software reliability engineering topics is outside the scope of this book; we will briefly outline the basics here. Musa highlights the essential steps of the software reliability engineering process as defining necessary reliability levels for a particular system, developing operational profiles, preparing for test, executing the test, and applying failure data to guide decisions that will influence reliability [Musa 1998].

The importance of studying software reliability arises to a large extent from the unique characteristics of software reliability as compared to hardware reliability. 'Hardware' not only denotes computer hardware but all other 'hard' ware such as nuts and bolts, cars, airplanes; in short any *physical* device of utility. We will now discuss how software reliability is different from hardware reliability.

17.10.1 Difference between Software and Hardware Reliability

The study of software reliability brings to focus some unique characteristics of software vis-à-vis hardware.

Software is perhaps the only engineering artefact which is customarily applied for purposes widely different from the intents of its original Design. The impossibility of using a standard automobile as a Mars rover is too obvious to be dwelt upon. But similar range of plasticity is often demanded from software systems. The Y2K confusion towards the end of the last century brought to the fore the fact that 'legacy' software written decades earlier were still in active service, successfully supporting vital military, business and healthcare services, way beyond their originally envisaged context and capacity. This embodies both the greatest facility of software and its even greater challenges. From the time software is released to the time it is retired, it undergoes a continual process of change. How software handles such change is the subject of much scrutiny and outside the scope of current discussion (see Chapter 18). But the stimulus is invariably changing user Requirements, which necessitate redesign and modification of existing components [Datta and van Engelen 2006], [Datta 2006]. Requirements change due to a variety of factors. But once Requirements change, the software system needs to change too. Whenever software components are modified in response to changed Requirements, chances of introducing inadvertent faults arise. The reasons for these faults are varied: misunderstood Requirements, developer ineptitude, cost and time pressures, mere oversight etc.

The unique features of software make the formulation and understanding of its reliability models more complicated. While other engineering products show trends of degraded reliability with continued use, due to *physical* wear and tear of their components, software does not *wear out* in this sense. So can we expect software to run at constant levels of reliability for an indefinite period of time? Unfortunately, we cannot. Assuming environmental factors remain the same, variations of software reliability occur due to user-initiated Requirement changes. Figure 17.7 illustrates the characteristics of hardware and software reliability.

17.10.2 Some Useful Software Reliability Relations

To give a flavour of the mathematical nature of software reliability studies, we present some useful relations below. We do not get into the details of the derivations.

Taking the reliability of a system to the probability of success P_s, which is the probability of failure P_f subtracted from 1, we present the following mathematical treatment of some key ideas of reliability [Shooman 2001].

Assuming t as the time of operation which starts at $t = 0$, and t_f denoting the time to failure, reliability as a function of time $R(t)$ is expressed as:

$$R(t) = P_s = P(t_f \geq t) = 1 - P_f = 1 - P(0 \leq t_f \leq t)$$

We may also express reliability in terms of the cumulative probability distribution function for the random variable time to failure, $F(t)$, and the probability density function $f(t)$. Now the density function is related to the distribution function by the relation $f(t) = \dfrac{dF(t)}{dt}$, and $F(t) = 1 - \int f(t)dt$. Since by definition $F(t) = P(0 \leq t_f < t)$, we can rewrite the previous equation as the following [Shooman 2001]:

$$R(t) = 1 - F(t) = 1 - \int f(t)dt$$

By introducing a *hazard function* $z(t)$—also called a *failure rate function*— which is defined as the probability of failure in the interval dt, given no failure has occurred in the interval 0 to t, we can arrive at the useful relation below [Shooman 2001] (after several steps of the derivation)

$$R(t) = e^{-\int_0^t z(x)\, d(x)}$$

The mean time to failure (MTTF) is often an important parameter in reliability studies, especially when complete information on failure trends is not known. It the can be shown that the MTTF is given by the following relation [Shooman 2001]:

$$MTTF = E(t) = \int_0^\infty t f(t)dt$$

Assuming a constant failure rate, by taking $z(t) = \lambda$ the following useful relations are arrived at [Shooman 2001]:

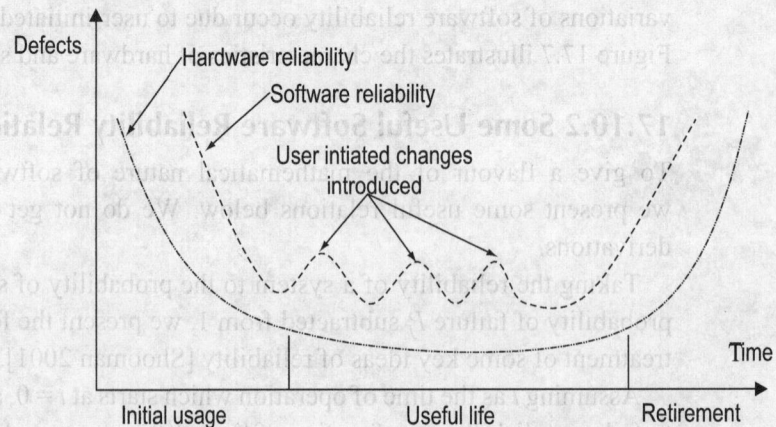

Figure 17.7 Hardware reliability versus software reliability

$$R(t) = e^{-\int_0^t \lambda dx} = e^{-\lambda t}$$

$$MTTF = E(t) = \int_0^\infty e^{-\lambda t} dt = \frac{1}{\lambda}$$

So, by knowing the failure rate we are able to calculate the reliability and the MTTF. We will use these relations in the Worked-Out Examples section and the numerical problems in the exercises.

17.11 TOWARDS SOFTWARE QUALITY

The aim of Testing and studying software reliability is to improve the quality of the software product. While the spirit of this statement is widely agreed upon, there is little consensus on what 'quality' means within the software context. 'Quality' is often used to suggest inherent worth or goodness, as in 'a quality movie', or spending 'quality time'. But in the industrial context, 'quality' has a more dry and down-to-earth meaning; often it just denotes conformance to pre-defined standards. In engineering college, one of our professors explained the idea of quality with great insight. He took a pen out of his pocket and said, 'Can I say the pen has quality if it writes well?' All of us said yes. He then corrected us by saying, 'It can only be said the pen meets quality standards if it has to be jerked *exactly* twice every time for the ink to flow!'

Whichever way you see quality, it has strong connections to the value a user gets out of an industrial product. To ensure this value is derived consistently, quality assurance aims to make sure the product matches pre-defined standards. Any organization that produces an industrial artefact has to fulfill the following quality criteria: Meet a well-defined need, use or purpose; satisfy customers' expectations; comply with applicable standards and specifications; comply with statutory (and other) Requirements of society, make the artefact available at competitive prices; and at a cost which will yield a profit [Datta 1998].

Quality in general and software quality in particular is a dynamic collection of ideas, best practices and procedures whose detailed discussion is beyond the scope of this book. We will very briefly touch upon three topics related to ensuring quality of a product. This is by no means an exhaustive list, and there are many other important ideas related to software quality.

17.11.1 ISO 9000 Series of Standards

The International Organization for Standardization (ISO) 9000-series standards represent an important initiative towards standardizing software quality. 'ISO 9000 series is a generic standard for management systems first introduced by ISO in 1987' [Datta 1998] and revised in 1994 and 2000. ISO 9000 standards are not exclusively for software; these are a set of five related standards relevant to a range of industrial processes and products. The standard ISO 9001 for quality systems is considered most relevant to software development. Specific guidelines for applying ISO 9001 to software development are contained in ISO 9000-3. ISO 9000 builds around the tenet that merely sticking to standards will not automatically ensure improved quality for a product, but it will bring down the chances of ending up with a poor quality product [Schach 2005].

17.11.2 Capability Maturity Model

The Capability Maturity Model (CMM), developed by the Software Engineering Institute at Carnegie Mellon University (CMU-SEI) 'is a conceptual framework that represents process management of software development' [Raynus 1998]. CMM specifies five *maturity levels* for an organization:

- Level 1, *initial*: Software process is unorganized and chaotic. Projects success depends on individual heroics.
- Level 2, *repeatable*: Basic processes are in place to control cost, scheduling etc. Software engineering methods and techniques have been formally introduced.
- Level 3, *defined*: Software processes for management and engineering are standardized and documented across the whole organization.
- Level 4, *managed*: Detailed measurements of process and quality are collected. Software process and products are understood and managed in quantitative terms.

- Level 5, *optimized*: Continuous process improvement across the organization is undertaken through feedback and innovations.

The CMM offers a scale for measuring the maturity of an organization's processes and its capability of delivering quality software.

The People Capability Maturity Model or P-CMM is a set of best practices for the management and development of the workforce of an organization. Like the CMM, it is a framework rather than a set of guidelines for improving an organization's people [Schach 2005].

Capability Maturity Model Integration or CMMI is a process improvement approach that provides organizations with the essential elements of effective process integration.

17.11.3 Six Sigma

In a preceding section, we had a flavour of how mathematical techniques can be applied to the study of software reliability. *Six Sigma*, invented in the 1980s in Motorola is a multifaceted approach—based on statistics, but encompassing many other disciplines—towards enhancing customer satisfaction through the reduction of defects. Today it is one of the most popular quality improvement paradigms. The 'sigma' in Six Sigma refers to the symbol of standard deviation in statistics. 'A centred six-sigma process has a normal distribution with a mean at the desired performance level and specifications placed six standard deviations to either side of the mean' [Siviy et al. 2007]. On an intuitive level, Six Sigma is almost zero defects; to be precise 3.4 defects per million opportunities (DPMO). Six Sigma has several methodologies which help reach such high levels of freedom from defects, like Define-Measure-Analyse-Improve-Control (DMAIC), Design for Six Sigma (DFSS), Lean etc. Six Sigma can be seen as a combination of a philosophy, collection of performance measures, multiple improvement frameworks and an analytical toolkit [Siviy et al. 2007]. Although originally invented for the manufacturing sector, Six Sigma is now widely applied to software development.

SUMMARY AND TAKE-AWAYS

In this chapter, we have discussed software testing, outlined the basic ideas of software reliability and touched upon some topics related to software quality. Here are the salient points from our discussion:

- Software testing, reliability, and quality are related though distinct areas that influence software development.
- It can be proved that even for a very simple program, exhaustive Testing is practically impossible.

- Black-box Testing and white-box Testing are two major Testing philosophies.
- Sound debugging methods are essential to rectifying the errors discovered through Testing.
- Inspections, walkthroughs and reviews are sometime useful methods for detecting errors without running a piece of code; however they can never substitute Testing.
- Effective test case Design aims to ensure Testing needs are fulfilled within the constraints of time and effort.
- Software reliability is inherently different from hardware reliability; It is strongly influenced by user initiated Requirement changes.
- ISO 9000, Capability Maturity Model and Six Sigma are few of the approaches towards addressing software quality at the organizational level.

WHERE TO LOOK FOR MORE

- Definitive reference to the Capability Maturity Model: http://www.sei.cmu. edu/cmm/
- International Organization for Standardization: http://www.iso.org/
- What is Six Sigma? http://www.isixsigma.com/sixsigma/six_sigma.asp

WORKED-OUT EXAMPLES

1. 100 items are tested for 1,000 hours and 4 failures are reported. Assuming a constant failure rate, what is the value of λ? What is the MTTF? What is the reliability for 5,000 hours? [Shooman 2001].

   Answer: Evidently, $\lambda = 4 / (100 * 1,000) = 4 * 10^{-5}$. Thus MTTF = $1 / \lambda$ = 25,000 hours. Now reliability for 5,000 hours $R(5,000) = e^{-4 * 10-5 * 5000} = 0.82$.

2. We have an examination grading system, which was used for grading the work of 51 students. Their mean score was correctly reported as 73.2, but the median was reported as 26 instead of the correct value of 82. The following observations or 'clues' are given:

   (a) The median is *only* incorrect for report number 3; it is correct for all other reports.

   (b) There is nothing wrong with the formula that calculates the median.

   (c) The error occurred for 51 students but *did not* occur for test cases with 2 or 200 students.

   (d) The median printed for a test run with one student is 1.
   What are the steps for debugging by induction when applied to this situation? [Myers et al. 2004]

Answer: Reviewing the clues for patterns, we find that the error appears to come only for odd number of numbers. This observation is important since medians are calculated differently for even and odd number of cases. Another pattern is that the reported median is always less than or equal to the number of students ($26 \leq 51$ and $1 \leq 1$). From the given clues it appears that the median being reported is equal to half the number of students, rounded up to the next integer. As the median formula is given to be correct, one plausible hypothesis is that if the grades are thought to being stored in a sorted table, the program is printing out the entry number of the middle student instead of the corresponding grade (which would have been the median). This hypothesis will now have to be proved by examining the code or running more focused test cases. [Myers et al. 2004].

EXERCISES

Review Questions

Review Questions test your understanding of the key concepts presented in this chapter.

1. Software testing, reliability, and quality
   (a) are connected by cause-and-effect relationships
   (b) are related ideas whose interconnections can be readily seen
   (c) are related but distinct groups of ideas
   (d) are unrelated

2. According Dijkstra, software testing can show
   (a) the absence but not the presence of errors
   (b) presence but not the absence of errors
   (c) both the presence and absence of errors
   (d) whatever the testers want to be shown

3. Which of the following statements can be said about faults and human mistakes?
   (a) The latter results in the former
   (b) The former results in the latter
   (c) Each can cause the other
   (d) None of the above

4. According to the IEEE Standard 610.12 1990, a failure involves a system or component
   (a) not meeting its performance targets
   (b) not merely performing its functions
   (c) performing wrong functions
   (d) not performing its functions within performance Requirements

5. Defect is a
   (a) specific kind of error
   (b) a specific kind of fault
   (c) a specific kind of failure
   (d) a term encompassing faults, failures, and errors

6. To denote unacceptable behaviour of a system, the word 'bug'
   (a) was first used in the software context
   (b) is never used in the software context
   (c) was used in a similar sense long before software systems were built
   (d) is used exclusively in the hardware context

7. Key objective(s) of Testing is to
   (a) find errors

(b) prove a system is running well

(c) predict the occurrence of errors

(d) all of the above

8. Planning and allocation of resources for Testing

(a) should be done at the beginning of Testing

(b) should be done at the beginning of the project

(c) should be done at the end of the project

(d) can be done at any time

9. Different types of software testing

(a) can be classified into rigid hierarchies

(b) have no classification

(c) can be classified into different categories which are largely perceptional

(d) none of the above

10. The basic assumption behind black-box Testing is that

(a) internal structure and functioning of the system being tested is fully visible to the tester

(b) internal structure and functioning of the system being tested is partly visible to the tester

(c) internal structure and functioning of the system being tested is not visible to the tester

(d) no assumption is made about the internal structure and functioning of the system

11. Black-box Testing is also known as

(a) data-driven Testing

(b) input-output Testing

(c) both (a) and (b)

(d) none of the above

12. Exhaustive input Testing is

(a) not recommended

(b) not practicable

(c) recommended and practicable

(d) recommended but not practicable

13. A more apt name for black-box Testing would be

(a) grey-box Testing

(b) opaque-box Testing

(c) out-of-the-box Testing

(d) inside-the-box Testing

14. White-box Testing assumes the tester

(a) can see the inner workings of the system being tested

(b) can analyse the inner workings of the system being tested

(c) either (a) or (b) depending on the system

(d) (a) and (b)

15. White-box Testing may also be called

(a) grey-box Testing

(b) open-box Testing

(c) logic-driven Testing

(d) closed-box Testing

16. Exhaustive-path Testing mandates

(a) all unique paths of the control flow through a program be tested

(b) a carefully chosen selection of unique paths of the control flow through a program be tested

(c) none of the unique paths of the control flow through a program be tested

(d) extent of path coverage be decided one a case by case basis

17. If exhaustive path Testing can be performed to the full,

(a) it would ensure the system is free from defects

(b) it would not ensure the system adhered to its specifications

(c) it would ensure the system has a manageable number of defects

(d) none of the above

18. If unit Testing is done thoroughly and all detected errors fixed, it can be reasonably expected that integration Testing will not uncover any more bugs.
    (a) This is a true statement.
    (b) This is a false statement
    (c) Whether the statement is true or false depends on the system being tested.
    (d) Unit Testing has no bearing on integration Testing.

19. Which of the following statements is true about integration Testing?
    (a) It is done before integration.
    (b) It is done after Testing.
    (c) Integration is followed by Testing.
    (d) Depends on the system being tested.

20. Integration Testing strategies are
    (a) always incremental
    (b) always non-incremental
    (c) may be incremental or non-incremental
    (d) none of the above

21. Top-down integration Testing is helpful for Testing
    (a) low-level control points early
    (b) middle-level control points early
    (c) high-level control points early
    (d) control points at all levels early

22. Bottom-up Testing needs low-level components to be
    (a) combined together in builds or clusters
    (b) tested at the end
    (c) tested at the beginning
    (d) (a) and (b)

23. Sandwich Testing is a form of
    (a) unit Testing
    (b) integration Testing
    (c) black-box Testing
    (d) white-box Testing

24. Regression Testing checks whether
    (a) introduction of some bugs have fixed others
    (b) fixing some new bugs have fixed other existing ones
    (c) fixing some bugs have introduced others
    (d) the total number of bugs has increased from one iteration to another

25. On practical considerations, regression Testing needs
    (a) all test cases must be run
    (b) a critical set of test cases must be run
    (c) no test cases must be run
    (d) depends on the specific situation

26. Stress Testing is done to
    (a) check if the development team is operating under too much stress
    (b) test the software system under abnormal operating conditions
    (c) (a) and (b)
    (d) none of the above

27. Which of the following is not a probable stress Testing situation?
    (a) millions of simultaneous users when the system expects hundreds such users
    (b) tens of hits on a website in a given time when the system expects thousands such hits
    (c) thousands of database transactions in a minute when the system expects hundreds such transactions
    (d) all of these are valid stress Testing scenarios

28. User Acceptance Testing must be done
    (a) immediately before a system's release
    (b) immediately after a system's release
    (c) immediately after the system development starts
    (d) anytime in the development process

29. Inspections, walkthroughs and reviews
    (a) can be substitutes for Testing
    (b) cannot be substitutes for Testing
    (c) are special categories of regression Testing
    (d) none of the above

30. Test case Design techniques can be classified according to
    (a) different types of debugging techniques
    (b) different types of inspection techniques
    (c) different Testing philosophies such as black-box and white-box Testing
    (d) all of the above

31. Decision or branch coverage Testing has
    (a) a wider scope than statement coverage Testing
    (b) a narrower scope than condition coverage Testing
    (c) either (a) or (b) depending on the system
    (d) both (a) and (b)

32. Through equivalence partitioning, we try to
    (a) reduce the number of inputs to be tested for
    (b) increase the number of inputs to be tested for
    (c) refine the inputs so that test results are better
    (d) partition the system to be tested into more manageable chunks

33. Boundary value Analysis mandates,
    (a) we check the statements lying at the boundary of the program more closely
    (b) we attach more value to the boundary variables
    (c) we analyse the program variables with a special technique
    (d) we looks more closely at the edges of the equivalence classes, as they are most prone to throwing up unexpected errors

34. Which of the following types of test is most likely to be done by automated Testing?
    (a) Black-box test
    (b) White-box test
    (c) Integration test
    (d) Stress test

35. JUnit is a
    (a) special type of unit tests
    (b) a method for Testing units of code
    (c) a framework to write and run repeatable tests
    (d) an automated-Testing technique

36. In the context of automated Testing, an oracle is
    (a) a database
    (b) a program that can forecast the output of a test
    (c) someone who manages the test
    (d) may be any of the above depending on the system being tested

37. Correctness proofs are
    (a) mathematical techniques for proving a program will always give correct output
    (b) intuitive methods to test correctness
    (c) mathematical techniques for proving a program's specification is correct
    (d) mathematical techniques for proving a program adheres to its specifications

38. Software reliability is defined in terms of
    (a) the probability of failure
    (b) the probability of execution without failure for an indefinite amount of time
    (c) the probability of execution without failure for a specified interval of time
    (d) the probability of finding faults

39. Unlike hardware reliability, software reliability is not influenced by
    (a) the evolution of software systems
    (b) changing user Requirements

(c) physical wear and tear

(d) all of the above

40. One of the main reasons for the difference between the plots of hardware and software reliability versus time is

(a) injection of user initiated changes in software

(b) the inherent difference between software and hardware

(c) lack of software reliability data

(d) none of the above

41. The study of software reliability involves

(a) looking at individual defects

(b) looking at collection of defects

(c) both (a) and (b)

(d) none of the above

42. An important parameter in reliability studies is the

(a) Mean Failure Rate

(b) Mean Failure Detection Rate

(c) Mean Failure Resolution Rate

(d) Mean Time to Failure

43. In the software engineering context, 'quality' can be said to refer to

(a) the excellence of the software product

(b) the goodness of the software product

(c) the level of adherence of the product to predefined standards

(d) all of the above

44. ISO 9000 series of standards are

(a) specific to software systems

(b) generic to management systems

(c) meant only for manufacturing organizations

(d) open to interpretation

45. As per the CMM, the highest level of maturity for an organization is

(a) managed

(b) defined

(c) optimized

(d) repeatable

46. Six Sigma is

(a) exclusively based on statistics

(b) a combination of a philosophy, collection of performance measures, a set of frameworks and an analytical toolkit

(c) a software development methodology

(d) all of the above

47. In debugging by brute force

(a) large number of people must be deployed for Testing

(b) symptoms of errors are forced to occur

(c) execution trace logs and outputs of print statement may be used

(d) all of the above

48. Identifying specific symptoms of errors and then extending the observation into a general hypothesis about the cause of the error is the basis for debugging by

(a) backtracking

(b) brute force

(c) deduction

(d) induction

49. Debugging by deduction needs to start with

(a) the hope of finding an error

(b) a single hypothesis that has to be validated

(c) a list of probable causes of the error

(d) there is no specific start or end to deduction by hypothesis, it is an ongoing process

50. Debugging by backtracking is most likely to work for

(a) small systems

(b) medium systems

(c) large systems

(d) all of the above

Reflective Questions

Reflective Questions require you to think more deeply about some of the ideas and come up with your own interpretations and answers.

1. We mentioned a successful software test as one that detects errors and an unsuccessful one that does not. If you go to a doctor with some symptoms and the doctor runs a diagnostic test, what would you regard as the successful outcome of such a test? What are the similarities and/or dissimilarities between software testing and diagnostic tests for our health?

2. 'The probability of finding more errors in a program is proportional to the number of errors already found in that program'—What do you think is the reason this empirical observation is true? Based on the validity of this statement, we may make the claim: Since more Testing will lead to the detection of more bugs, which in turn will increase the probability of finding more bugs, Testing effort should be minimal. Do you support or oppose this claim? Explain your position.

3. Between black-box and white-box Testing, which do you think is likely to detect more bugs—from the developer's point of view, as well as the user's point of view?

4. Top-down and bottom-up approaches represent two contrasting but complementary ways of designing software. Can you give some examples of systems (other than software) that were designed in the top-down way and those designed in the bottom-up way? Do you think systems which were designed by the top-down approach will be more amenable to top-down integration Testing, and vice-versa?

5. Beta releases are very common these days, especially for open source systems. How should it be decided how long to continue beta Testing of a particular system?

6. A difficulty with correctness proofs is the problem of ensuring specifications actually reflect the program's intent. Can you think of a mathematical framework for expressing software specifications? Does any such framework already exist?

7. Software reliability as a function of time is given by the relation, $R(t) = e^{-\lambda t}$. How does this function look when plotted? Compare this with Figure 17.7 for software reliability; how do you explain the difference? Can you derive a relation for the software reliability curve shown in Figure 17.7 as a function of time?

8. While discussing some of the debugging techniques, we used the metaphor of a detective solving a mystery. Note that we did not draw a similar comparison for debugging by brute force. What do you think would be a brute force debugging equivalent of solving a mystery? Are there any difficulties in trying to solve a mystery this way?

9. You are the technical lead of a team of software engineers who have developed a banking website. At 3 AM one morning, you get a phone call from the customer manager reporting a critical error that has led to the website being 'down' and huge loss of revenue for the customer organization. Which debugging technique will you apply to
    (a) get the error fixed as fast as possible?
    (b) establish the root cause of the error?

10. With reference to the example discussed in the Case Study section, do you notice any difference between Figure 17.6 and

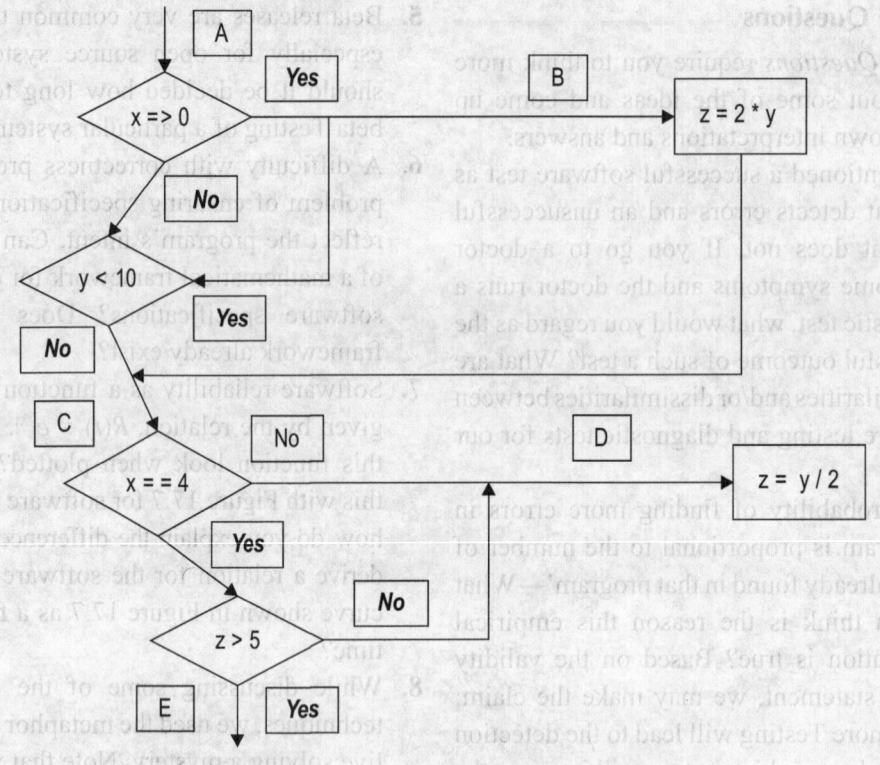

Figure 17.8 Flow chart for reflective question 10

Figure 17.8 (other than the physical orientation of the boxes, arrows etc.)? If we want to cover *all* the conditions in the program segment of Figure 17.8, what ranges of data will we need? How many more test cases will we need than those needed for condition coverage Testing of Figure 17.6?

Numerical Problems

Numerical Problems require you to do quick 'back-of-the-envelope' kind of calculations to arrive at answers to some simple but insightful problems.

1. Assuming it takes 10 minutes for one tester to write, run and verify each test case, how much time will a team of 4 testers need to fully perform exhaustive path Testing of a program that has a set of 8 independent paths which are being iterated over 4 times.

2. 300 items are tested for 10,000 hours and 17 failures are reported. Assuming a constant failure rate, what is the value of λ? What is the MTTF? What is the reliability for 1,000 hours?

3. We have discussed how software reliability differs from hardware reliability. Let us assume we have a special type of software system whose reliability mimics hardware reliability to some extent, such that the hazard function $z(t) = kt$ where k is a constant. What would the expressions be for the reliability $R(t)$ and MTTF for this system?

4. In the above problem, if $k = 2$, what is the reliability for 1,000 hours and the MTTF?

5. A program segment has two decision blocks, each with three conditions, and each decision block has two branches going out of it. How may test cases do we need to exercise all conditions, decisions, and branches? If a program segment has x decision blocks, each with y conditions, and each decision block has z branches going out of it, can you derive a general formula for the number of test cases needed to exercise all conditions, decisions, and branches. (An example of a decision block with two branches is *if-else*. The Java *switch* statement is an example of a decision block with multiple branches.)

Programming Examples

Programming Examples require you to analyse or write a program or a program segment to understand a specific problem.

1. Write a program in Java which accepts the three sides of a triangle from the user; and reports whether such a triangle exists, and if it does, whether it is an equilateral, isosceles or scalene triangle. How will you test this program? Classify the bugs you detect into faults, failures, and errors.

2. Write a JUnit test for the Java program of the above problem.

3. The article *JUnit Test Infected: Programmers Love Writing Tests* available at http://junit. sourceforge.net/doc/testinfected/Testing. htm describes the use of JUnit through a simple example. Using that example, write a JUnit test for a Java class *SetPlay* that offers the *union* operation for two sets. If we want *SetPlay* to also offer the *intersection* operation, how will you write the unit test before you write the actual code?

4. How many test cases do you need to Design to perform statement, decision and condition coverage Testing for a Java program which accepts the three sides of a triangle from the user; and decides whether such a triangle exists?

5. Write the above program (which accepts the three sides of a triangle from the user; and decides whether such a triangle exists) in a language other than Java. Is there any difference in the number of test cases required to conduct statement, decision, and condition coverage Testing between this program and the one you wrote earlier in Java?

REFERENCES

Bruegge, B., and Dutoit, A.H. (2003), *Object-Oriented Software Engineering: Using UML, Patterns and Java,* Prentice Hall, 2nd edition.

Datta, J.P. (1998), 'ISO 9000: A roadmap for Design, installation and Implementation of quality management systems', Manuscript.

Datta, S. (2006), 'Master of science thesis: A mechanism for tracking the effects of

requirement changes in enterprise software systems', http://etd.lib.fsu.edu/theses/available/etd-07052006-183531/. last accessed on May 19, 2010.

Datta, S. and van Engelen, R. (2006), 'Effects of changing Requirements: a tracking mechanism for the Analysis workflow', In *SAC '06: Proceedings of the 2006 ACM symposium on Applied computing*, pp. 1739–1744, ACM Press, New York

Datta, S., van Engelen, R., Gaitros, D., and Jammigumpula, N (2007), 'Experiences with tracking the effects of changing Requirements on Morphbank: a Web-based bioinformatics application', In *ACM-SE 45: Proceedings of the 45th annual southeast regional conference*, pp. 413–418, ACM Press, New York.

Musa, J.D. (1998), *Software Reliability Engineering*, Osborne/McGraw-Hill.

Myers, G.J., Sandler, C., Badgett, T., and Thomas, T.M. (2004), *The Art of Software Testing*, 2nd ed., Wiley.

Pan, J. (1999), 'Software reliability', http://www.ece.cmu.edu/~koopman/des_s99/sw_reliability/, last accessed on May 19, 2010.

Peters, J.F. and Pedrycz, W. (1999), *Software Engineering: An Engineering Approach*, Wiley.

Pressman, R.S. (2000), *Software Engineering: A Practitioner's Approach*, McGraw Hill.

Raynus, J. (1998), *Software Process Improvement with CMM*, Artech House Publishers.

Schach, S. (2005), *Object-oriented and Classical Software Development*, Sixth Edition, McGraw-Hill International Edition.

Shooman, M.L. (2001), *Reliability of Computer Systems and Networks: Fault Tolerance, Analysis, and Design*, Wiley-Interscience.

Siviy, J.M., Penn, M.L., and Stoddard, R.W. (2007), *CMMI and Six Sigma: Partners in Process Improvement*, Addison-Wesley Professional.

Sommerville, I. (2004), *Software Engineering*, Prentice Hall, 7th edition.

Torvalds, L. and Diamond, D., (2001), *Just for Fun: The Story of an Accidental Revolutionary*, HarperBusiness.

Towards Software Evolution

Learning Objectives

In this chapter, we explore what happens to a software system at the end of the development life cycle. Specifically, we will cover:
- Life of a software system after the development life cycle
- Maintenance and modification
- Software configuration management
- Software entropy
- Software evolution

18.1 MOTIVATION

In fairy tales, the story usually ends with everyone living happily *ever after*. But software development is no fairy tale and the story of a software system does not end at the end of the software development life cycle with everyone living happily ever after. In the real world, the most challenging aspects of software development usually arise after the system has been delivered to users. Every software system that is used is subjected to change, and ongoing modifications and maintenance are essential instruments to keep pace with that change. Software development theory usually pays lip service to these post-delivery concerns. But software development practice, to be successful, needs to be concerned with life after the life cycle. In this chapter, we will outline life after the life cycle for a software system, the issues of maintenance and modification, software configuration management and the concepts of software entropy and software evolution.

18.2 LIFE AFTER THE LIFE CYCLE

As discussed in detail in Chapter 14 and elsewhere in this book, the software development life cycle (SDLC) stretches across workflows and phases. The SDLC is primarily concerned with the *making* of a software system. It is in a way akin to the period of gestation for living beings. Accordingly, releasing a software system to the user domain is analogous to birth; thus the 'real life' for software starts post release. The SDLC sheds very little light on life after the life cycle. One is left with the impression that after delivering a software system to its users, developers can wash their hands off and move on to another project. In the real world of paying customers, this is almost never the case. Most software development teams are bound by contractual obligations that require continuous involvement in support, maintenance and enhancement activities after the final release of the system to the users. Even when there is no legal binding, professional ethics demand software engineers take responsibility for what they have delivered to users.

Life after the life cycle presents unique challenges to the developers and maintainers of a software system. As has been established in Chapter 17, not *every* scenario can be tested due to the practical constraints on time and effort. While a software system is being used, the untested and seemingly obscure scenarios have a habit of messing things up every now and then, causing much consternation. Additionally, there is the phenomenon of 'piecemeal growth' [Gabriel 1998] which is sometimes described by other phrases such as continual development, ongoing modifications etc. Gabriel defines piecemeal growth as '... the process of Design and Implementation in which software is embellished, modified, reduced, enlarged and improved through a process of repair rather than replacement' [Gabriel 1998]. 'Repair rather than replacement' is the key idea behind piecemeal growth. A software system is rarely removed to make way for a new system; the older system is tweaked incrementally to incarnate the new one. Piecemeal growth embodies an essential fact of life for software development.

But piecemeal growth comes with its price. We cannot hope to continually change a system in subtle but significant ways, yet expect no concomitant cost. As we shall see in a later section, the ideas of software entropy and evolution addresses the effects of piecemeal growth.

18.3 MAINTENANCE AND MODIFICATION

In the dawn of the computing era, maintenance was a relatively insignificant aspect of development activity. With each passing decade, it has become more and more demanding in terms of time and effort. Software maintenance is defined as '...the modification of a software product after delivery to correct faults, to improve performance or other attributes, or to adapt the product to a changed environment' [Yang and Ward 2003]. Maintenance activities are classified into the following categories [Yang and Ward 2003]:

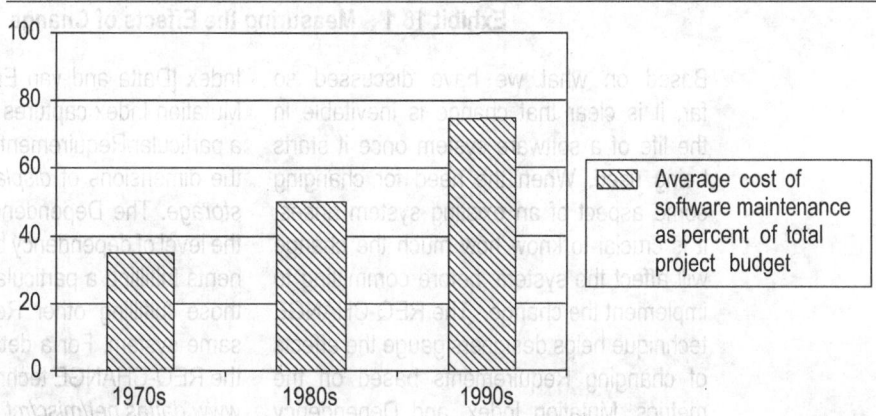

Figure 18.1   The rising cost of software maintenance (inspired by Yang and Ward 2003)

- *Corrective maintenance*: Undertaken to rectify faults that have manifested while the system is running.
- *Adaptive maintenance*: Undertaken to prepare the system to function smoothly on the face of changes in external environment.
- *Perfective maintenance*: Undertaken to address changes in user Requirements of a system.
- *Preventive maintenance*: Undertaken in anticipation of future problems.

Software maintenance has its own life cycle, different from the SDLC. It is an iterative three stage process, consisting of *request control*, *change control* and *release control*. Request control consists of collecting information about the maintenance request, analysing impact and prioritizing it. Change control involves identifying the request to be worked on, reproducing the problem scenario, analysing code, designing, documenting changes and creating tests, modifying the code and following quality assurance procedures. Release control is all about determining and building the new release, Testing and deploying it, and subjecting it to user-acceptance Testing [Yang and Ward 2003].

The rising cost of software maintenance is portrayed in Figure 18.1.

18.4 SOFTWARE ENTROPY

The idea of *entropy* resonates in disciplines as diverse as thermodynamics and information theory. Entropy is usually taken to be antithetical to energy, and to represent chaos, disorder and the lack of potential for an organized outcome. According to the Second Law of Thermodynamics, a 'closed-system's' disorder will only increase, or remain unchanged; it can never be reduced. Jacobson introduces an analogous concept for software systems, calling it 'software entropy'

Exhibit 18.1   Measuring the Effects of Change

Based on what we have discussed so far, it is clear that change is inevitable in the life of a software system once it starts being used. When the need for changing some aspect of an existing system arises, it is crucial to know how much the change will affect the system, before committing to implement the change. The REQ-CHANGE technique helps designers gauge the effects of changing Requirements based on the metrics, Mutation Index, and Dependency Index [Datta and van Engelen 2006]. The Mutation Index captures the extent to which a particular Requirement has changed along the dimensions of *display*, *processing*, and *storage*. The Dependency Index measures the level of dependency between the components fulfilling a particular Requirement and those fulfilling other Requirements of the same system. For a detailed discuss ion of the REQ-CHANGE technique, refer to *http://www.dattas.net/misc/rct.pdf*.

[Jacobson 1992]. Basing his arguments on the laws of software evolution (discussed in the next section), Jacobson gives the following mathematical treatment of software entropy.

On the assumption that a system has some initial software entropy, it has been empirically observed that *the increase in software entropy is proportional to the entropy of the software when the modification started* [Jacobson 1992]. This matches with our intuitive understanding that it is easier to impose further order on an already ordered system than a disordered one. Mathematically, it is expressed as ΔE proportional to E, that is

$$\Delta E \propto E$$

Or, in the language of differential calculus, taking k as the constant of proportionality,

$$\frac{dE}{dt} = kE$$

The solution to this equation is plotted in Figure 18.2. From the figure, it is apparent that how long a system will be serviceable depends on 'how well-structured' the system is initially. After the critical software entropy limit is reached, modifying the system further becomes unviable. This may be an opportune time to re-engineer the system to bring down its entropy so that it can continue to be maintained. So, as a thumb rule, the longer we want to have a system in service, the lesser initial entropy the system needs to start with. Even then, there will inevitably come a time when the limiting entropy is reached and it will become infeasible to maintain the system further.

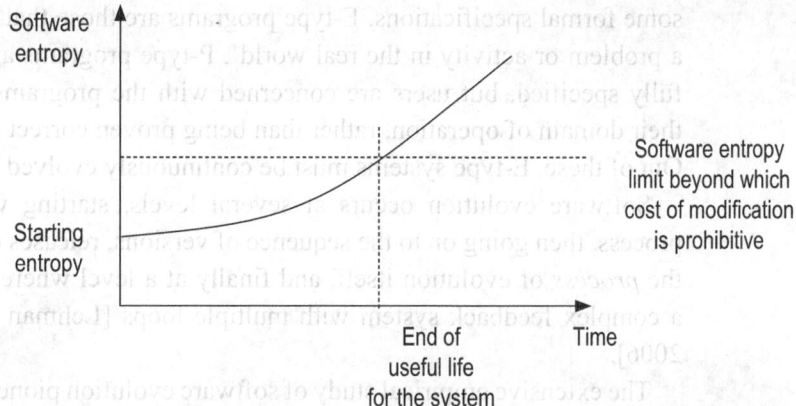

Figure 18.2   Characteristics of a software system's entropy (inspired by Jacobson 1992)

Jacobson's treatment does introduce an elegant analogy between thermodynamic and software entropy. However it is far from complete. As he himself says, the main difficulty with this line of reasoning is that 'people are involved' [Jacobson 1992]. This implies that mathematical equations can hardly capture the diversity, variation and inter-personal dynamics of software development to the fullest extent. Moreover, note how software entropy is never formally defined, its similarity to thermodynamic entropy is only loosely highlighted. Also, how do we know if a system is 'well structured' or not, initially? In spite of these lacunae, software entropy is a useful and interesting concept. Those who have built software for real users in the real world, know how difficult it progressively becomes to modify and maintain a system, and how this task is easier if the initial Design is clean and flexible. Software entropy is a formalization of these observations.

18.5 SOFTWARE EVOLUTION

An integrated view that ties together all that we have been discussing in this chapter centres around the idea of *software evolution*. Our common sense understanding of 'evolution' reflects on change over time across a wide variety of systems. Evolution in the software context concerns itself with the study of *how* software changes over time as well as the *why* and *what* of the change. The former has been called the *verbal* view of software evolution, while the latter is the *nounal* view [Nazim H. Madhavji 2006] .

To better analyse their evolution, software programs have been classified into *S-type*, *P-type* and *E-type*. An S-type program is one which can demonstrably satisfy the necessary and sufficient conditions of mathematical correctness against some formal specifications. E-type programs are those that 'operate in or address a problem or activity in the real world'. P-type programs are those which can be fully specified, but users are concerned with the programs running correctly in their domain of operation, rather than being proven correct against specifications. Out of these, E-type systems must be continuously evolved to continue in use.

Software evolution occurs at several levels, starting with the development process, then going on to the sequence of versions, releases or upgrades, then onto the *process* of evolution itself, and finally at a level where software evolution is a complex feedback system with multiple loops [Lehman and Fernandez-Ramil 2006].

The extensive empirical study of software evolution pioneered and continued by Lehman, Ramil and others, have resulted in the following *laws of software evolution* [Lehman and Fernandez-Ramil 2006] which we briefly mentioned in Chapter 7. In the context of our discussion regarding laws of software development in Chapter 3, we may point out here that most of these laws are more of the *observational* kind. They illustrate the dynamics of software evolution but do not readily give rise to mathematical relations such as the Ohm's law or Newton's laws of motion. Even then the laws given below have a major influence in building the theory of software evolution.

1. *Continual Change*: An E-type system needs to be continuously adapted, otherwise progressively less satisfaction is derived from its use.
2. *Increasing Complexity*: With change, an E-type system's complexity increases which makes its evolution more difficult unless work is done to maintain or diminish its complexity.
3. *Self-regulation*: Feedback regulates the evolution of global E-type systems.
4. *Conservation of Organizational Stability*: The work-rate of an organization in which an E-type system evolves tends to be constant over the operational lifetime of that system or phases of that lifetime.
5. *Conservation of Familiarity*: In general, the need to maintain familiarity constrains the incremental growth (growth rate trend) of E-type systems.
6. *Continuing Growth*: To maintain user satisfaction over system lifetime, E-type system's functional capability must be continually improved.
7. *Declining Quality*: The quality of an E-type system will appear to be declining, unless it adapts and evolves to address changes in the operational environment.
8. *Feedback System*: The processes of E-type evolution are multi-level, multi-loop, multi-agent feedback systems.

Summary and Take-Aways

In this chapter, we have discussed what happens to a software system after it is delivered to users. In summary, the points discussed are:

- In the real world, the life of a software system begins after it has been delivered to the users.
- The conventional software development life cycle does not throw much light on post-delivery concerns.
- There are several types of maintenance activities and software maintenance has a life cycle different from software development.
- Software configuration management outlines a set of techniques to address change in software systems.
- Piecemeal growth is a fact of life for real-world software systems; it involves the continued modification of a system to meet changing user needs through a process of repair rather than replacement.
- Software entropy, inspired by the concept of thermodynamic entropy, tries to express the consequences of continual modifications on a software system. As a thumb rule, the longer we want to have a system in service, the lesser initial entropy the system needs to start with.
- The study of software evolution seeks to take a holistic view of the continual change and modifications that a software system undergoes after delivery. Based on empirical data, eight laws of software evolution have been suggested. A key insight from the study of software evolution is that the complexity of a software system that is in use will increase unless specific effort is directed towards containing the complexity.

Where to Look for More

- A definitive collection of software evolution resources can be found at Feedback, Evolution And Software Technology: http://www.doc.ic.ac.uk/~mml/feast/

EXERCISES

Review Questions

Review Questions test your understanding of the key concepts presented in this chapter.

1. Post-delivery concerns are addressed within the software development life cycle.
   (a) True
   (b) False
   (c) Depends on the type of the life cycle
   (d) Depends on the kind of concerns

2. The key concept of piecemeal growth is
   (a) continual maintenance
   (b) bug fixing

(c) repair rather than replacement

(d) involves all of the above

3. Which of the following is not a type of maintenance?

   (a) Defective maintenance

   (b) Adaptive maintenance

   (c) Corrective maintenance

   (d) Preventive maintenance

4. Which of the following seems to have inspired the idea of software entropy most closely

   (a) Joule's heating

   (b) Newton's Second Law of Motion

   (c) Thermodynamic entropy

   (d) Information entropy

5. Which of the following is a type of software system as classified in the study of software evolution?

   (a) B-type

   (b) C-type

   (c) D-type

   (d) E-type

Reflective Questions

Reflective Questions require you to think more deeply about some of the ideas and come up with your own interpretations and answers.

1. Is the so-called piecemeal growth unique to software? Can you think of any other domain where it may be relevant?

2. What do you think is the biggest benefit and the biggest cost of piecemeal growth?

3. Out of the types of maintenance, which one(s) do you think is/are most important in ensuring end-user satisfaction?

4. We have reflected on the rising cost of maintenance with each passing decade since the early computing era. Why do you think this has happened? Do you see a reversal of this trend in the near future?

5. The formulation of software entropy is based on the assumption—increase in software entropy is proportional to the entropy of the software system when the modification started. Can you point out any particular kind of software systems where this assumption is not valid? Are there any other hidden assumptions behind the formulation of software entropy?

6. What is the key difference between the so-called verbal view and the nounal view of software evolution?

7. Why do you think the laws of software evolution are relevant to a particular type of software systems?

8. The idea of software entropy can be inferred from which law of software evolution?

9. Why do you think software maintenance has a different life cycle than the SDLC?

Numerical Problems

Numerical Problems require you to do quick 'back-of-the-envelope' kind of calculations to arrive at answers to some simple but insightful problems.

1. We have a software system, whose entropy is given by the relation

$$\frac{dE}{dt} = kE^2$$

How would the entropic characteristics of this system vary from the one discussed with the relation $\frac{dE}{dt} = kE$?

Given the same initial entropy, which of the two systems will be in service longer?

2. Let us consider a hypothetical software system. Between the second and third iterations of development, its Requirements changed in the following dimensions:

- R_1 – D, P, S
- R_2 – D
- R_3 – No change
- R_4 – P, S

In the third iteration, the Component Sets for the Requirements were:

- R_1 – {C_2, C_3, C_5, C_6}
- R_2 – {C_1 C_2, C_4, C_7}
- R_3 – {C_5, C_6, C_7}
- R_4 – {C_1, C_6}

Calculate the Mutation Index and Dependency Index for the Requirements of the third iteration. Using the REQ-CHANGE technique, predict the extent to which change in each Requirement will affect interaction of the system's components.

Programming Examples

Programming Examples require you to analyse or write a program or a program segment to understand a specific problem.

1. Write a program to calculate the area of a rectangle. Then over several iterations of piecemeal growth, modify the same program to additionally calculate the area of a circle, a triangle and a trapezoid. How would you measure the entropy? Do you notice any increase in the entropy of the program based on your measure? What could you have done to start with a low-level of entropy?

   Based on your experience with this program, criticize the mathematical formulation of software entropy.

REFERENCES

Datta, S. and van Engelen, R. (2006), 'Effects of changing requirements: a tracking mechanism for the analysis workflow', In *SAC '06: Proceedings of the 2006 ACM symposium on Applied computing*, pp. 1739–1744, ACM Press, New York, USA.

Gabriel, R.P. (1998), *Patterns of Software: Tales from the Software Community*, Oxford University Press, USA.

Jacobson, I. (1992), *Object-Oriented Software Engineering: A Use Case Driven Approach*, Addison-Wesley.

Jacobson, I., Booch, G., and Rumbaugh, J. (1999), *The Unified Software Development Process*, Addison-Wesley.

Lehman, M. and Fernandez-Ramil, J.C. (2006), *Software Evolution and Feedback: Theory and Practice*, chapter Software Evolution, Wiley.

Nazim H. Madhavji (Editor), Juan Fernandez-Ramil (Editor), D.P.E. (2006). *Software Evolution and Feedback: Theory and Practice*, Wiley.

Yang, H. and Ward, M. (2003), *Successful Evolution of Software Systems*, Artech House Publishers.

PART V

LATEST TRENDS OF
SOFTWARE DEVELOPMENT

Software Engineering and the World Wide Web

Learning Objectives

In this chapter, we will discuss the relevance of the World Wide Web (WWW) for software development. Specifically, we will focus on:
- The Web vis-à-vis the Internet
- How software engineering has evolved with the coming of the Web
- The architectural characteristics of Web-based software systems
- Salient features of Web-based software systems
- The potential of the Web as a software development medium

This is not a chapter on the technology behind the Web or the Internet. Such material is more suited to a course on networking and thus beyond the scope of this book. This is a chapter on how the Web has transformed software engineering over the past decade and half and what lies ahead.

19.1 MOTIVATION

The World Wide Web—*WWW* or merely *the Web,* in common usage—is regarded as one of the most transformational technologies of the modern age. Every day our lives are touched by the Web in many different ways. We are increasingly attached to it for information, utility, and entertainment. It is thus no wonder that software engineering, from what it produces to how it is practiced, has been strongly influenced by the coming of the Web. The Web itself is fascinating in its technological innovations, as well as academic, commercial and social impact. The invention of the Web (Exhibit 19.1) is a great saga of human ingenuity.

To a great extent, the Web has affected how large-scale software is deployed, used and built. This influence has grown significantly over the last decade, with increasing popularity of the Web. Web technologies, which concern themselves with the ways and means of purveying, processing and presenting information, have undergone a sea change in recent years. Many practicing software engineers today need to Design, build and maintain software systems for the Web.

Exhibit 19.1   Weaving the Web

The first website went online on August 6, 1991. Isn't it wondrous that the Web—where many of us work, play, idle, and practically live on—is less than two decades old! The story of the Web's birth is told insightfully in the book *Weaving the Web* [Berners-Lee 1999] by Tim Berners-Lee, who is credited with inventing the World Wide Web. According to Berners-Lee, while working as a software engineer at CERN —the European Organization for Nuclear Research—he married the idea of hypertext (that allows documents to be 'linked' to one another) with the Transmission Control Protocol and domain name system of the Internet; and there was the Web. In reality however, it called for a lot more ingenuity and engineering acumen to create what was to become one of the greatest inventions of the modern age. At the heart of Web's conception lies 'hypertext'; the word was coined by Ted Nelson in 1965. Even before

that, in a 1945 article in *The Atlantic Monthly* called 'As We May Think' Vannevar Bush wrote about a device he called 'Memex', which sought to create an elaborate network of information across microfilms and books, using electromechanical linkages. The Web is thus a culmination of decades of thinking into connecting diverse sources of information into a seamless knowledge network. In his book, Berners-Lee outlines the genesis of ideas leading up to the Web as well as his and his colleagues' work in bringing the Web to life. Berners-Lee now heads the World Wide Web Consortium (W3C), an international body which seeks to guide the future directions of the Web. He is also an active proponent of Web 2.0—the next generation of the World Wide Web (see Chapter 23 for more details).

Check out the world's first website that Berners-Lee created: http://info.cern.ch/.

Very often, Web development skills are confused with the expertise for engineering software applications for the Web. The familiarity with Hyper Text Markup Language (HTML), Cascading Style Sheets (CSS) and client-side scripting languages such as JavaScript are important elements in *presenting* attractive web pages. But software engineering concerns for Web applications centre more on the *processing* of the information supplied by and provided to users.

As anyone who has surfed the Web is aware, Web pages come in a large variety of content and 'cleverness'. Skills for building such web pages is very helpful in today's world, where creating and maintaining websites is essential to a wide range of professional and amateur Web users. However, we will not go into the discussion of those skills here. Instead, we will emphasize on things which are essential for any software engineer handling a Web application—some of the foundational concepts dealing with the exchange of information over the Web.

In casual conversation, 'Internet' and 'Web' are used interchangeably. However, there are significant differences, which we highlight in the following section.

Next we discuss how coming of the Web has changed software applications, followed by a discussion of the architecture of typical Web-based software systems. Subsequently, we will highlight some of the salient features of software systems on the Web and conclude the chapter with a discussion of the burgeoning role of the Web as a software development medium.

Before getting deeper into the chapter, we would like to clarify the meanings of the phrases 'Web applications', 'Web-based software systems' and 'software systems for the Web' which will be used interchangeably in this chapter. These phrases will denote business applications with a Web interface—such as an online banking application whose Design and development was illustrated in Chapter 15.

19.2 INTERNET AND THE WWW

In common parlance, 'network' suggests *connection* between entities that help easy transfer of commodities between them. A computer network is a collection of machines connected together to facilitate information exchange between them. The Internet is a network of networks—a system of interconnected computer networks that allows for the interchange of data using a set of standardized protocols. Next to the actual connections between the networks and the computers within those networks, the most important aspect of Internet are the protocols—universally accepted rules for transmitting and receiving data. The Internet involves a host of protocols, primary among which are the *Transmission Control Protocol* (TCP)—which breaks up a piece of data to be transmitted into *packets*, and reassembles the packets when they are delivered—and the *Internet Protocol* (IP), which governs the delivery of packets between *IP addresses*. An 'IP address' is a 32-bit number (subsequently enhanced to 128-bits in the new addressing system. IPv6) identifying devices in a computer network that use the IP for communicating between its nodes.

With reference to Figures 19.1 and 19.2 network architectures come in two major varieties, *client/server* and *peer-to-peer*. In client/server architecture, all the processing and storage of information takes place at the centralized server, which the clients access to send and retrieve information from. As we shall discuss in a following section, software applications for the Web are mainly based on the client/server architecture. In a peer-to-peer architecture, every peer is able to connect to other peers directly, and no peer has primacy over another. Most Internet file sharing applications operate on a peer-to-peer basis.

Table 19.1 shows the major milestones in the history of the Internet (The information for this table has been gathered from http://www.zakon.org/robert/Internet/timeline/) As is evident, the Internet existed at least two decades before the Web came into being. So, how are the Internet and the Web related?

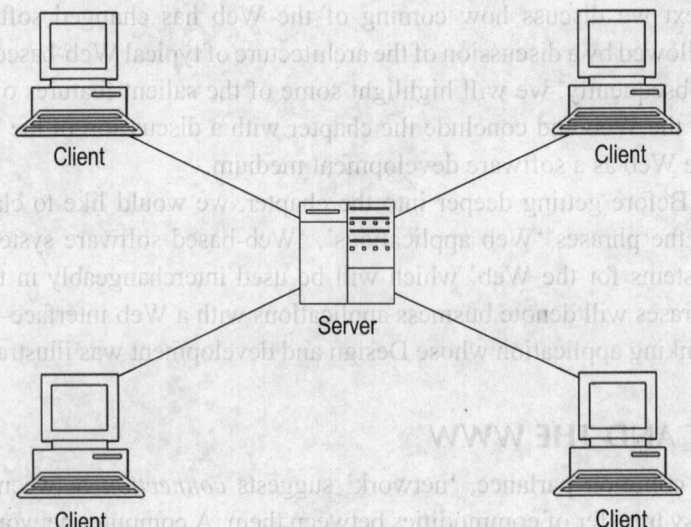

Figure 19.1   Client/server network architecture

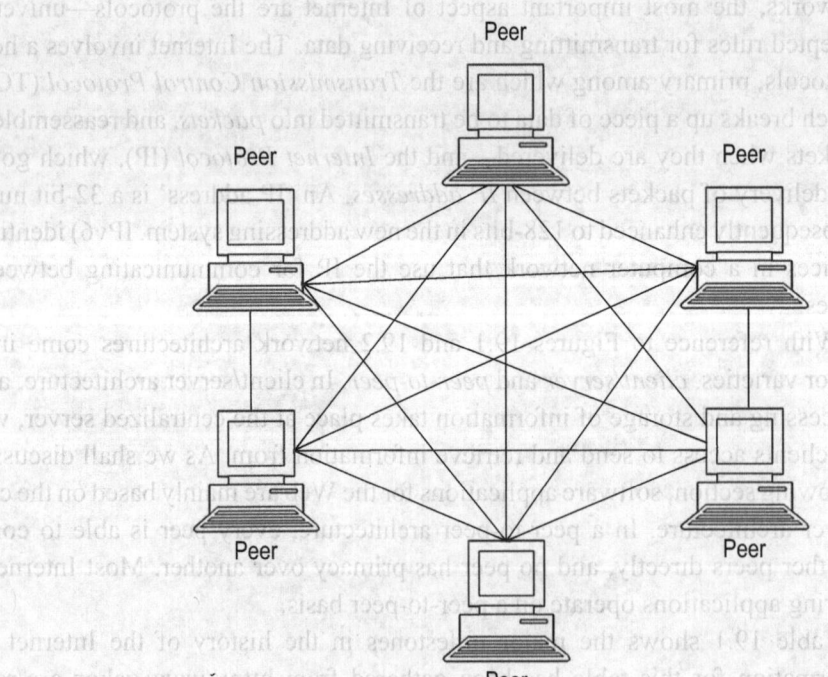

Figure 19.2   Peer-to-peer network architecture

Table 19.1 The Internet: A timeline

Year	Notable event
1970	AT&T installs first cross country link to connect networks across USA
1973	The ARPANET (the first successful packet switching network in the world, built by the *Advanced Research Projects Agency* of USA) goes international by extending to the University College in London
1984	ARPANET is divided into MILNET—for military purposes and ARPANET—for research
1990	The ARPANET project ends and the Internet formally turned over to the public
1991	The Commercial Internet Exchange (CIX) Association established to facilitate businesses use of the Internet
1993	The first Web browser, Mosaic launched

The Web can be seen to be a service based on the infrastructure of the Internet, allowing for the interconnection of information resources through *hyperlinks*. The most important aspect of the Web is existence of hyperlinks—'clickable' parts of a web page that let us go from one page to another. Hypertext Transfer Protocol or HTTP governs the transfer of information between Web servers and clients. A Web server may be thought to be a computer that generates web pages and a client can be a Web browser on another computer that accesses the pages over the Internet's infrastructure. A *Uniform Resource Locator* (URL) uniquely identifies a resource on the Web. For all practical purposes it is the Web address we type into our browsers address field when we want to go to a web page. Two other significant characteristics of the Web, which impart great power and utility, are its scalability and lack of centralized control. Scalability ensures theoretically there is no upper limit on the amount of information that can be put on the Web; any one is free to have his or her own website. Lack of centralized control means there is no authority that can claim oversight on the Web's functioning.

Figure 19.3 depicts one aspect of the relationship between the Internet and the Web. Very simply, the Internet and the Web can be said to share a superset-subset relationship. Some facilities, such as email or file transfer can both be Internet-based or Web-based. Though a large majority of email is now Web-based, those accustomed to using Unix-based operating systems will know the use of emails that is *not* Web-based. However, the superset-subset stereotyping of the Internet-Web relationship hides certain subtleties. The Web is not just a subset of the Internet; it also embodies a higher layer of abstraction towards end-user utility and usability. With reference to Figure 19.4, we note that at the lower most layer lies the network of individual computers. To effectively exchange information across this 'physical'

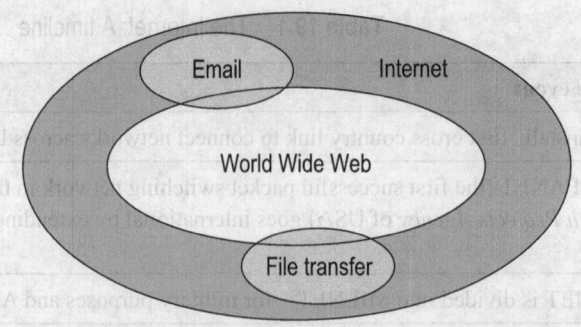

Figure 19.3 The Internet and the Web

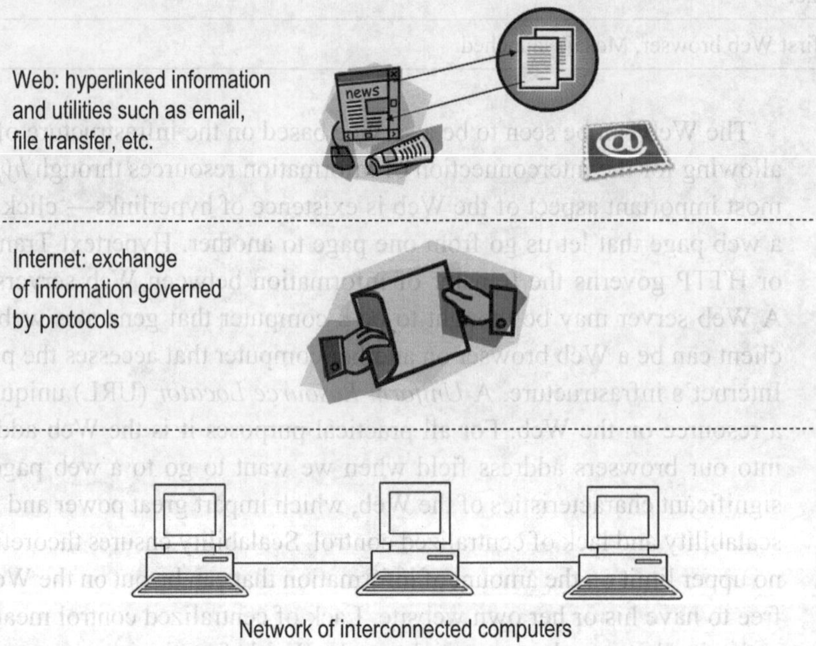

Web: hyperlinked information and utilities such as email, file transfer, etc.

Internet: exchange of information governed by protocols

Network of interconnected computers

Figure 19.4 Network, Internet, and Web: layers of abstraction

infrastructure, we need the Internet, with its protocols, at one level higher. On top of the Internet lies the Web, with its hyperlinks for seamless traversal between information sources and utilities such as emailing (which, as discussed earlier, can also be non-Web-based).

In Table 19.1, we have shown the major milestones in Internet's history. How fast has the Web grown? Figure 19.5 shows the growth of the number of websites from 1990 to 1995 (The information for this figure is gathered from http://royal. pingdom.com/2008/04/04/how-we-got-from-1-to-162-million-websites-on-the-

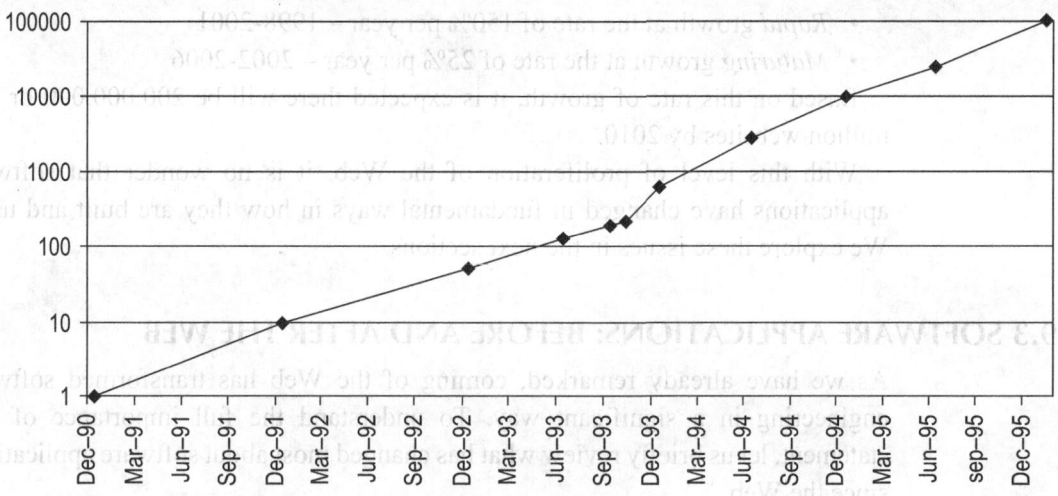

Figure 19.5   Number of websites versus time (1990–1995)

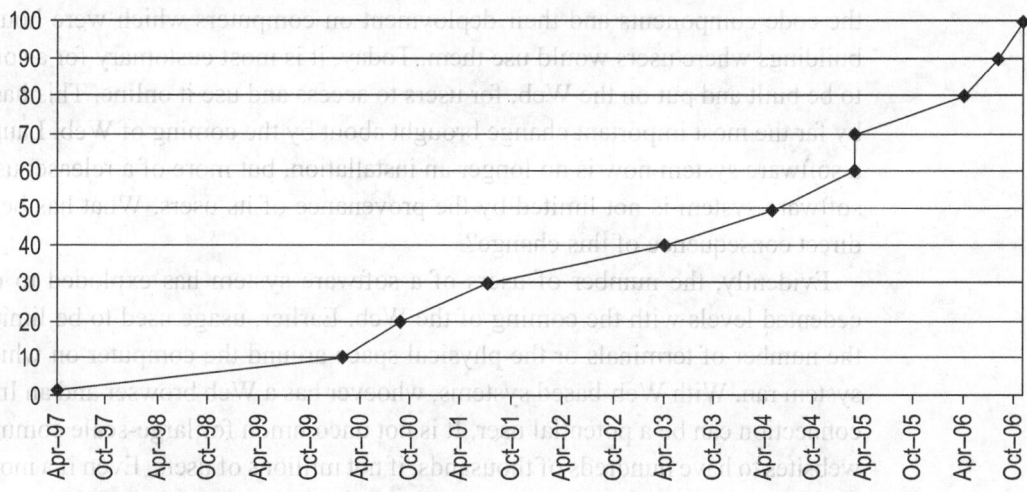

Figure 19.6   Number of websites in millions versus time (1997 – 2006)

Internet/). Figure 19.6 depicts the growth of the number of websites from 1997 to 2006 (The information for this figure is gathered from http://www.metrics2.com/blog/2006/11/03/world_wide_web_sites_crossed_100_million_milestone.html).

According to Nielsen, the Web's growth can be segregated into three major stages (http://www.useit.com/alertbox/web-growth.html):

• *Explosive* growth at the rate of 850% per year – 1991-1997

- *Rapid* growth at the rate of 150% per year – 1998-2001
- *Maturing* growth at the rate of 25% per year – 2002-2006

Based on this rate of growth, it is expected there will be 200,000,000 or 200 million websites by 2010.

With this level of proliferation of the Web, it is no wonder that software applications have changed in fundamental ways in how they are built and used. We explore these issues in the next sections.

19.3 SOFTWARE APPLICATIONS: BEFORE AND AFTER THE WEB

As we have already remarked, coming of the Web has transformed software engineering in a significant way. To understand the full importance of this statement, let us briefly review what has changed most about software applications since the Web.

Before the Web, development teams usually built a software system, which was *installed* at the customer premises. 'Installation' meant the physical delivery of the code components and their deployment on computers which were housed in buildings where users would use them. Today, it is most customary for a software to be built and put on the Web, for users to access and use it online. This has been by far the most important change brought about by the coming of Web. Launch of a software system now is no longer an installation, but more of a release; use of a software system is not limited by the provenance of its users. What has been the direct consequence of this change?

Evidently, the number of users of a software system has exploded to unprecedented levels with the coming of the Web. Earlier, usage used to be limited by the number of terminals or the physical space around the computer on which the system ran. With Web-based systems, whoever has a Web browser and an Internet connection can be a potential user. It is not uncommon for large-scale commercial websites to have hundreds of thousands, if not millions of users. Even if a moderate percentage of this user base is using the system concurrently, it leads to Design and deployment challenges wholly unknown in the pre-Web days. The change is not exclusively one of quantities, with merely *more* users in the post-Web era. The Web is a highly *interactive* medium which allows users to take a variety of actions and the system is expected to respond appropriately. Web-based systems no longer merely present data to users; users are allowed to do much more with the data they are presented with. With increased interactivity has arisen the need for enhanced event handling—a user click on a web page is an event which must be detected and responded to by the system. For a large Web-based system, millions of such events are being generated at any given time. Large number of users also means much higher chances of many of them using the system simultaneously, so Design considerations have to include issues of concurrency and load balancing.

Thus, with the coming of the Web, the ways users use a software system have undergone a radical change. The change is not only confined to the user interface—after all, it is quite possible to have a sophisticated user interface without a system being Web based. The change is most significant at the architectural framework of the system, which we discuss in the next section.

19.4 ARCHITECTURE OF WEB-BASED SOFTWARE SYSTEMS

In Chapter 11, we reviewed the role architecture plays in the effectiveness and long term health of a software system. What are the special characteristics of the architecture of Web-based software systems? This is a question we will examine now, with reference to Figure 19.7.

Web-based systems usually have a three-tier architecture. Users access the system through their Web browsers, the middle tier consists of the Web server (responsible for generating the web pages) and the application server (which gives the environment for the components containing the business logic of the application to run) and the database lies at the back-end. This may be compared with the model-view-controller architectural pattern discussed in Chapter 11.

Given this typical architectural framework, what are the important characteristics that set apart Web-based software systems?

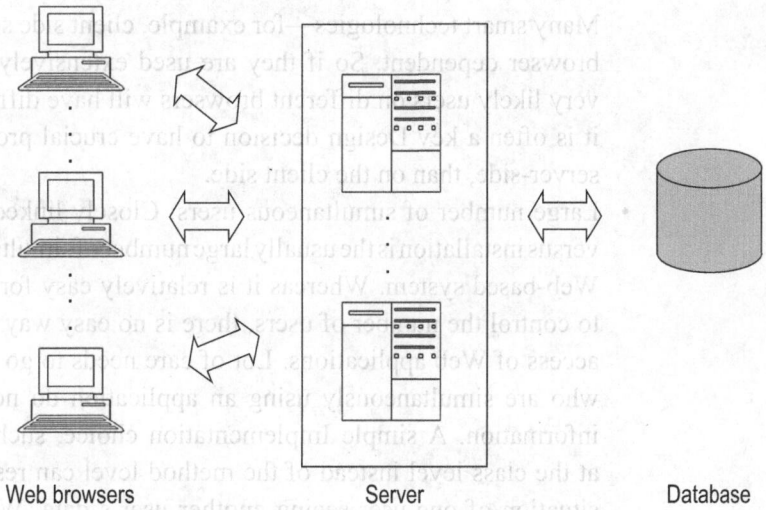

Web browsers   Server   Database

Figure 19.7 Typical architecture for a Web application

19.5 SOFTWARE SYSTEMS ON THE WEB: SALIENT FEATURES

We can identify the following salient features of Web-based software systems. These are the most generic features; specific systems may have other unique characteristics.

- Release versus installation: As discussed earlier, Web-based systems do not need to be installed on client premises, they are released to users over the Web. This leads to a quicker turnaround time for getting systems to users. Additionally, it opens up the possibility of alpha and beta releases (see Chapter 17)—sometimes called *soft launches*—which allow a larger pool of would-be users to work with and test the system before it is formally released to the entire user base. This facilitates vital feedback to the development team from real-world users. Thus, Web-based systems are specially suited to be developed through the iterative and incremental methodology, with time boxed development cycles and frequent releases of parts of the final system's functionality to users.

- Ensuring uniformity of user experience: Using the Web as a user interface brings its own set of challenges. The Web is accessed by end-users through a variety of browsers, and over widely varying connection types, and access speeds, and computing platfroms. Ensuring a uniform user experience for the entire range of users is tricky. Although many websites come with common disclaimers to the effect 'this is best viewed by so and so browsers, so and so version', there is no guarantee such directions will be followed. This often results in user interface related bugs that developers find extremely difficult to reproduce, let alone fix. There is another more insidious problem. Many smart technologies—for example, client side scripting languages—are browser dependent. So if they are used extensively in an application, it is very likely users on different browsers will have different experiences. Thus it is often a key Design decision to have crucial processing be done on the server-side, than on the client side.

- Large number of simultaneous users: Closely linked to the issue of release versus installation is the usually large number of simultaneous users for a typical Web-based system. Whereas it is relatively easy for a non-Web application to control the number of users, there is no easy way to restrict simultaneous access of Web applications. Lot of care needs to go into ensuring that users who are simultaneously using an application do not access one another's information. A simple Implementation choice, such as declaring variables at the class level instead of the method level can result in the inappropriate situation of one user seeing another user's data. Web applications, if they are popular, have a tendency to have their user bases scale up very rapidly.

So, in developing such applications, current and projected loads need to be reviewed carefully while making Design choices.

- Load balancing: What happens when we type 'a URL' into our browser and hit enter? If the same request is made by 10 different people it is very unlikely the same server will service all the requests. There is a piece of software, which balances the load—distributing requests across many of the servers that support the website in a way transparent to the end-user—and ensures one server does not get clogged by an overdose of user requests or the application can remain functional even when one or more of the servers are temporarily non-functional. Load balancing is a key aspect of any successful large-scale Web application. It ensures the user experience remains unaffected even when pieces of the physical hardware supporting the website may be under maintenance. Although we have been loosely referring to the underlying hardware supporting a Web application as a 'server', as discussed earlier, there are several categories of them, such as Web servers, which are responsible for the generation and presentation of web pages versus the application servers, which provides the executing environment for the software components embodying the business logic. Load balancing is necessary both for the Web server and the application server.

The Web has also emerged as a great enabler of software development, which we will discuss next.

19.6 WEB AS A SOFTWARE DEVELOPMENT MEDIUM

From the discussion so far, it is clear the Web has transformed the way software systems are built and used. However, in addition to the specific characteristics of Web applications we have discussed so far, it is important to understand the Web's impact on how software is developed now and will be in the future. Over the last couple of years, the Web has emerged as a versatile, powerful and flexible software development medium. Discussing the genesis and implications—technological, cultural, social and political—of this phenomenon in detail is beyond the scope of this book. We will briefly try to understand what the Web as a software development medium means for today's software engineers.

What is by far the largest and most widely used open source software? Arguably, it is the Linux operating system (for more discussion on open source software development, see Chapter 22). Would Linux have been possible in a pre-Internet and pre-Web world? Most likely, no. The collaboration amongst countless developers, working on their own towards a shared objective with minimal centralized supervision, is strongly contingent upon a robust networking infrastructure, such as, the Internet, and subsequently, the Web. Raymomd has even suggested that the biggest contribution of Linus Torvalds—the founder of Linux—lies not in creating the open source operating system but in creating the Linux development model [Raymond 2001].

How has the Web revolutionized software development? Most significantly, it has facilitated the formation and functioning of distributed teams, whose members may be located far apart but are still able to work together in building a software system. The whole paradigm of global software development (discussed in detail in Chapter 21) has arisen due to the easy accessibility of the Web. The implications of global software development are many. Most importantly, the flow of software development, enhancement and maintenance activities to organizations and geographies that offer value and cost advantages. There are also open source distributed development platforms allowing remote collaboration amongst members of a software development team.

On a less apparent but equally consequential level, the Web has opened up possibilities for large-scale knowledge sharing amongst software engineering practitioners. Whenever stuck with an unusual error message, software engineers now instinctively search for a solution on the Web. Today, quick answers to technical questions can be easily accessed at online forums. The Web is getting positioned to address one of the holy grails of software development, building systems by the combinations of standardized and off-the-shelf components.

The proliferation of the Web as a development medium has further possibilities with the maturing of the Web itself, as we discuss in Chapter 23.

SUMMARY AND TAKE-AWAYS

In this chapter, we have discussed how the World Wide Web has influenced the development of software systems. Our discussion can be summarized as:

- The Web can be said to be a service based on the infrastructure of the Internet, allowing the interconnection of information resources through hyperlinks.
- The Web serves as a powerful user interface for many software applications.
- The typical architecture of Web-based systems has a Web-based front-end, a middle tier with the Web and application servers, and back-end database.
- Some of the important concerns of Web applications are: release versus installation, ensuring uniformity of user experience, large number of simultaneous users, load balancing, etc.
- The Web serves as an effective software development medium by facilitating collaboration and knowledge sharing.

WHERE TO LOOK FOR MORE

- The World Wide Web Consortium is the definitive guide to the latest directions of Web technology—www.w3.org
- The Internet Society: 'An independent international non-profit organization … to provide leadership in Internet related standards, education, and policy around the world.'—www.isoc.org

EXERCISES

Review Questions

Review Questions test your understanding of the key concepts presented in this chapter.

1. Most recently, software engineering has been strongly influenced by the
   (a) coming of the Internet
   (b) coming of the Web
   (c) invention of the digital computer
   (d) discovery of the electron

2. A key concern for software engineers who build applications for the Web is
   (a) to build attractive web pages
   (b) learn HTML
   (c) increase the number of 'hits' to their website
   (d) understand how information is exchanged over the Web

3. 'Web applications' are
   (a) ways to apply the Web towards business needs
   (b) Web-based word processing software
   (c) business applications with a Web interface
   (d) none of the above

4. Which of the following is a true statement?
   (a) The Internet came before the Web.
   (b) The Web came before the Internet.
   (c) Both of them came together.
   (d) All the statements are false.

5. The Transmission Control Protocol (TCP)
   (a) governs the delivery of packets across the Internet
   (b) transmits signals across a network
   (c) breaks up a piece of data to be transmitted into packets and reassembles the packets when they are delivered
   (d) all of the above

6. The Internet Protocol (IP)
   (a) governs the delivery of packets between IP addresses
   (b) defines the infrastructure of the Internet
   (c) defines the infrastructure of the Web
   (d) none of the above

7. The IP address
   (a) is an address of a website
   (b) is the same as a URL
   (c) identifies a device on a network
   (d) defines the Internet Protocol (IP)

8. In a peer-to-peer network,
   (a) peers can only communicate through a centralized server
   (b) peers can communicate directly amongst themselves
   (c) peers cannot communicate amongst themselves
   (d) one peer is a server while others are clients

9. Software applications for the Web are
   (a) usually based on a peer-to-peer architecture
   (b) usually based on a client/server architecture
   (c) usually based on a peer-server architecture
   (d) none of the above

10. The ARPANET was a
    (a) predecessor to the Internet
    (b) successor to the Internet
    (c) contemporary of the Internet
    (d) none of the above

11. Which of the following is one the most important characteristic of the Web
    (a) Hyperlinks
    (b) TCP/IP
    (c) Client/server
    (d) All of the above

12. A uniform resource locator (URL)
    (a) uniquely identifies resources on the Internet
    (b) uniquely identifies resources on the Web
    (c) uniquely identifies resources on any computer
    (d) uniquely identifies resources on a network

13. An emailing service
    (a) can only be based on the Web
    (b) can only be based on the Internet
    (c) can either be Internet or Web-based
    (d) can neither be Internet or Web-based

14. With the coming of the Web, which of the following regarding a software application can be said to have undergone a significant change?
    (a) The way information is stored
    (b) The way users access an application
    (c) The way a system is performance tuned
    (d) The way the system is advertised

15. Which of the following are important concerns software engineers must address for Web applications?
    (a) Release versus installation
    (b) Ensuring uniformity of user experience
    (c) Large number of simultaneous users
    (d) All of the above

16. The power of the Web as a development medium comes primarily as it
    (a) allows a centralized control of development activities
    (b) allows quick bug-fixing
    (c) offers a fast and reliable collaboration medium
    (d) helps generate beautiful web pages

Reflective Questions

Reflective Questions require you to think more deeply about some of the ideas and come up with your own interpretations and answers.

1. Evolutionary development embodies the integration of exiting ideas and practices in a new and insightful way, whereas a revolutionary change breaks away from the past and brings forward something radically new. Was the invention of the Web evolutionary or revolutionary? Justify your answer.

2. We have outlined the client/server and peer-to-peer architectures. Which one do you think is more resilient to failure? Is that the architecture that is more widely used for Web applications? Justify your answer.

3. Have you ever used the Internet without the Web? Explain your usage scenario.

4. A superset-subset relationship has been depicted between the Internet and the Web. Does this representation fully capture the relationship between the Internet and the Web? Justify your answer.

5. What are the advantages of having a software system released over the Web versus installed at the customer premises? What are the disadvantages?

6. We have discussed a typical architecture of Web-based software systems. Can you think of some Web application that may not have a similar architecture?

7. For an online auction, what kind of challenges might a large number of simultaneous users pose?

8. What can be the potential problems with using the Web as a development medium?

Numerical Problems

Numerical Problems require you to do quick 'back-of-the-envelope' kind of calculations to arrive at answers to some simple but insightful problems.

1. Figures 19.5 and 19.6 shows the Web's growth over the years. Look at the respective y-axes of the graphs; one is log scale, while the other is not. Can you justify the choice of the scale in the light of the data presented in the two graphs. Can you model the Web's growth using mathematical equation? Using your model, predict the number of websites in the year 2020. What the assumptions underlying your model?

Programming Examples

Programming Examples require you to analyse or write a program or a program segment to under stand a specific problem.

1. One of the biggest utilities of the Internet as well as the Web lies in allowing the transparent exchange of information. Write two different programs in Java that exchange information, when
   (a) they are running on the same computer
   (b) running on different computers.
   What are the challenges you faced?

REFERENCES

Berners-Lee, T. (1999), *Weaving the Web: The Original Design and Ultimate Destiny of the World Wide Web by its Inventor*, Harper San Francisco, 1st edition.

Raymond, E.S. (2001), *The Cathedral and the Bazaar: Musings on Linux and Open Source by an Accidental Revolutionary*, O'Reilly.

Towards Enterprise Software Development

20.1 MOTIVATION

'We come across a young man, saturated with a mass of theoreticals like integrals and vectors, entropy and critical point, gyration and attenuation, eutectics and S-curves, resonance and rectification, endothermic and mass transfer et seq. He has been asked to accept on faith that all of these somehow have their application in the world of engineering outside his campus.' [Datta 1974]

The above quote describes a fresh engineering graduate. It is as relevant to software engineers of today as it was to conventional engineering graduates at the time of writing in 1974. How does 'what has been asked to be accepted on faith' compare with what the mint fresh software engineer actually faces in the *real* world? There is a world of difference, and bridging that divide is what this chapter is all about.

So far, in earlier chapters of this book we have studied the basic principles of software engineering. We have seen how software development is planned and managed, and how software is made, maintained and modified. What we studied is how circumstances of software development should be in an ideal world. It is always helpful to think of ideal situations—like the frictionless plane, zero resistance wire etc.—when we first get acquainted with new ideas. But once we have imbibed the concepts in their full extent and scope, it is time to explore how they fit in the real, *non-ideal* world. Enterprise software development is the *real* world of software engineering for a vast majority of those in the software engineering profession.

In the real world, the force of friction is ever present and it cannot be assumed away. It must instead be accounted for and utilized in clever ways—a car can only move forward because the road offers friction to its tires. Similarly, the realistic challenges of enterprise software development must be addressed if software engineers have to succeed individually or collectively. What are some of these challenges? Let us see a couple of examples.

Every software engineering student learns in the classroom that software Requirements must *freeze* before the actual building of software can begin. (We have also made such sweeping assumptions—sometimes implicitly, sometimes tempered by caveats—in earlier chapters) And the very first thing every software engineer learns on the job is that freezing of Requirements never occurs in a project with real customers and real users. Real customers are those who have paid to have the software built for them and real users are the ones who will be pleased if the software does for them what they want and angry if it does not. In the real world, software has to be built in a continuously dynamic space of changing needs, shifting business goals, technological advancement and constraints of time and resources.

As another example, in every textbook, a software engineering project is assumed to be *greenfield*—building a system from scratch where customers come with new Requirements. In reality, almost every major software project that is commissioned now has to integrate legacy applications (and what is more insidious, legacy thinking) into new development. The technical, managerial and political implications of such situations is something hardly ever discussed in the theory of software engineering.

These and many other issues make the practice of software engineering so different from its theory. And in this difference lie the charm and challenge of software development as well as the scope for a particular engineer's originality to make a difference.

In this chapter, we discuss how software engineering in the real world differs from software engineering in the ideal. We will see how we can connect the theory learnt in class with the practice permeating the world at large. In the next few sections, we discuss how enterprise software differs from other kinds of software. We then highlight some challenges typical of enterprise software development. The following section describes how software engineering principles and techniques can be oriented to enterprise software development. The Case Study section illustrates the use of the ideas of the chapter in a practical scenario.

20.2 HOW IS ENTERPRISE SOFTWARE DEVELOPMENT DIFFERENT?

Enterprise software systems are one of those 'things' whose size and variety makes it so difficult to have a single all-pervading definition. It is easier to look at their characteristics separately; and from the parts try and conceive the whole. In my book *Metrics-Driven Enterprise Software Development,* I have described these systems as:

> 'Enterprise software systems usually support business processes, they need to respond to changing user needs, they are bound by business and technological constraints, and their soundness of Design and Implementation is of material interest to different groups of stakeholders ... their scope of operation ranges across a diverse spectrum of geography, nationality, and culture' [Datta 2007].

According to Fowler, enterprise applications need to support persistent data, concurrent access to data, lots of 'user interface screens', integration with other enterprise applications, bridging conceptual dissonance between diverse business processes, and reconciling complex business 'illogic' [Fowler 2003].

The interplay of these features gives rise to the unique characteristics of enterprise software systems. In the above descriptions the word 'business' occurs frequently. 'Enterprise' has many meanings in many contexts. For example, it is sometimes taken to mean boldness or readiness in an undertaking as in 'an enterprising youth'. It may also mean a business initiative or venture. 'Enterprise' in 'enterprise software systems' relates closely to the business angle. (Boldness and readiness are also vital ingredients in the success of enterprise software projects!) Enterprise software is that particular kind of software that is commissioned, developed, maintained and enhanced with a view to supporting businesses, providing value to users and increasing profits. As success in business is about catching the right opportunity, enterprise software projects have to operate under many constraints, usually tight.

In addition to the need to operate under business constraints, enterprise systems also require to position themselves to leverage the latest technology. With the rapid pace of advance that is typical of software technology, the ability to adopt a new solution earlier than competition often leads to large enhancements to the value delivered to end-users. For example, the coming of AJAX (http://www.w3schools.com/ajax/default.asp) has radically changed the look and feel of web pages within a short span of time; any organization that uses this new technology can take user experience to a new level. The soundness of Design and Implementation of enterprise systems is also of material interest to its stakeholders. Enterprise applications have to make their mark in the marketplace; if they do well, it would mean increased business for the customer organization, and accolades for the software engineers who developed it. And the converse is also equally and painfully true; elegance of Design, solidity of Implementation mean nothing if they do not translate to added value to the end-user. This is sometimes a bitter pill to swallow for the software engineer fresh out of college, brimming with the urge to deliver the most optimal solution, always. Many a time, innovative ideas need to be moderated by the pressures of time and cost. This does not mean the good software engineer has to turn away from innovation, it means innovation has to help deliver better solutions *within constraints*. Over the last few years, enterprise software development is increasingly taking on a global colour, stretching across continents and cultures. We have discussed at length the implications and opportunities (as well as threats) of such situations in Chapter 21. Enterprise software systems are primarily concerned with persisting data, expedient access to such data and some manipulations on the data. The operations on data are not complex—for example, it would be highly unlikely for an enterprise application to solve differential equations—but the sheer amount of data that needs to be stored, fetched, updated and displayed makes it a daunting task. The copious user interface screens Fowler talks about also indicate the multiple paths of user input that can affect the data.

Fowler makes three other very important points about enterprise software systems which need some Elaboration. Enterprise applications seldom, if ever, are able to function all by themselves. Let us consider a Web-based shopping mall which allows users to purchase books (and other goods) by using their credit cards and delivers the merchandize to a physical address. How many enterprise applications are in play when we buy a book from this online store and have it sent to our home? One, the online shopping mall itself; two, the credit card company (and probably an authentication company too) three, the original seller of the book; four, the courier company which delivers the book. So, four enterprise applications in the least have to seamlessly integrate to fulfill an order starting with an user click on the Web. What does 'integration' mean in this context? Evidently it does not mean human intervention at every point to keep things together. Integration here is the sum total of all the interfaces and interactions between diverse systems, the countless automatic transactions that must go right if the whole system is to function smoothly. Much of enterprise software development is about making diverse systems talk (and make sense out of the conversation!) that were never meant to communicate with one another. This brings us to what Fowler has called 'conceptual dissonance'; the disparity in the intents and interests of entities that enterprise applications must serve. An enterprise system is often the meeting point of many business processes, none of which is obligated to be aligned with others. This is more of an *internal* challenge as compared to the *external* demands of integration discussed earlier. 'Business illogic' is indeed a very pithy description of the seemingly unreasonable and frequently irrational nature of what ironically gets called business *logic*. Business logic is entirely about business and not about logic; I prefer the phrase 'business rules' to describe the set of guidelines which business imposes on enterprise software. But enterprise software systems have to support business (il)logic effectively, as their existence is often the reason why the system needs to be built in the first place. That timeless line from Tennyson's *Charge of the Light Brigade* 'Their's not to reason why ...' could be the *mantra* for enterprise software developers when thinking about business logic.

But by far the most important feature as well as the most important challenge of enterprise software systems is the need to address changing user needs. We discuss it in detail in a later section.

From the above discussion, we should be have a fair idea by now of what one can expect in the development of enterprise software systems. But what are the other software systems enterprise applications are distinct from? Earlier in the book we have talked about programming projects for class assignments; these are certainly not enterprise applications. Software programs that support large-scale scientific computing are often of much complexity and criticality, but they are not enterprise systems either. Often software is written to understand or experiment with an idea, producing a 'proof of concept' system—these are also not enterprise systems. All the systems mentioned above are useful in their own ways, and their importance should never be undermined. But they do not exhibit the following symptoms of an enterprise system: Is someone paying to have the system built and hope to make a profit out of running of the system? Are those who build the system different from those who will be using it? Are there similar systems already running which the system has to compete with and improve upon? This is by no means an exhaustive list, but whenever we meet a new software system, either in conception or operation, running through these questions can help us understand the nature of the system.

From what we have seen so far, enterprise software systems seem unpredictable, inconsistent and complex; in short, difficult to Design, develop and maintain. Why still, should we be concerned with them? Àla George Mallory's reply when asked why he wanted to climb Mount Everest, we could say, 'Because it is there'. But such an answer, though of much romance, will not serve our purpose. We need to understand why enterprise systems are important after all. That is what we attempt in the next section.

20.3 IMPORTANCE OF ENTERPRISE SOFTWARE

The importance of enterprise software systems in our lives has risen many times in the last decade. Today, we rely on enterprise software for many of our needs; financial, educational, communicational, healthcare, entertainment etc. The list is long and varied. Indeed, in almost every realm computers have entered our lives to make a qualitative change, there is one or many enterprise applications working in the background. What has transformed in the last ten years or so to lead to this pervasion of enterprise software systems? An easy answer would be to say the Internet, but this is not fully accurate. Internet and the World Wide Web—though almost synonymous now—are different from one another (see Chapter 19). It is primarily the Web as we know it today that has led to the rapid rise in the use of enterprise applications.

The Web by itself is not enough. The increasing availability of high bandwidth network connectivity—the veritable road through which we can connect to the Web—is also making the access of information and services from the Web faster, easier and cheaper. Note how we mentioned information and services in the last sentence. The Web is no longer a place where we go to *know* things (or let things about us be known) but also to *do* things. Increasingly, typical desktop applications such as word processing or working with spreadsheets are moving to be on the Web. There is also initiatives to move the operating system, on to the Web (http://www.cs.duke.edu/ari/issg/webos/).

So the importance of enterprise software systems arises to a large extent from the explosion of Web-based content and services in the past few years. As was discussed earlier, at the heart of enterprise applications lie integration amongst diverse business processes and systems; and this integration is one of the foundations of the Web's growing presence in our work, play and the overall experience of living. So enterprise software is no longer solely about large businesses and corporations, it is something which we use in our every action on the Web, from checking emails to booking tickets or merely searching for information. It is no wonder enterprise software systems also have a set of unique challenges. We discuss these in the next section.

20.4 CHALLENGES UNIQUE TO ENTERPRISE SOFTWARE DEVELOPMENT

As we have discussed, enterprise software systems need to deliver business value. Business value ultimately comes down to providing information or service to the end-user. End-users frequently do not know what they want until they see it; and when they see it they usually want something else. This is often described as the IKIWISI syndrome—I Know It When I See It. The IKIWISI syndrome is especially endemic in users of software systems due to a cognitive gap that widely exists in our understanding of and expectations from software as a medium [Datta 2007]. To deliver value to users we need to remain sensitive to *what* users want without worrying too much about *why* they might possibly want it. Whenever users change their mind or want something different, enterprise applications have to respond appropriately. As we have already seen, the process of software-making is driven by Requirements. Thus, when Requirements change, all of the downstream activities are influenced. As we mentioned earlier in the chapter, the need to respond to changing user needs is a key expectation from enterprise software systems. It is natural for software systems to change as new bugs are unearthed and resolved; but enterprise applications change due to a wholly different reason—user initiated changes in Requirements. The Requirements that change due to users asking for something more, or something done differently are usually *functional* Requirements. As we have seen in Chapters 14 and 15, functional Requirements are those that deal with the functionality of the system, that is, what the system does for the user. In addition to changes in functional Requirements asked for by users, *non-functional* Requirements also pose a major challenge to enterprise software systems.

Consider this scenario: You are trying to log in to your email account on the Web. You enter your user id and password and hit the 'Sign In' button. Nothing happens for the next ten minutes. Even if you find the system has let you in (if the user id/password combination was correct) after those ten minutes, this will certainly be a frustrating experience. What has gone wrong here? The system is delivering the functional Requirement alright—granting access upon verifying credentials; but it is violating the crucial non-functional parameter of delivering the functionality within an acceptable amount of response time. As we have already seen, the non-functional parameters of usability, reliability, performance, supportability etc. (URPS+) encapsulate the needs a system must fulfill in addition to functioning as designed. Non-functional Requirements are often swept under the carpet, and remain there, until very near the deadline, something is noticed about the system that users will dislike. Non-functional Requirements can make or break or enterprise software systems. They are insidious; people tend to ignore them until there is something very wrong with them. And then it is usually too late for anything to be done. Usability and performance are by far the most influential parameters for affecting the user satisfaction of an enterprise application; but reliability and supportability can cause more pain on a longer term to the development team. For a non-enterprise system, performance etc. are usually hardly any concern.

Related to the above points is the need for enterprise software systems to continually function in a state of flux. One of the reasons for this flux is certainly changing user needs. Additionally, an enterprise application is forever at the intersection of technological enhancements and business innovation. The only way organizations can stay in business (and make profit) is to either harness new technology (eg. a financial company letting its customers transact online) or outwit competition by coming up with a new business model (eg. outsourcing work to provide cost and quality benefits). These two prerequisites for an enterprise's survival are closely linked in today's world. Being able to offer better value to customers means outsmarting completion in the use of technology; and this comes from a sound sense of both business and technological trends. As we have highlighted earlier, to be a good software engineer one needs to be more than just a good programmer. To this we may add now, effective enterprise software development needs more than just software development skills; it is essential to be aware of business trends on a global scale. Enterprise software built today is meant to deliver value tomorrow; so a continual focus on the future is necessary for those who Design and deliver enterprise software solutions. However, a degree of discretion is necessary when building for the future. A common pitfall of software development is too much *future proofing*—investing time and attention into planning for future features which may never be used. But an appreciation of how business and technology will develop in the near future is a key ingredient in the success of enterprise applications.

Based on what we have discussed so far, it is natural to ask, are there development methodologies specially suited to enterprise software systems? The answer is not a straight yes; neither it is a straight no. The twilight zone between yes and no is usually very interesting; and we take it up in the next section.

See Exhibit 20.1 for a flavour of two real-life enterprise software systems.

20.5 ENTERPRISE-ORIENTED SOFTWARE ENGINEERING

In a field so ready to spawn new abbreviations, let us be clear at the very beginning that Enterprise-Oriented Software Engineering—EOSE—is not a sub discipline; at least, not yet. And that is why we need to learn and know about it; to stay ahead of competition!

In Chapter 1, we gave various definitions of software engineering. We also discussed how the unique characteristics of software engineering make it almost impossible to come up with one-definition-fits-all. On a similar vein, we may give a *working definition* of enterprise-oriented software engineering as: the study and practice of software engineering principles in the context of the development, maintenance, and enhancement of enterprise applications. We identify some of the features of enterprise software systems below.

Exhibit 20.1   Enterprise Applications in Flesh and Blood

It is said that 'the largest system ever used by us humans may be a system with close to 1 billion users, namely the global telecommunications network' [Jacobson et al. 1999]. For example, making a telephone call from San Francisco to Stockholm will probably go through some 20 systems such as switches, satellite and transmission systems. Each such type of system required approximately 1,000 person-years to develop, and a majority of the effort related to software development.

We have another major enterprise system much closer at hand.

In the late 1980s and early 1990s, the online passenger reservation system for the Indian Railways was built. As we know, Indian Railways mimics India in its size and diversity, carrying more than a million ton of freight traffic and about 14 million passengers daily across 6,856 stations (http://www.indianrail.gov.in/). No computerized system for reservation existed before; all bookings were done across the counter, manually. This was a greenfield projects in a true sense of the term— projects do not get that green any more. This was also an enterprise software project of the most gigantic proportions: involving not just one enterprise (Indian Railways is quite an enterprise in its own right, a separate government department presenting its own annual Budget to Parliament), but touching the lives of millions of people in a direct way. The software framework built by this project was later enhanced with a Web interface, which now allows booking of Indian railway tickets from anywhere in the world.

But the online passenger reservation system did something more than providing an online passenger reservation system. (Interestingly, when the system was commissioned, online did not necessarily mean Web-based —it merely referred to something that was available in a way such that time and place were of no constraint to the user.) It brought in a radical change in legacy thinking—the Indian Railways was no longer just a behemoth of a public transport system, it was seen more as a modern and flexible mass transit infrastructure. This is the kind of change great enterprise software systems usher in. They not merely change the way of doing, they also change the way of thinking [Datta 2007].

20.5.1 Identifying and Understanding Stakeholders' Needs

Stakeholders are those individuals or groups who have a material interest in the functioning of an enterprise application. Some stakeholders are easy to identify: those who have commissioned the project, those who will be using the system when it is ready, the development team which is building the system etc. It is also clear what the 'material interest' is for each of these stakeholders. There may be other, more covert, stakeholders also. What about a competing company whose business will be affected when the system moves into production? Or departments within the customer organization whose personnel will be affected by the new system?

When an enterprise software project starts, one of the earliest tasks for the development team is to identify who all the likely stakeholders are, and make as good an estimate as possible on the respective stakes. Guessing on the stakeholders and their stakes may neither be easy nor accurate. But with experience, the guesswork becomes more refined. Throughout the development life cycle of enterprise systems, one needs to keep track of how each stakeholder is being affected, for better or for worse. Stakeholder needs may or may not map directly to the formal Requirements the system is supposed to implement. And this is why special care needs to be taken to address them. It is best to try and capture all the relevant needs as Requirements so that their tracking becomes easier; but all enterprise projects will have new needs coming up or old ones changing. The development team has to remain sensitive to these, if the final product they deliver will satisfy customers, and not just be a software meeting all written specifications. This volatility and mutability of Requirements brings us to the very important issue of choosing a development methodology.

20.5.2 Choice of a Methodology

In Chapter 4, we have studied how the evolution of software engineering as a discipline has given rise to a number of development methodologies. We had also made a strong case for iterative and incremental development (IID) being specially suited to tackling many of the challenges of 'real-world' software development. Real-world software development is very much what we have been discussing in this chapter and it will influence our discussion regarding the choice of a methodology.

IID is not so much a single methodology, but the guiding *spirit* for any methodology which seeks to develop a software system through iterations and release it in increments. We may recollect that an *iteration* can be thought to be a mini-Waterfall; a time-boxed set of life cycle activities focusing on a sub-set of the system's Requirements towards the release of a part of the final system—an *increment*—for the *real* users to test and provide feedback. Many process frameworks or processes such as the Unified Software Development Process (UP) [Jacobson et al. 1999] or Rational Unified Process (RUP) [Krutchen 2004] use the spirit of IID.

The most important criterion for any methodology seeking to support enterprise software development is its facility to utilize continual feedback from *real* users. 'Real' is emphasized because users who will finally use the system need to get their hands on to it as early as possible. IID offers that easily—as new increments are released, the development team comes to know what the users like about the system, what they do not, or what they want done differently.

We have discussed different methodologies in some depth in Chapter 4. Whichever methodology is actually chosen for a project, sensitivity to continual user involvement and feedback is crucial to the success of enterprise software development. Is the Waterfall model never suitable for enterprise software development? This is a question often asked and seldom answered in software engineering textbooks.

As we saw earlier, there is nothing wrong with the Waterfall model as such. Indeed, it provides a streamlined and intuitive approach for software building. We have also seen how each iteration in IID in a way mimics Waterfall. The Waterfall model can deliver quality software in systems of well-defined scope, where the problem domain is well-understood and there are not very many unknowns in technological, environmental and other parameters. From the symptoms of enterprise applications described earlier of development, it is apparent they are different from the ones best-suited to the Waterfall model. Besides, Waterfall model is often beset with too much *ceremony*, which means there is a strong chance developers swamp themselves with documents and diagrams, instead of creating the most important artefact of a software project—running code. So the Waterfall model is usually unsuitable for enterprise software development, barring the exceptional cases of very small systems. The main reason for Waterfall's unsuitability is its tendency to encourage developers to blinker themselves from user feedback while the system is being built.

As we discuss in the next section, user involvement and feedback is a key driver for success in enterprise software projects. There is no sure shot formula for choosing one methodology over another in enterprise software projects, but the use of metrics can instill more discipline in the decision process, (See Exhibit 20.2).

20.5.3 User Involvement and Feedback

Any system which seeks to continually monitor and moderate its behaviour must leverage the power of feedback. Let us for a moment, stop thinking about the software product, and take the development process as our 'system'. How do we ensure this system has the *intelligence* to know which way it is heading, how that differs from where it needs to go, and take corrective action if necessary? The only way to do it is to hear what the users are saying.

But this is much easier said than done. What should we let the users see to give feedback on? Involving users at every step of the development process may create unnecessary noise that will benefit neither the developers nor the users. IID answers this question by providing periodic incremental releases—chunks of functionality that constitute parts of the final system—for the users to try out, test and comment upon. For feedback to work most effectively, the users have

Exhibit 20.2   Metric for the Crossroads of Software Development Methodologies

Deciding on a particular development methodology is often one of the earliest and most important decisions facing a software development project. The metric *Agility Measurement Index* can help us understand and weigh the different factors involved in reaching that decision. Calculation of the Agility Measurement Index considers the duration, risk, novelty, projected effort, and interaction levels expected in a project the metric can help, recommend whether the Waterfall model, Unified Process, or Extreme Programming is best suited for that project. For a detailed discussion of the Agility Measurement Index metric, refer to *http://www.dattas.net/misc/ami.pdf*.

to be 'real'; in the sense that they will be the ones who will use the system when it is finally launched into production. The worst thing to happen to an enterprise application during development is to be tested by the development team itself—there is no surer recipe that the most serious defects will remain hidden. Tracking and acting upon user feedback is by no means straightforward; it takes a lot of communicational finesse to extract the gist of what the users say and translate that to executable tasks in the development context. This is again one of those skills which can be learnt only by experience; but it is very helpful to be aware of such learning needs when entering the real world of enterprise software development. Figure 1.3 shows the mechanism of feedback.

In the next section, we conclude the discussion on enterprise software development by highlighting another important aspect.

20.5.4 Continual Development

Whenever we think of a *project*—in software development context, or otherwise—we are accustomed to think in terms of a clear beginning and an end to a set of coordinated activities. This stereotype hardly ever fits the enterprise software development scenario. An enterprise software project may have a clear beginning—when the development team is hired by the customer organization. But it may not end with delivering the system the customer had asked for. In addition to the maintenance and enhancement issues examined in Chapter 18, there is also another angle to the continual development of an enterprise software project. If the project delivers value to the end-users, it will most likely bring more business to the customer, who will want the system be tuned (or at times radically modified) to meet the latest business needs. Continual development means continual change and continual change means breaking what was done before to make something new. It is difficult to predict how the system will need to change next as business situations are often as fickle as the weather. But for enterprise software developers,

Figure 20.1 Feedback in software development

it is vital to have the mindset to live with continual change and deliver the best value in spite of it.

CASE STUDY

Our protagonist Preeti is leading an enterprise software project. Given the scope of the project, the unknowns in the business domain and the fair chance of changing Requirements, the iterative and incremental development methodology has been chosen. To be able to estimate the project's duration and resource needs, Preeti will need to know the approximate number of iterations of development. How does she do it; is there a ready formula to calculate the number of iterations?

The answer is no, simply because ready formulas often do not work well in engineering—and software engineering is no exception. But there are *heuristics* though—rules of thumb that combine empiricism and intuition – to help practitioners make decisions. Kruchten discusses such a rule of thumb [Krutchen 2004] which we will examine now, and help Preeti solve the problem at hand.

According to Kruchten, an iteration 'ideally' should be from between two to six weeks, but there are several project-specific parameters which influence the actual duration, such as the size of the project team, how familiar the organization is with iterative and incremental development, the stability and maturity of the organization, and how much the development team can leverage automation in the life cycle activities such as Testing etc. Kruchten points to an empirical formula by Joe Marasco: $D_{weeks} = \sqrt{S}_{ksloc}$, where D_{weeks} is the duration of an iteration in weeks and $\sqrt{S}_{ksloc}$ is the size of the project in thousand lines of source code. Kruchten also advises this formula should be taken with a 'grain of salt'!

In Chapter 14, we have introduced the life cycle phases of Inception (I), Elaboration (E), Construction (C), Transition (T). Kruchten gives his 'six plus or minus three' rule of thumb [Krutchen 2004] saying that 'normal' projects have 6±3 iterations. This is broken up per phase as:

- Low: three iterations – zero for I, one each for E, C, T
- Typical: six iterations – one for I, two each for E, C, and one for T
- High: nine iterations – one for I, three each for E, C, and two for T

Low, typical, high denotes *levels* which indicate scope of the project in terms of its novelty, complexity, size etc.

So, Preeti has a rule of thumb now to start with her estimate of the number of iterations. The beauty of rules of thumb is that they are certainly not cast in stone, and can be customized as necessary, to fit a particular case. Preeti's project should have qualified for the 'typical' level; but the system her team is looking to develop will need to interact with a legacy system developed and maintained by another organization. Initial communication overheads are most likely to be there, so Preeti keeps two iterations for Inception, two each for Elaboration and Construction, and one for Transition, leading to a total of seven iterations.

Now, how should Preeti decide on the number of weeks for each iteration? Can she use Marasco's formula given above? What are the assumptions and approximations that need to be made? How should she heed Kruchten's warning for taking the formula with a grain of salt? We examine these crucial issues in one of the Numerical Problems of the Exercises section.

WORKED-OUT EXAMPLES

1. Using the formula, $D_{weeks} = \sqrt{S}_{ksloc}$ calculate the length of an iteration for a project with 200,000 lines of code. Are there any assumptions?

   Answer: Now, $D_{weeks} = \sqrt{200} = 14$ weeks (approximately). This seems a slightly long time for an iteration; We may consider what a real life 200,000 lines of code system might be like. Chances are very high that such a system will actually consist of a number of sub-systems, instead of being a single monolithic system. So assuming there are 8 subsystems of equal size, each of 200,000 / 8 = 25,000 lines of code, then the duration of each iteration for each subsystem is 5 weeks, which is a more manageable time frame.

2. How suitable is the metric *Mutation Score* (defined below) to compare how much one Requirement changes across iterations?

   Mutation Score is a measure of the number of times Requirements change over the number of iterations. The software system undergoes [Datta 2007].

Let the *Mutation Score* for Requirement R_n be denoted by $MS(R_n)$. Let *m* be the number of iterations for a project.

1. At iteration 1, set $MS(R_n) = 0$.

2. For iteration 2 to m

if R_n has *changed*,

$MS(R_n) = MS(R_n) + 1;$

Repeat steps 1 and 2 for all Requirements, R_1 to R_n

Answer: The key step in the above algorithm to calculate the *Mutation Score* is to decide if R_n has changed. A quick way to decide will be to do a textual comparison (through some automated tool) of the versions of the Requirement across the iterations. This will tell us whether the description of the Requirement has changed. Under the assumption the descriptions of the Requirements would have changed only if the underlying intent also has changed, the *Mutation Score* would be a helpful metric. But not all changes in Requirements have the same potential to impact development; this metric does not help us differentiate between the levels of Requirement change.

SUMMARY AND TAKE-AWAYS

In this chapter, we have highlighted some of the key elements of enterprise software development. The importance of enterprise software systems lies in the fact that a large majority of professional software engineers need to Design, develop, maintain and enhance these systems as a part of their job. Given below is a summary of the discussion in this chapter:

- Enterprise software systems support business processes. They need to respond to changing user needs, are bound by business and technological constraints, and their soundness of Design and Implementation is of material interest to different groups of stakeholders [Datta 2007].
- Additionally, enterprise applications are characterized by persistent data, concurrent access to data, lots of 'user interface screens'. They may need to integrate with other enterprise applications, and to bridge conceptual dissonance between diverse business processes, and reconcile complex business 'illogic' [Fowler 2003].
- The increasing importance of enterprise software systems in our daily lives is closely linked to the increasing penetration of the Web since mid 1990s.
- Major challenges in enterprise software development come from frequent user-initiated changes to functional Requirements as well as stringent demands on non-functional Requirements.

• Key elements of enterprise-oriented software engineering include identifying stakeholders' needs, choice of a methodology, user involvement and feedback and continual development.

WHERE TO LOOK FOR MORE

• *Metrics-Driven Enterprise Software Development* [Datta 2007] illustrates the use of simple, intuitive metrics to guide the software development life cycle for enterprise applications—www.dattas.net

EXERCISES

Review Questions

Review Questions test your understanding of the key concepts presented in this chapter.

1. Enterprise software systems
   (a) are software applications to support business processes
   (b) are software applications developed by enterprising people
   (c) are software developed with the latest technology
   (d) none of the above

2. In real life software projects,
   (a) Requirements freeze before development starts
   (b) there are no clearly defined Requirements
   (c) Requirements are frozen from the start of the project
   (d) Requirements never freeze

3. According to Fowler, enterprise applications need to support
   (a) persistent data
   (b) concurrent access to data
   (c) conceptual dissonance between diverse business processes
   (d) all of the above

4. Enterprise applications commonly require
   (a) to interface with other enterprise applications

(b) to solve differential equations
(c) to have many users
(d) all of the above

5. A programming exercise for a course
   (a) can be an enterprise application
   (b) is always an enterprise application
   (c) is never an enterprise application
   (d) none of the above

6. The rise in the importance of enterprise applications
   (a) is strongly influenced by the penetration of the Web
   (b) has been primarily due to increasing computer literacy
   (c) is a direct result of outsourcing
   (d) all of the above

7. Functional Requirements
   (a) are not relevant to enterprise applications
   (b) change frequently in enterprise applications
   (c) coincide with non-functional Requirements in enterprise applications
   (d) remain unchanged once they have been agreed upon

8. Non-functional Requirements
   (a) are easy to track in enterprise applications
   (b) do not matter in enterprise applications

(c) can be addressed after delivering a product

(d) can affect the user experience of enterprise applications to a very large extent

9. In an enterprise software development project,

   (a) stakeholders' every need is given in writing to the developers

   (b) stakeholders' every need has to be carefully understood by the developers

   (c) there is unlikely to be any stakeholders

   (d) (a) and (b)

10. In choosing a methodology for an enterprise application

    (a) one can rely on established formulas

    (b) the choice is immaterial

    (c) one can not rely on any established methodology

    (d) one has to rely to a large extent on intuition and experience

11. Iterative and incremental development is

    (a) suitable for enterprise applications

    (b) unsuitable for enterprise applications

    (c) an inefficient development paradigm

    (d) none of the above

Reflective Questions

Reflective Questions require you to think more deeply about some of the ideas and come up with your own interpretations and answers.

1. In the Case Study section, we mentioned the formula $D_{weeks} = \sqrt{S}_{ksloc}$ for calculating the duration of an iteration in weeks. What are the assumption(s) implicit in the formula?

2. Create a checklist of five questions to find out whether a software system is an enterprise application or not. What are the details you need about a software system to answer the questions? Who is the best person to have such information?

3. Write three points for and three points against the statement: Enterprise software systems are the only important software systems.

Numerical Problems

Numerical Problems require you to do quick 'back-of-the-envelope' kind of calculations to arrive at answers to some simple but insightful problems.

1. Using the formula, $D_{weeks} = \sqrt{S}_{ksloc}$ calculate the size of a software system (in thousand lines of source code) that has a team size of 4 developers and uses 1600 resource-hours of effort for an iteration of development. State any assumptions you need to make.

2. A project has 7 iterations, but no Inception phase. In terms of Kruchten's rule of thumb, which *level* should this project be categorized in?

3. With reference to Exhibit 20.2, calculate the *AMI* given the *Actual Score* (*A*) for the dimensions D = 2, R = 4, N = 2, E = 3, I = 5. (Assume same *Max and Min Scores* as in the exhibit.) Which methodology would you recommend?

Programming Examples

Programming Examples require you to analyse or write a program or part of a program to understand a specific problem.

1. We mentioned that one of the non-functional Requirements, performance, is often of great importance in enterprise software systems. One indication of performance is how quickly a program segment can perform its task. Suppose we want to fetch 10,000 items from a data structure. Write a Java program using the Hashtable class to accomplish this task; and report how long it takes to fetch the items. Can you substitute Hashtable with any

other Java supported data structure to have better performance? If there is a difference in the performance, what do you think the reason is?

2. Write a Java program to calculate the *AMI* metric and give you a recommendation on what methodology to choose for a project. A simple way to embed the *intelligence* of recommending a methodology may be to use *if-else* statement(s) whose condition(s) relate to the value of the metric. Can you think of other ways to embed such intelligence?

3. Integrated Development Environments or IDEs can help you write, compile, run, debug and test programs quickly. Write a Java program between 1-100 lines of code using any text editor (say Notepad) and compile and run it from the command prompt. Now download a Java IDE such as NetBeans (www.netbeans.org) or Eclipse (www.eclipse.org) and repeat the same activity. How much do you save in time and effort? Would the saving be more if the program was between 100-10,000 lines? Can you formulate an empirical relation between the lines of code and the resource hours to make it correctly run with and without using an IDE?

REFERENCES

Berners-Lee, T. (1999), *Weaving the Web: The Original Design and Ultimate Destiny of the World Wide Web by its Inventor*, Harper San Francisco, 1st edition.

Boehm, B. and Turner, R. (2003), 'Observations on balancing discipline and agility', In *ADC '03: Proceedings of the Conference on Agile Development*, p. 32, Washington, DC, USA. IEEE Computer Society.

Datta, J.P. (1974), 'The fresh engineer entering indian industry—"a babe in the woods"?', *ISTD REVIEW (Journal of the Indian Society for Training and Development)*, IV(4).

Datta, S. (2005), 'Integrating the furps+ model with use cases—a metrics driven approach', In *Supplementary Proceedings of the 16th IEEE International Symposium on Software Reliability Engineering (ISSRE2005)*, Chicago, IL, November 7–11, 2005, pp. 4–51—4–52.

Datta, S. (2006), 'Agility measurement index: a metric for the crossroads of software development methodologies', In *ACM-SE 44: Proceedings of the 44th annual southeast regional conference*, pp. 271–273, ACM Press, New York, USA.

Datta, S. (2007), '*Metrics-Driven Enterprise Software Development: Effectively Meeting Evolving Business Needs*, J. Ross Publishing.

Fowler, M. (2003), *Patterns of Enterprise Application Architecture*, Addison-Wesley.

Jacobson, I., Booch, G., and Rumbaugh, J. (1999), *The Unified Software Development Process*, Addison-Wesley.

Krutchen, P. (2004), *The Rational Unified Process: An Introduction,* 3rd ed., Addison-Wesley.

Global Software Development

21.1 MOTIVATION

As we have discussed at length in the earlier chapters of this book, software as an engineering artefact differs fundamentally from other engineering artefacts. Without reiterating the details, we may note that software functions on the substrate of the digital medium, which has widely proliferated over the past two decades. While it remains virtually impossible to build a bridge without the engineers' physical presence at its site; it is customary these days for software systems to be built remotely, without many of those commissioning, developing and using the system ever coming into face-to-face contact. The benefits of this situation are many and the challenges no less numerous. As the world becomes increasingly wired, global software development will only grow in expanse and importance.

Global software development represents a convergence of a number of technological, social, economic and political trends of our times. The study of these trends is illuminating and entertaining, and the subject matter of much deliberation. We do not have the scope to get into their details here; we will outline as much is necessary as background to global software development.

21.2 WHAT IS SO SPECIAL ABOUT GLOBAL SOFTWARE DEVELOPMENT?

Global software development represents a shifting paradigm in the dynamics of the industrial world. Its impact has been profound and is likely to be long lasting. But perhaps the most significant point about global software development is that it is very much a *work-in-progress*;—we are all actors in a drama whose plot is just unfolding. Never before in the history of human enterprise has one system of industrial production been so influential in shaping circumstances at the individual as well as national and international levels.

The key element in the functioning of any software system is flow of information. Information flow has been vital in the functioning of human societies as well, since antiquity. Ways of transferring information across distances started with messenger pigeons and evolved into the postal system. The information exchange for software systems takes place through electromagnetic waves. With the maturing of the global telecommunication network, transfer of information has increasingly become quick, easy and cheap. Global software development leverages this communication boom in two ways. On one hand, the network infrastructure supports the flow of information that is the life-blood of the software system. On the other hand, the information flow amongst the stakeholders of a software system—customers, users, designers, developers—also utilizes the superior networking facilities available now. Figure 21.1 illustrates the two levels of information flow.

These two levels of information flow associated with software systems makes them so uniquely positioned to embrace a global development paradigm. To build a bridge, we need information flow between the stakeholders—how high and wide the bridge needs be, how much traffic will pass over it, how much wind load it will have to endure etc. This information exchange can happen over distances, using telecommunication networks. But we need steel, mortar and bolts to actually *build* a bridge. These commodities still cannot be transported near instantaneously across great expanses as information can be. Thus we can think, plan and talk about building a bridge remotely; but to actually build one, we need to have all the material and people who will process the materials be physically present at the site of the bridge. May be in future, we will either be able to teleport bridge building materials, or remotely control bridge building robots. But we are not there yet; and till that time, software will remain by far the only large-scale industrial artefact whose production and consumption can happen remotely.

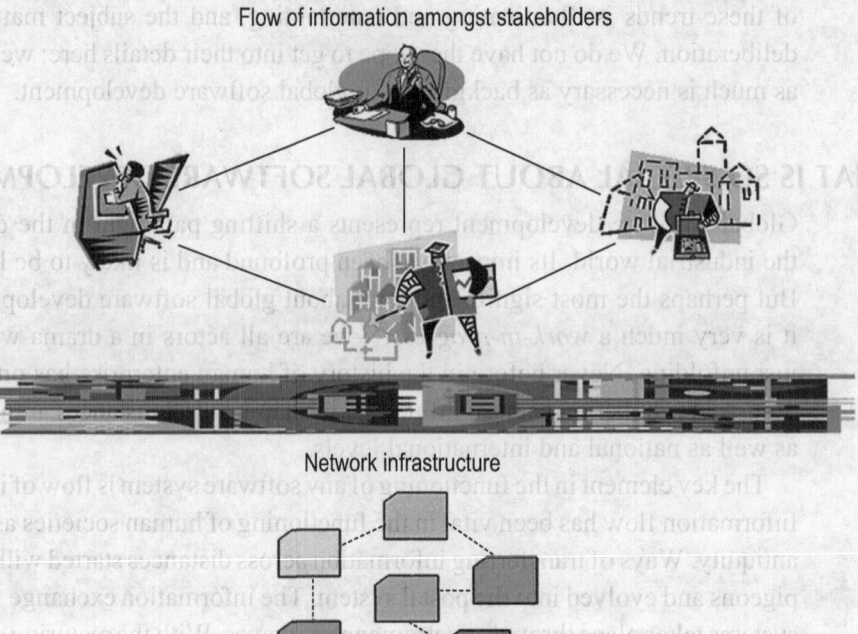

Flow of information amongst stakeholders

Network infrastructure

Flow of information within the software system

Figure 21.1   Global software development: two levels of information exchange

21.3 GENESIS OF GLOBAL SOFTWARE DEVELOPMENT

As we remarked earlier, global software development is a recent phenomenon. The Year 2000 problem—or Y2K, as it is popularly called—was a watershed in the genesis of global software development. As the year 2000 approached, it was discovered that many programs mainly written in COBOL decades earlier, were still supporting large-scale business applications. This just goes to show how resilient and long lasting software can be. But the problem lay elsewhere. These programs customarily used two digits in the date format (DD-MM-YY or MM-DD-YY) to record the year. With the coming of the year 2000, '00' in the year field would have led these programs into assuming the year as 1900, instead of 2000. That had potentially disruptive implications for many business processes. For example, a banking application calculating interest could have ended up with negative accrued interest, or other even more serious errors. By the end of the 1990s, businesses all over the world had become significantly dependent on software, and the Y2K problem was certainly something that could have affected their business continuity. So, there was a widespread scramble to rectify the date format in these legacy applications and make them year 2000 compliant. Much of this work was done by Indian software engineers. In preparation for addressing the Y2K problem, large-scale investment was made in communication infrastructure to integrate India into the global supply chain. As Friedman has pointed out, development of this infrastructural support allowed India to take global software development to the next level beyond the Y2K problem [Friedman 2007]. With the successful solution to the Y2K problem, businesses all over the world realized the great potential of having software remotely designed, developed, enhanced and maintained by those best skilled to do it, irrespective of where they are on the globe. Global software development had started in earnest. The current decade of the 2000s has seen the reach and penetration of global software development grow by leaps and bounds. And many experts agree it has yet to reach its full potential.

21.4 DISTRIBUTED TEAMS AND REMOTE CUSTOMERS

Global software development essentially means distributed teams and remote customers. In recent years, significant work has been done in trying to understand how these factors influence software development: Collaboration platforms for offshore software development are evaluated in [Rodriguez et al. 2007], Shami et al. simulate distributed development scenarios in [Shami et al. 2004] and a research agenda for this new way of software building is presented in [Sengupta et al. 2006]. Herbsleb and Grinter in their papers [Herbsleb and Grinter 1999a], [Herbsleb and Grinter 1999b] have taken a more *social* view of distributed software development. In terms of Conway's Law—*organizations which Design systems are constrained to produce designs which are copies of the communication structures of these organizations* [Conway 1968]—Herbsleb and Grinter seek to establish the importance of the match between how software components collaborate and how the members of the teams that develop the software components collaborate. This collaboration is a key aspect for the success of global software development.

The Agile Manifesto lists the principles behind agile software development—methodologies being increasingly adopted for delivering quality software in large and small projects in the industry (see Chapter 2 for more details) including those utilizing dispersed development [Kornstadt and Sauer 2007]. The Manifesto mentions the following among a set of credos: 'The most efficient and effective method of conveying information to and within a development team is face-to-face conversation', and 'Business people and developers must work together daily throughout the project' (http://agilemanifesto.org/principles.html). Evidently, the very nature of distributed development precludes this kind of interaction between the project stakeholders. How can some of the effects of global software development on software Design be understood better?

[Datta and van Engelen 2008] identify the key drivers of the effects of global software development on software Design as *locational asynchrony* (LA) and *perceptual asynchrony* (PA). LA and PA may exist between customers and developers, or within the development team itself. Locational asynchrony arises from factors like differences in geography and time zones. An example of LA would be the difficulty in explaining a simple architectural block diagram over email or telephone, which can be easily accomplished with a white board and markers in a room of people (something similar to the *consequence of distance* highlighted in Herbsleb and Grinter 1999a). Perceptual asynchrony tends to be more subtle, and is caused by the complex interplay of stakeholder interests that global software development essentially entails. For example, in distributed development scenarios, developers who have no direct interaction with the customer often find it hard to visualize the relevance of the module they are working on, in the overall business context of the application—this is a manifestation of PA. [Datta and van Engelen 2008] examines the effects of LA and PA on the Design of several projects of varying degrees of distributed development.

21.5 OUTSOURCING: A QUICK REFLECTION

No discussion of global software development can be complete without broaching the (often uncomfortable) topic of outsourcing. Very simply, outsourcing can be said to be the transfer of the production of some goods and services essential to the functioning of a particular organization, to other organization(s) offering benefits of cost, quality, flexibility and any other parameter influencing the success of the former organization.

Unfortunately, the most discussed, and much maligned aspect of outsourcing is *cost*. Outsourcing is often portrayed as an arrangement to take away jobs from the so-called developed economies and give them to the so-called developing ones. The only motivation for outsourcing is said to be the surfeit of cheap labour that the latter economies offer, willing to work for a fraction of the remuneration of their counterparts in the former. Without getting into the moral, ethical, or political implications of such a situation, let us briefly review what makes software development so amenable to outsourcing.

For any industrial production other than software, there is a need for large-scale investment of capital. Cars cannot be built without an intricate network of factories and ancillary plants: Building airplanes requires such a huge capital outlay, there are only few companies in the world that can build them. In comparison, what do we need to build software? We merely need computers and connectivity, both of which are now cheap, reliable and readily available. This has contributed to taking software development close to a level-playing field in terms of the availability of resources for fulfilling business needs irrespective of economic or geographic constraints.

However, what makes outsourcing work is not exclusively cost benefits. Elements of quality and flexibility are inherent in the value delivered through outsourcing. Certain geographies—most significantly India—have become preferred outsourcing destinations due to specific skills offered at competitive prices here; skills which are in short supply at locations from which the work is being outsourced. Flexibility, in terms of tailoring solutions to fit a diverse range of business problems, as well as developing wide variety of skills also play a crucial role in ensuring high value from outsourcing.

A clear understanding of outsourcing—its drivers and consequences—leads to a recognition of its dynamic nature. As a particular geography becomes a preferred outsourcing location, other regions try to culture the specific skills that made it a preferred location in the first place. Soon, newer regions emerge as preferred locations, and start exerting competitive pressures. Thus every region, organization or group which seek to reap outsourcing's benefits need to engage in a continual cycle of self-improvement.

21.6 GLOBAL SOFTWARE ENGINEER

Today, every successful software engineer needs to be a global software engineer. In addition to software engineering skills, global software engineers have to inculcate some traits that will enable them to practice their profession in diverse working environments. Being a global software engineer is much more a process of *becoming* than learning; one can only become through experience and a conscious effort. Given below are some of the characteristics of a global software engineer. This is by no means an exhaustive list.

- *Ability to adapt to diverse professional cultures*: It may come as a surprise that this list begins with 'culture' and not 'technology'. This is deliberate. Whether we are building a system for a client based in India, or the United States, or Philippines, sequence diagrams, inner classes, third normal forms and such technicalities stay the same. What varies are how a business situation is perceived and communicated, how people work, how they are expected to receive and respond to feedback etc. The global software engineer needs to quickly observe, absorb and imbibe these cultural differences. Sometimes it may be as simple and straightforward as when to report to work in the morning. At other times, it may involve subtle and sensitive issues like how best to criticize someone's plan or approach, without sounding offensive.

- *Ability to communicate at different levels*: For a global software development project, stakeholders are spread out not only geographically but also across languages, time zones and cultures. A global software engineer has to maintain simultaneous communication channels across a wide range of stakeholders, who may be speaking very different languages, even when all of them are speaking English! Successful communication is to be able to share ideas amongst people with different backgrounds and preparation. It is unfortunate that conventional wisdom often regards communication as essentially 'soft' and outside the purview of the so-called 'hard' disciplines of science and engineering. In software engineering, very often the soft stuff is the hard stuff.

- *Sensitivity towards social and political issues:* The primary objective of the global software engineer is to develop software systems. He or she is very likely to work with project stakeholders from different countries, often while located *at* different countries. It is important to develop a close working relationship with these stakeholders. However, it is equally important to remain sensitive to social and political issues that may be associated with global software development. Outsourcing is a case in point. It is not unusual now for the global software engineer to be working with someone whose job is under threat due to outsourcing. In such cases, it is essential to remain sensitive to these issues, and not act or speak in a way that will increase any grievance. Failure to do so will only result in lack of amicability that makes the project environment more difficult.

SUMMARY AND TAKE-AWAYS

The discussion of this chapter can be summarized as follows:

- Global software development represents a convergence of a number of technological, social, economic and political trends of our times.
- Due to some unique characteristics, software remains by far the only large-scale industrial artefact whose production and consumption can happen remotely.
- The Y2K problem has played a key role in the proliferation of global software development.
- Locational asynchrony and perceptual asynchrony are important factors that can affect distributed software development.
- Outsourcing is an important element of global software development. The benefits of outsourcing come from reduced cost, enhanced quality and greater flexibility. Outsourcing calls for continual improvement to stay competitive.

- Through experience and conscious effort, global software engineers need to develop traits that make them suitable for functioning in diverse work environments.

WHERE TO LOOK FOR MORE

- Information-Seeking in Global Software Teams—www.cs.cmu.edu/~sfussell/CHI2007/MilewskiSlides.pdf.
- Global software development and delivery: Trends and challenges—http://www.ibm.com/developerworks/rational/library/edge/08/jan08/fryer_gothe/index.html.

EXERCISES

Review Questions

Review Questions test your understanding of the key concepts presented in this chapter.

1. Global software development
   (a) is a recent phenomenon
   (b) has been around since the early era of software engineering
   (c) will be realized in future
   (d) none of the above
2. Global software development is influenced by which of the following trends?
   (a) Technological
   (b) Social
   (c) Political
   (d) All of the above
3. Global software development is *most* like which of the following?
   (a) An epic
   (b) A comic strip
   (c) A drama being played out
   (d) A short story
4. The Y2K problem is
   (a) a software bug
   (b) a problem related to global software development

(c) a problem with the date field of legacy code
(d) a problem related to connecting to remote servers

5. Locational asynchrony arises from factors like
   (a) physical location of servers
   (b) differences in geographies and time zones
   (c) differences in the ages of team members
   (d) all of the above
6. The benefits of outsourcing derive from the parameters of
   (a) cost
   (b) quality
   (c) flexibility
   (d) all of the above

Reflective Questions

Reflective Questions require you to think more deeply about some of the ideas and come up with your own interpretations and answers.

1. We have remarked 'two levels of information flow associated with software systems makes them so uniquely positioned to embrace a

global development paradigm'. Analyse this statement with two points in favour and two against.

2. If there was no Y2K problem, where do you think global software development would be today?

3. Perceptional asynchrony is not necessarily a problem associated with distributed teams and remote customers. Do you support this statement? Give reasons for your answer.

4. What do you think are the advantages India offers as an outsourcing location? What are the disadvantages?

5. 'In software engineering, very often the soft stuff is the hard stuff.' What is being referred to as 'soft' and 'hard' here? Do you agree with this statement?

Programming Examples

Programming Examples require you to analyse or write a program or a program segment to understand a specific problem.

1. While discussing the Y2K problem, we mentioned the implications of how the year is recorded in a date field. Write a program in Java that uses the language's in-built features for working with dates (such as the Calendar class) to calculate the number of days between two dates given by the user. Write the same program without using the language's in-built features for working with dates. Are both programs Y2K compliant? If not, what did you do to make your program Y2K compliant?

REFERENCES

Conway, M. (1968), 'How do committees invent?', *Datamation Journal*, pp. 28–31.

Datta, S. and van Engelen, R. (2008), 'An examination of the effects of offshore and outsourced development on the delegation of responsibilities to software components', In the *Second International Conference on Software Engineering Approaches for Offshore and Outsourced Development* held at ETH Zurich, Switzerland, July 3–4, 2008, Proceedings published in Springer LNBIP, Vol. 16.

Friedman, T.L. (2007), *The World Is Flat: A Brief History of the Twenty-first Century*, Picador.

Herbsleb, J.D. and Grinter, R.E. (1999a), 'Architectures, coordination, and distance: Conway's law and beyond', *IEEE Softw.*, 16(5):63–70.

Herbsleb, J.D. and Grinter, R.E. (1999b), 'Splitting the organization and integrating the code: Conway's law revisited', In *ICSE '99: Proceedings of the 21st international conference on Software engineering*, pp. 85–95, IEEE Computer Society Press, Los Alamitos, CA, USA.

Kornstadt, A. and Sauer, J. (2007), 'Mastering dual-shore development—the tools and materials approach adapted to agile offshoring', In Meyer, B. and Joseph, M., editors, *SEAFOOD*, volume 4716 of *Lecture Notes in Computer Science*, pp. 83–95, Springer.

Rodriguez, F., Geisser, M., Berkling, K., and Hildenbrand, T. (2007), 'Evaluating collaboration platforms for offshore software development scenarios', In Meyer, B. and Joseph, M. (Eds), *SEAFOOD*, volume 4716 of *Lecture Notes in Computer Science*, pp. 96–108, Springer.

Sengupta, B., Chandra, S., and Sinha, V. (2006), 'A research agenda for distributed software development', In *ICSE '06: Proceeding of the 28th international conference on Software engineering*, pp. 731–740, ACM, New York, USA.

Shami, N.S., Bos, N., Wright, Z., Hoch, S., Kuan, K.Y., Olson, J., and Olson, G. (2004), 'An experimental simulation of multi-site software development', In *CASCON '04: Proceedings of the 2004 conference of the Centre for Advanced Studies on Collaborative research*, pp. 255–266. IBM Press.

Open Source Software Development

Learning Objectives

Over the last decade, open source software has emerged as a useful alternative to proprietary software. In simplest terms, proprietary software is a commodity users have to pay for to use; whereas open source software is free of charge. However, open source software is beyond being merely free of charge; it represents a trend in the evolution of software development that is very important in the present and very consequential for the future. In this chapter, we will review some of the significant characteristics of open source software and its impact on the software engineering profession. Specifically, we will discuss:

- What is open source software?
- The evolution of open source software
- The range and limitations of open source software
- The implications of open source software on the software engineering profession

22.1 MOTIVATION

What has had the most profound impact on software engineering in the recent past? There is no universally acceptable answer to this question. However, it is quite likely that the open source paradigm of building software will be regarded by many as the factor of most consequence. It is interesting that the biggest impact came not from a new language, methodology or hardware breakthrough—indeed, all of these have had significant effects—but from a whole new way of building software. Recognizing the risk of over simplification, it may be said that the open source paradigm is about developers building a software system and sharing the source code (along with the executable) with users without charging any money. Users are allowed to use or modify the software as they desire. Even from this overly simplified picture, we are hit with the *why* question. If there is nothing called a free lunch in this world, why will software be given away for free?

This is a complex question, deals as it does with the issues of intellectual property rights as well as human ingenuity and motivation. Discussing the open source phenomenon in totality, even when limited to the software context, is beyond the scope of this book. In this chapter, we will only have the opportunity of outlining its major contours, and while doing so, address the question posed above. Readers will note there are numerous exhibits in this chapter; this is deliberately so. The open source odyssey is full of anecdotes, many of which reflect on basic software engineering credos. The exhibits try to capture some of these.

22.2 WHAT IS OPEN SOURCE SOFTWARE?

Although many aspiring software engineers have heard about open source software, and very many have used them—usually by downloading from the Web—confusion prevails as to what exactly is open source software. Does mere access to the source code and/or the permission to use and distribute a piece of software without paying any fees define open source software? The Open Source Initiative, an organization founded by Bruce Perens and Eric S. Raymond—one of the earlier pioneers of the open source software movement—defines open source in terms of the following key characteristics (http://www.opensource.org/docs/osd):

1. The rights of free distribution of the software without the need to pay a royalty or other fee.
2. Availability of the source code to the user.
3. Permission to modify and derive other software from the original software and distribute them under same license as the original software.
4. Maintenance of the integrity of the author's source code and assigning a different name or version number than the original work to any derived works.
5. Non-discrimination against persons or groups in the distribution of the software.
6. Non-discrimination in allowing the use of the software in any endeavor.
7. Applicability of the rights related to the software to all whom the software is redistributed.
8. Recognition that the software's license cannot be specific to a product.
9. Recognition that the software's license cannot restrict other software.
10. Recognition that the software's license has to be technology neutral.

In a nutshell, open source software is one which can be accessed and used without the need to pay any royalties. The source code may be accessed and modified; however if the modified code is redistributed, it must be identified as different from the original software.

22.3 EVOLUTION OF OPEN SOURCE SOFTWARE

Although open source software has gained popularity in the recent years, it is by no means a recent phenomenon. Very interestingly, in the 1960s even large corporations—subsequently seen to be the *bete noires* of the open source movement—gave away free software with their large-scale commercial computers. The hardware was being 'sold' and not the software, so the source code was available to be used and modified. Source code sharing was also common in the academic community. Stallman reflects that when he joined the Massachusetts Institute of Technology (MIT) Artificial Intelligence Lab in 1971, he became a part of the software sharing community that had existed ever since computers did, 'just as sharing of recipes is as old as cooking' [Muffatto 2006]. However, the situation changed by the mid 1970s as companies recognized the commercial potential of software. Earlier software was intimately tied to the hardware it ran on. But with gradual de-coupling of software and hardware—a piece of software could run on diverse hardware platforms and vice versa—software began to be *sold* as a proprietary product that could not be redistributed or modified. From then to now presents a fascinating story of the evolution of open source. From Muffatto's account of the history of open source [Muffatto 2006], the following phases can be recognized.

22.3.1 From Free to Proprietary

In the 1960s and 70s, the use of software was wholly confined to the academic and research community. Only the very 'geeky' took an active interest in software, and true to the knowledge sharing traditions of the academia, software was freely distributed amongst the members of the small group who were interested in it. The communication between software researchers were supported by the ARPANET, a predecessor to the Internet. This era was marked by the lack of standardization and software was hardware specific. Compatibility issues prevented one software system being run on different machines. The most significant attempt at breaking this bottleneck resulted in the development of the C programming language in 1972 by Dennis Ritchie. C was a multi-platform language and could run independent of the underlying machine. It was quickly adopted as a language of choice and still remains popular due to its great versatility. The Unix operating system was developed between 1969 to 1974 by Ken Thompson and his team at Bell Labs, using C. Unix became the first machine independent operating system. The source code of Unix was freely available for modification and distribution and in 1979 the first version of BSDUnix was released. BSDUnix was developed at the University of California-Berkley in parallel with Bell Lab versions of Unix; BSD stands for the Berkley Software Distribution. The release of BSDUnix marks the first 'open' license giving rights of use, modification and redistribution of source code. In the mid 1980s, AT&T decided to 'commercialize' the production, distribution and sales of Unix. From then onwards, Unix became a proprietary software and was not available for free. The members of the community who had originally participated in the development of Unix were dismayed and angered.

Few other developments also marked the pioneering era. Successful collaboration needs a cheap, fast and reliable mode of communication. In 1973, Vinton Cerf and Bob Kahn proposed the TCP/IP (Transmission Control Protocol/Internet Protocol—see Chapter 19 for related discussion), which facilitated information flow amongst connected computers and laid the framework for the future Internet. In 1979, the Sendmail program was developed by Eric Allman, a student at the University of California, Berkley. It was distributed freely and became the virtual standard for electronic mails over the ARPANET (emails were first known as 'electronic mails' and then 'e-mails'). In the first half of the 1980s, personal computers or PCs became widespread. A large majority of PC users did not have a programming background and looked to use the PC as an utility in their business or personal needs. They were willing to pay a price for the software that ran their PCs. The age of proprietary software had arrived.

22.3.2 Open Source Response

As the commercial potential of software became clear, large and small companies started to hire the pioneers of the early software systems from the academia into corporate laboratories. The intellectual property rights of software begun to be zealously guarded. Any unauthorized use of software was taken as an act copyright infringement. In 1984, Stallman started the GNU project (G-N-U is a recursive acronym—an abbreviation that refers to itself —for 'GNU's not Unix') to develop an open-source alternative for Unix. The Free software foundation (FSF) was founded in 1985 to support GNU (see Exhibit 22.1). In 1988, the first version of the General public license (GPL), to protect the freedom to copy, distribute and change a product was mooted. GPL had a significant role in rallying together the open source community by establishing a framework distinct from the commercial model, as well as protecting the rights of free software. However, there was still the need for a viable open source alternative to Unix, both as a practical necessity, as well as an affirmation that large-scale open source software could work.

This void was filled by Linus Torvalds' Linux operating system, which started with the modest goal of modifying the Minix clone of Unix in 1991. Linux was not developed by Torvalds alone. In fact, he later remarked that had he known about the enormity of the task of building an operating system, he may not have started. Linux was and is being built by an unprecedented scale of collaboration, by a community of developers widely dispersed in geography and background, but united in the common cause of open source software. The first official version of Linux was released in 1994. As a testimony to how much Linux caught the imagination of the software community, its number of lines of code increased from 10,000 to 1.5 million by 1998 and had 12 million users across 120 countries. Linux had shown to the world what was believed for several decades by a small group of free spirited individuals; open source works!

In the next phase of the evolution of open source, we find the development of a unique business model. In 1994, Bob Young and Marc Ewing founded a company called Red Hat. Red Hat's objective was to make Linux more user-friendly and add utility programs that could be used with the operating system. Red Hat operates as a so-called *pure player*, a company whose business model aims at deploying an open source product and related services. This is a very significant idea in the software business; a company making profit not exclusively by selling a software product, but by helping users get the maximum benefit from its use. Following Red Hat, many other companies have emerged who do business based on this model, many of them built around Linux distributions. Other than Linux, another pioneering large-scale software product was released in 1995—the Apache Web server, which soon became the most popular HTTP server on the Internet.

22.3.3 Spread of the Mantra

Till the point we have traced the history of open source so far, 'free software' was how open source was known and referred to. Well, free is a very loaded word. The sense that it was being perceived in many circles was free as in 'free of charge', thus indicating there could not be any business interests associated with it. However, this was far from the connotation the Free Software Foundation had in mind. With the famous slogan, 'free speech, not free beer', they tried to drive home the point that the most important attribute of the so-called free software was intellectual freedom (see Exhibit 22.1). To take the open source movement to the next level, it was imperative to get companies interested in it. In the 1997 Linux Congress, Raymond presented the seminal paper, *Cathedral and the Bazaar*, where he analysed the commercial software development model with the open source one, through the respective metaphors of the cathedral and the bazaar (See Exhibit 22.2). Raymond

Exhibit 22.1   Free as in Freedom

The Free Software Foundation (FSF) was founded in 1985 to support the GNU project, and take forward the message of open source software to a larger community. However, the founders of GNU and FSF, most notably Richard Stallman, had to soon clarify that the *free* in 'free software' meant free as in freedom rather than free as in gratis. Stallman's collection of essays, *Free Software, Free Society* argues how the philosophy of open source software extends the traditions of freedom of expression. In Stallman's words, 'The term free software has nothing to do with price. It is about freedom.' [Muffatto 2006].

A major step in establishing the free software philosophy was the publication of the first version of the General Public License (GPL) in 1988. GPL introduces the concept of *copyleft*, opposing everything that *copyright* stands for. Copyleft guarantees the prerogative to copy, distribute or change a product. The preamble to the GPL states '... the GNU General Public License is intended to guarantee your freedom to share and change free software—to make sure the software is free for all its users' [Muffatto 2006].

also proposed the term 'open source' to remove many of the ambiguities associated with the meaning of 'free'.

22.3.4 Open Source as an Institution

In 1998, two events proved decisive in the history of open source. The first was the adoption of the 'open source', name and the founding of the Open source Initiative. The second was the announcement by Netscape that the source code of their Web browsing software Netscape Navigator was being made public. This was the first time a mainstream software company had adopted open source.

The next ten years after Netscape's announcement has on one hand seen a large proliferation of open source, both in terms of the size of user and developer pools, as well as support from many large corporations, government and non-government agencies. However, there are many corporations—some very large and influential ones— which see open source as a fundamental threat to their business model, question the very validity of the GPL, and have embroiled the open source community in countless law suits.

However, the open source movement thrives and will continue to play a significant role in the world of software development in the coming years.

22.4 RANGE AND LIMITATIONS OF OPEN SOURCE SOFTWARE

Today, if we can think of something we want to do with software, chances are very high we will find some open source software that does it for us. One only

Exhibit 22.2	**Linus' Law**

There is a popular adage, 'Too many cooks, spoil the broth'. Any large scale open source system—characterized best by Linux—has hundreds, if not thousands, or tens of thousands of developers. Not only are there too many cooks, the cooks range across a variety of skills, backgrounds, cultures and locations; and there is hardly any hierarchical control. How then does the broth remain unspoiled?

This question was addressed with great insight in 1997, when Eric S. Raymond presented the paper 'The Cathedral and The Bazaar' [Raymond 2001] as mentioned earlier. This is a must read for any one trying to understand the underpinnings of open source. Raymond compared the model of the classical and commercial software development with the new model of dispersed development and free distribution of code through the metaphors of the cathedral and the bazaar. According to Raymond, the cathedral stands for a structure built through immense planning and forethought; whereas the bazaar represents the spontaneous, pulsating, and emergent crucible of open source development. Among much of the enduring wisdom the paper resonates with, Linus' Law stands out—'Given a large enough beta-tester and co-developer base, almost every problem will be characterized quickly and the fix obvious to someone' or 'Given enough eyeballs, all bugs are shallow'. Linus' Law seems to fly in the face of Brooks' Law, which mandates that adding more developers will only slow down a project further. Raymond has argued there are indeed exceptions to Brooks' Law citing Linux as the most notable exception [Raymond 2001]. Evidently a necessary condition for the exceptions is the availability of a fast and reliable communication medium. Thus it is no surprise that the coming of age of open source has coincided with the popularization of the Internet and then the Web.

needs to visit websites such as http://sourceforge.net/ to know the range of open source systems. There are systems varying widely in utility and technology, and of course, quality. Perhaps, by far the most widely and critically used open source software is the Linux operating system. The sheer range of open source systems available and in use today, establishes the power and versatility of the open source paradigm. A small group of dedicated developers can create a large and complex system; a large group of dedicated developers can surely create a very large and very complex system, as is proved by the ubiquity and popularity of Linux. There is however a common concern—mainly amongst those who question the very credibility of anything worthwhile being free in this world—on the quality and reliability of open source systems.

As we discussed at length in Chapter 17, concepts of software reliability and quality can be subtle and notoriously difficult to quantify. There is no empirical evidence to suggest open source software is less reliable or of poorer quality than proprietary software that is sold against a fee. In fact, it has been widely established that the more *eyeballs* impinge on a software system, the better it becomes, due to continuous bug finding and fixing (See Exhibit 22.2). The open source world has no dearth of eyeballs, and this is precisely why many large-scale proprietary software systems are now launched via series of beta releases. These seek to leverage the potentially limitless eyeballs of a band of early users to detect and report bugs. Open source software does not come with guarantees or warranties; they are usually useless either way. Users want to be able to use a software system with satisfaction, not be cushioned by legal safeguards. Even for the proprietary software that we buy, what do we do when something does not work as expected? Most frequently, we post a query on an online forum, and more often than not get a helpful response. In any case, faith in unknown developers' skill and judgement supports our confidence in a piece of software. The open source paradigm operates on this faith; and not on legally binding promises of support when things go wrong. The engine behind the success of open source is the hacker's pride (see Exhibit 22.3).

22.5 OPENS SOURCE SOFTWARE AND THE PROFESSIONAL SOFTWARE ENGINEER

In the preceding sections, we have outlined various facets of the open source software phenomenon, from its evolution to its range and utility. However, the question of significant import to professional software engineers is what the open source paradigm means for the future of their profession. Will the fact that open source software can be used without paying any money, adversely affect the opportunity for software engineers to earn their living? This is a question of both philosophical and practical implications.

What would happen if there was a band of volunteers who built bridges for free, working in their spare time and not charging for their labours? In such a situation, professional bridge builders may soon be out of work; those needing the service of a bridge would expectedly wait for it to be built for free.

What drives our band of hypothetical bridge builders? Surely, the creative pleasure from building beautiful bridges over the most difficult of chasms. This kind of motivation dreads drudgery and revels in newer challenges. But it is unlikely that all the bridges that need to be built needs novelty of the highest order. In fact, as with other engineering artefacts, the economies of scale—both in terms of materials and human expertise—for bridge building can be best leveraged with repetitive Implementation of a standardized Design. Once the Design has crystallized following initial innovation, there is hardly scope for radical ingenuity; although there may be place for incremental improvement. Routine Implementation is not something that drives our bridge building aficionados. Thus there is likely to be little interest in building all those standard bridges the world needs, out of a solely creative urge. And this is where bridge builders who build bridges for a living—irrespective of whether each individual bridge has scope for great originality—operate. And the bridge building profession survives.

Bridge building in many respects is similar to software building and the band of volunteers for software development come from the open source community. Open source developers revel in creating novel software systems in fundamentally new ways, and do a great job of it. But for routine applications like a billing system for a utility company or an online banking system, it would be difficult if not impossible, to find open source developers to build these systems for the fun of it. So we need professional software engineers who will build, enhance and maintain them for a living. What we have discussed so far connects with a larger philosophical question of what human beings are willing to do for pleasure versus what they are willing to do for profit. Both can be very strong motivations, and each has its place and relevance. It is unlikely one will ever come to replace the other. And consequently, the software engineering profession is safe!

However, there is another, far deeper reason why the world will continue to need professional software engineers. It pertains to the basic way how a bridge differs from a software product. We have underscored the unique characteristics of software as an engineering artefact at several places throughout this book. Users of software as compared to those of other conventional engineering products need to bridge a significant cognitive gap. We know, almost instinctively, what to do with a bridge (cross it), or a house (live in it). The use of software for its best benefit is not that deeply and widely understood, or that easily and instinctively. The task of professional software engineers is not merely to build and maintain software system; it is also essentially to help users address their business problems using the software. For the professional software engineer, a software system, in spite of all the technical ingenuity is finally means to serve users' ends. So, understanding the business problem and tailoring a software solution to best fit the problem is the most important point in a software engineer's charter. While the open source movement produces exciting and robust software products, the professional software engineer needs to utilize their potential in the proper business context. The open source paradigm is not in conflict with professional software engineering; it is complementary to it.

SUMMARY AND TAKE-AWAYS

The open source paradigm of software development is arguably one of the factors that have impacted software engineering the most in recent years. In this chapter, we have briefly discussed the salient features of open source software development. Our discussion can be summarized as:

- Very simply, open source software development involves granting the user rights to use and modify a software system, without paying any fees.
- The Open Source Initiative has defined open source software in terms of 10 key characteristics.
- The release of BSDUnix in 1979 marked the marks the first 'open' license giving rights of use, modification and redistribution of source code.
- The arrival of the personal computer in the early 1980s led to an explosion in the number of computer users, many of whom were willing to pay for the software they used.
- In 1988, the first version of the General Public License (GPL) which protects the freedom to copy, distribute and change a product was mooted.
- The business model based on the open source paradigm centres around the concept of a company making profit not exclusively selling a software product, but by helping users get the maximum benefit from its use.
- There is no empirical evidence to support open source software is of lesser quality or reliability than proprietary software.
- The open source paradigm is not in conflict with professional software engineering; it is complementary to it.

WHERE TO LOOK FOR MORE

It is no surprise that open source resources are widely available on the Web. Some of the notable websites are:

- Free Software Foundation—http://www.fsf.org/
- The Linux Foundation—http://www.linuxfoundation.org
- Free Software Free Society: selected essays of Richard M. Stallman—http://shop.fsf.org/product/free-software-free-society/
- The Cathedral and the Bazaar—http://www.catb.org/~esr/writings/cathedral-bazaar/

EXERCISES

Review Questions

Review Questions test your understanding of the key concepts presented in this chapter.

1. Which of the following does not directly relate to the concerns of open source software development?
   - (a) Software Design
   - (b) Motivation
   - (c) Ingenuity
   - (d) Hardware Design

2. The Open Source Initiative
   - (a) has developed the best open source system
   - (b) is an accreditation body for open source developers
   - (c) has defined open source in terms of some key characteristics
   - (d) sells open source software.

3. Copyleft
   - (a) is a political manifesto
   - (b) a credo opposing copyright
   - (c) an open source software product
   - (d) none of the above

4. The idea of open source software
   - (a) started around the 1990s
   - (b) has not gained popularity as yet
   - (c) was around since the 1960s

   - (d) none of the above

5. Which of the following was the first attempt at making software independent of the underlying hardware?
   - (a) Development of C
   - (b) Development of FORTRAN
   - (c) Development of Java
   - (d) Development of the Internet

6. Which of the following can be said to have indirectly, but strongly influenced the practical aspects of the open source paradigm?
   - (a) Object-oriented programming
   - (b) Procedural programming
   - (c) HTTP
   - (d) TCP/IP

7. The development of Linux started in 1991 with the aim of
   - (a) creating the largest open source system in the world
   - (b) creating a new software development paradigm
   - (c) modifying the Minix clone of Unix
   - (d) all of the above

8. The open source paradigm shifts the emphasis from software as a product to
   - (a) software as an utility
   - (b) software as a toy

(c) software as a paradigm

(d) software as a service

9. Other than Linux which of the following is another pioneering open source product released in the 1990s

    (a) Apache Web server

    (b) C

    (c) Java

    (d) Python

10. Linus' Law—as codified by Raymond,

    (a) contradicts Brooks' Law

    (b) illustrates an exception to Brooks' Law

    (c) supports Brooks' Law

    (d) is unrelated to Brooks' Law

11. In the open source community, 'hacker' means

    (a) criminal

    (b) a creative person inclined to problem solving

    (c) a whiz-kid

    (d) a programmer

Reflective Questions

Reflective Questions require you to think more deeply about some of the ideas and come up with your own interpretations and answers.

1. We have indicated that the open source phenomenon is not strictly confined to the world of software development. Where else do you see a similar paradigm? Do you feel something as far removed from software development, as say movie production, can function on open source principles?

2. The YouTube Symphony Orchestra is a unique effort at collaborative music production. Visit the website—http://www.youtube.com/symphony—to know more about the idea and its execution. Is there an analogy with open source software development?

3. We have discussed how the open source paradigm has evolved for software. Do you think a similar paradigm can work for hardware? Justify your answer.

REFERENCES

Graham, P. (2004), *Hackers and Painters: Big Ideas from the Computer Age*, OReilly Media, Inc.

Muffatto, M. (2006), *Open Source: A Multidisciplinary Approach*, Imperial College Press.

Raymond, E.S. (2001), *The Cathedral and the Bazaar: Musings on Linux and Open Source by an Accidental Revolutionary*, O'Reilly.

Future of Software Development

23.1 MOTIVATION

For a subject like software engineering, today's students will be tomorrow's practitioners. Thus, a textbook needs to prepare students for the future. But the only way we can see the future is through our lens of the past. So, predictions have an implicit assumption: What can happen in the future may be projected from past experiences. This is often a very questionable assumption, especially for a field as mercurial as software engineering. Here predictions have an uncanny knack of going wrong, even when backed by the best wisdom and experience. Those who predict do so at significant risk to their reputation!

Still, we need to predict if we are to learn a subject such that our learning stays current in the foreseeable future, and hopefully beyond. In this chapter, we will look ahead at the various ways software engineering *may* go in the short and medium term. Instead of arbitrary crystal ball gazing, we will reflect on where the topics discussed in each part of this book may lead us next. Then, in a final section, we will outline a software engineer's survival toolkit; a minimal set of tricks that are likely to give an individual software engineer a better chance of keeping obsolescence at bay during his or her career.

23.2 EVOLVING TRENDS IN SOFTWARE DEVELOPMENT

The material of this book is organized into five parts. In the following sections, we discuss how the material of each part can be expected to evolve in the coming years.

23.2.1 Understanding of Software Engineering

The perception of software engineering as a pursuit with a distinct name [Bauer et al. 1968] and different from computer programming is just about entering its fifth decade. Software engineering as a profession for a large group of people is still much younger; probably starting its third decade. As a discipline matures, it's very understanding amongst practitioners, academics and the general public changes. For example, in the beginning medicine had a quasi-religious aura; curing of human afflictions was considered a divine prerogative. Practice of medicine meandered for ages through quacks and faith healers and finally established itself as a modern science. There is still noticeable quackery and faith healing in software engineering—not that they are not effective at times—all of which are contributing to the maturity of the discipline.

Currently, a large majority of practicing software engineers have not studied the subject in a software engineering degree program. While practicing software engineers are usually good at 'learning on the job', there are clear advantages of being introduced to a subject first in the classroom. Dedicated software engineering programs have just begun to proliferate across university curricula in the world. In the coming years, we will have many more software engineers not merely by practice, but also by training. This is likely to have a significant impact on the general awareness of software engineers about the context of their profession. From my personal experience, I can say that today there are many software engineers doing their jobs with reasonable finesse who have never heard of Brooks' Law. Would we trust an electrical engineer who has no inkling of Ohm's Law?

The great demand for software engineers in the past decade and half have acted like vortex for sucking in people with very different backgrounds and preparations into the profession. While they have done a competent job within limitations, the new age of software engineering will come when those who have invested their university education in software engineering will hit the work force. With increased formalization of software education, licensing and accreditation Requirements for software engineers will become more widespread [Kruchten 2008]. As it exists in many other professions, software engineers may be expected to demonstrate their continuing grasp of the subject through periodic evaluations.

The understanding of software engineering, as well as the appreciation of its power and challenges is likely to change profoundly with a more organized approach to learning the subject in the coming years.

23.2.2 Planning and Managing Software Development

Till now, planning and managing software development has been largely an empirical pursuit. People start out with what they think would work, rejoice and refine it when it does; despair and look for something else when it does not. This does not reflect on something particularly clumsy about software engineering; every engineering rests heavily on empiricism. Even when there are laws of physics to swear by, we may have to use unspeakable hacks and shortcuts. Any one who has tried to experimentally verify Ohm's Law in a high school physics laboratory class knows how difficult it is to reproduce its pristine certainty in practice. However, laws do provide a convenient starting point in other engineerings; software engineering is yet to have that privilege.

In terms of development methodologies, it is likely agile methods will be more widely adopted in mainstream development. However, the effectiveness of agile methods for very large projects is still under scrutiny. Iterative and incremental development is likely to remain very relevant, with the Unified Software Development Process being refined further by its proponents as well as practitioners. But often the process that guided the choice of a methodology has a more immediate impact. Process maturity, which involves the discernment and flexibility to choose a methodology that is best fit for the particular problem at hand, will hopefully be enhanced in the coming years. More rigor in time and effort estimation is something that is long overdue; whether that will happen depends to a large extent on the systematic collection of software project data. Initiatives along this direction have already begun—https://www.isbsg.org—and they need to be taken forward. With the availability of statistically significant amounts of data, data mining techniques [CSC-NCSU 2009] can effectively address many of software engineerings most pressing problems. Software metrics have gone a long way in introducing valuable heuristics; to take them to the next level, deeper penetration of measurement theory ideas will need to come. The holy grail of software automation, programs writing programs, will probably not be around for large-scale software development in the next few years. However, if software engineering has to attain the level of consistency and precision that is commonplace in other engineerings, large-scale automation has to come sooner than later.

23.2.3 Designing and Building Software Systems

Analysing the problem domain for a software solution and designing the solution remain by far the most *reflective* of all software development activities. In the best interest of software engineering, it should remain that way for many years hence. Analysis and Design of software systems is best understood in the context of systems Analysis and Design [Weinberg 1975], [Weinberg 1988], [Weinberg and Weinberg 1988] and a dedicated software engineering curriculum is likely to introduce these topics together. A deeper understanding of software architecture will be facilitated by the discernment of newer architectural patterns and the appreciation that software is usually an important component in a larger system of stakeholder concerns [Booch, 2009]. There is always the opportunity for a radical new paradigm of software development to appear, but object orientation, and aspect orientation, individually, or when combined still have considerable untapped potential. The most durable paradigms are those that arise out of existing ones, and are more evolutionary than revolutionary in nature. Java has by far been the biggest new entrant in the programming language landscape in the past two decades. In the mid- 1990s, Java was rapidly gaining ground with the burgeoning Web, and posed as a serious contender to the existing clout of C and C++. At that time predictions flew thick and fast about the likely successor to Java. It seemed there would appear a smart, new language every couple of years. That has certainly not happened. Next generation languages will eventually come to address the shortcomings of the current ones or to realize new programming paradigms.

23.2.4 Testing, Maintenance, and Modifications

Testing, maintaining and continuously modifying a software system in response to user demands and environmental changes remains by far the most *messy* parts of software development. No one wants to do it, but every one has to do it. Change is endemic and every system that needs to survive and function must devise its own effective mechanism to react to it. Recently, there have been initial explorations in emulating such mechanisms of living organisms in software systems [Gabriel and Goldman 2006]. This is an interesting area and one that may fundamentally change the way software systems are built. Software has to be increasingly designed and developed based on the recognition that changing Requirements is a fact of life for software systems.

23.2.5 What will be the Next Big Thing?

Very often, the 'next big thing' attracts much more attention than a set of incremental changes that may have much deeper impact in the long run. This attention is unfortunate. As we have highlighted throughout this book, game-changing developments seldom appear as the 'big thing' when they first come in. Given all these caveats, if one has to identify the next big thing, chances are very high it will be *Web 2.0* or the *Semantic Web*. The next generation Web will cease being a mere repository or purveyor of information. It will be more aware of the 'meaning' of the information it houses and transmits, and be able to act intelligently on that awareness. It is a tall order and the work has only just begun. But it promises a deep transformation of the way we see and use not only the

Web, but information in general; and along with it a profound change in the way software systems are designed and built.

Exhibit 23.1   The Rate of Knowledge Growth

For those who earn their living as software engineers, caring about the future of software engineering centres around the fear of being swept aside by the rush of new technology. How long will all the knowledge and skills we painstakingly acquire, stay current? There is no way to know for sure, but perhaps some educated guesses may be made.

Hamming—whose pioneering work on error correcting codes remains relevant even today—makes an interesting case for the rate of knowledge growth and supports it with quick and clever 'back of the envelop' calculations [Hamming 1997]. Hamming asserts that the body of knowledge in any given field doubles in about 17 years. He supports his statement by noting that libraries have to double their stock of books on a subject in approximately 17 years to stay current. Also, he notes that during his tenure at the Bell Telephone Laboratories—a great research establishment in his time—the number of employees doubled about every 17 years. Hamming also states that the number of scientists has grown *exponentially* over time and 90% of scientists who ever lived are now alive.

Can one of these statements be verified by assuming the validity of another? Hamming assumes, the number of scientists at any time t is given by,

$$y(t) = ae^{bt}$$

and the knowledge produced by those scientists alive has the constant of proportionality k to the number of scientists at a given time. As knowledge doubles every 17 years, we have,

$$\frac{1}{2} = \frac{\int_{-\infty}^{t-17} kae^{bt} dt}{\int_{-\infty}^{t} kae^{bt} dt}$$

From this relationship, we can solve for b, as the other parameters cancel themselves out. Next, Hamming makes another assumption: The average working life of a scientist is 55 years, that is the number of years the individual is active in his or her field. Then the ratio of the number of scientists living now to the all those scientists who ever lived is given by

$$\frac{\int_{t-55}^{t} ae^{bt} dt}{\int_{-\infty}^{t} ae^{bt} dt} = \frac{e^{bt} - e^{\{bt - b(55)\}}}{e^{bt}} = 1 - e^{-55b}$$

In the above equation, if we substitue the value of b solved for earlier, we have $1 - e^{-55b} = 0.894$, which is close to 90%.

So assuming the knowledge in a field doubles every 17 years and an average scientist is active for 55 years, it can be derived from an exponential model describing the growth of scientists with time, that indeed 90% of scientists who ever lived are living now.

With a little bit of stretch, Hamming's argument can be extended from scientists to software engineers. So, we have to watch out for obsolescence (knowledge doubling at least twice in the span of a career!), while at the same time remaining aware of increased competition (a very large percent of those who practiced our profession are practicing it now!).

This is precisely why software engineers of today need to be cognizant of what may come tomorrow. And that is the whole point of this chapter.

23.3 SOFTWARE ENGINEER'S SURVIVAL TOOLKIT

In the classic paper, *No Silver Bullet: Essence and Accidents of Software Engineering*, Brooks' posits how a single development, either in technology or management, will not be able to bring about order of magnitude improvements within a decade in productivity, reliability or simplicity of software development [Brooks 1987]. The 'no silver bullet' rubric is often invoked in a negative sense, to question the validity or promise of any new approach. However, the argument can be turned on its head, to cast a more encouraging light. If there will not be anything new or hot enough to change rules of the software development suddenly, is there a set of skills that guarantees a software engineer will not suddenly become obsolete in the course of his or her career, which typically lasts several decades? This is an interesting question, and one of much importance to the professional software engineer. With so much change going around in our field of expertise, and change at an even accelerated rate about to come, how do we stay current and relevant? In the /, natural selection favours those species with characteristics most amenable to winning the struggle for existence. What are the corresponding traits for software engineers to win the battle against obsolescence?

From my experience, I have identified the following skills, which I will call the *software engineer's survival toolkit*. This is by no means an exhaustive list, but being aware of these skills will at least ensure a software engineer is able to stay tuned to the progression of his or her profession.

23.3.1 Virtuosity with at least One Programming Language

Although programming is not software engineering, programming does play a central role in software engineering. Beautiful, resilient and flexible software systems will not be possible without beautiful, resilient, and flexible programs. Every programming language has similar basic constructs—for example, ways of decision-based branching, and repetitive execution through looping. Thus close familiarity with one programming language facilitates the learning and using of other programming languages. A software engineer's survival toolkit includes virtuosity with at least one programming language. The language needs to be known intimately—its most powerful features as well as strongest limitations—to the level such that, given a problem, the solution can be *thought out* in that language. This degree of skill does not grow overnight; it has to be cultured through programming experience in real life projects with paying customer to satisfy. I am very chary of software engineers who are chary of programming. Those who are not ready to get their hands dirty with code will forever live in the fear of obsolescence.

23.3.2 In-depth Experience with at least One Development Methodology

Just being able to program, even with great virtuosity is not enough in the scramble to stay current. Beautiful, resilient and flexible software systems may have significant scientific and aesthetic value. But in commercial software development, if they cannot satisfy stakeholder interests, they do not justify the time and money invested in building them. Development methodologies delineate the context of building a software system; encompassing technology, management, sociology, and very often, politics. How these factors correlate and co-influence one another is the key theme of any worthwhile methodology. A software engineer needs to know the steps and nuances of at least one methodology in great depth. Additionally he or she must have seen the methodology in action from the perspective of a stakeholder in a real life project. This experience is essential in knowing where the methodology works and where it does not. Not all aspects of a methodology are equally relevant to all parts of a system or project; the skill lies in being able to identify the most important aspects. An effective methodology can act like a strong magnetic field in aligning a diverse set of individuals to a common purpose. Without the experience of applying such a methodology, a software engineer can hardly stay current.

23.3.3 Detailed Understanding of at least One Application Domain

The systems software engineers build must serve a particular application domain, addressing one or a set of related problems. With the commissioning of the system, customers and users are looking to do something they could not do earlier, or do something they could already do faster and/or cheaper. For example, a new Web-based system for a credit card company may be offering the user facilities to view their statements online, and additionally, contend a charge showing up on their statement. To build this system, software engineers will not only need to be proficient in a programming language and a methodology, they will also need to understand the credit card business. Each application domain has its own peculiarities and understanding of one does not suffice understanding another. However, detailed knowledge of at least one application domain, gained through developing a software system for a business problem within that domain, reveals an important insight. The most difficult problems in software engineering do not come from the complexity of algorithms or programming language constructs; nor from the inadequacy of methodologies. They arise at the intersection of technology and business, largely from the fact that 'business logic' has no obligation to be logical [Fowler 2003]. Detailed understanding of even one application domain makes us aware of this reality. This awareness is vital in asking the right questions and nosing out the most consequential issues when confronted with a new problem in a new domain. The problem and the domain may not have been encountered before. But having faced and resolved many intersectional issues in another domain prepares the software engineer better. Every day, software engineering is penetrating newer domains; this trend is only likely to accelerate in the future. And while it is impossible for an individual software engineer to know all domains in equal depth, it is imperative for him or her to know one in detail, to be able to stay current.

23.3.4 Sense of History

A large majority of those who take up science or engineering have a strong loathing of history. This is usually a hangover from high school history lessons, intense as they are with long lists of dates and events, and names of obscure kings. But history defines to a large extent what we are today and what we will be tomorrow. Thus, scientists and engineers have a lot to gain professionally from a sense of history. This is even more relevant for software engineering; its past, present and future are closely meshed across just a few decades. Software engineering's history is also very well-documented in the public domain (for example—http://www.computerhistory.org/). But hardly if ever, software engineers show the inclination to learn how their profession evolved. This does not call for burning the midnight oil to read big, fat history books; few Web searches during the lunch hour can give valuable information as well as insights. Culturing a sense of history does not need memorizing dates and events; it comes from a familiarity with the major trends that have shaped a discipline. Although 'history repeats itself' has become rather cliched with overuse, every arena of human endeavor has some repeating patterns over time. History is a good mirror of the past. Sometimes it is a good predictor of the future, sometimes it is not. Either way, it is a great teacher. To be prepared for the future in as fast-evolving an area as software engineering, a sense of history is must.

SUMMARY AND TAKE-AWAYS

In this chapter, we reflected on some of the trends that may have significant impact on software engineering in the foreseeable future. The discussion can be summarized as:

- Any prediction has the implicit assumption that future events can be projected from past experiences.
- Though without guarantees of accuracy, we need to make some predictions about the future, as today's students will be tomorrow's practitioners.

- As more and more universities offer dedicated software engineering programs, the understanding of software engineering amongst those who study and practice it will change fundamentally.
- The availability of statistically significant amounts of data will facilitate the application of data mining techniques to software engineering.
- Increased focus on studying software Analysis and Design in the context of systems Analysis and Design will foster a more holistic view of software engineering.
- Software has to be increasingly designed and developed based on the recognition that changing Requirements is a fact of life for software systems.
- Web 2.0 or the Semantic Web has the potential to be the biggest factor of impact on how software systems are designed and built in the coming years.
- The key components of a software engineer's survival toolkit are virtuosity with one programming language, in-depth experience with one development methodology, detailed understanding of one application domain, and a sense of history.

WHERE TO LOOK FOR MORE

The Web is full of prognostications about the future of software engineering, some grim, but most wildly enthusiastic. A series of articles that offers a balanced view and much insight by Watts S. Humphrey can be found at http://www.sei.cmu.edu/news-at-sei/columns/watts_new/2002/1q02/watts-new-1q02.htm.

EXERCISES

Review Questions

Review Questions test your understanding of the key concepts presented in this chapter.

1. Software engineering as a field with a distinct name is just about
   (a) 10 years old
   (b) 20 years old
   (c) 30 years old
   (d) 50 years old
2. The change in the perception of software engineering amongst its practitioners is likely to happen due to
   (a) more people becoming software engineers

   (b) licensing of software engineers
   (c) proliferation of dedicated software engineering programs in universities
   (d) all of the above
3. With the availability of statistically significant data on software projects, which of the following will be facilitated
   (a) Emergence of new methodologies
   (b) Emergence of new programming languages
   (c) Better fault prediction
   (d) Enhanced use of data mining techniques
4. Software Analysis and Design are best studied in the context of

(a) hardware Analysis and Design

(b) computer Analysis and Design

(c) systems Analysis and Design

(d) none of the above

5. In recent studies, which of the following attributes of living organisms is being sought to be emulated in software systems?

(a) Mechanisms of reacting to change

(b) Reproducibility

(c) Embryonic development

(d) All of the above

6. The main point in Brooks' classic *No Silver Bullet* paper is

(a) software is soft

(b) software development can be improved by automation

(c) high-level programming languages can speed up development to a very large extent

(d) none of the above

Reflective Questions

Reflective Questions require you to think more deeply about some of the ideas and come up with your own interpretations and answers.

1. We have mentioned the journey of medicine from the initial days to its establishment as a modern science. Assuming software engineering is charting out a similar path; where are we now? How long do you think it will take for the so-called 'quackery and faith healing in software engineering' to disappear?

2. Evaluate the following statement with three points for and against the claim: While practicing software engineers are usually good at 'learning on the job', there are clear advantages of being introduced to a subject first in the classroom.

3. Why do you think a new programming language has not gained wide popularity in the last decade since Java?

4. Increasingly global software development is becoming common, where the stakeholders of a software system are geographically separated by large distances. In this context, do you think a 'sense of geography' should be added to the software engineer's survival toolkit? What factors should such a sense of geography address?

5. With reference to Exhibit 23.1 it may be noted that such back-of-the-envelope calculations give quick insights, but rest heavily on assumptions, some of which may bias the result. Can you detect such unstated assumption(s) in this case? What additional assumptions would you make in this case?

6. The figure of 17 years for the doubling of knowledge was arrived at from Hamming's own experience and observation. What is *your* personalized rate of knowledge growth?

Hint: You may think of counting the number of text books you had to study each year, from middle school onwards to the university level.

Numerical Problems

Numerical Problems require you to do quick 'back-of-the-envelope' kind of calculations to arrive at answers to some simple but insightful problems.

1. With reference to Exhibit 23.1, let the equation relating number of scientists to time be taken to be linear, rather than exponential: $y(t) = at+b$. Assuming 90% of all scientists who lived are now living and the working life of a scientist is 55 years, what is the number of years in which

knowledge would double with this model? If you take the your personalized rate of knowledge growth calculated in the last of the Reflective Questions, and keep other assumptions unchanged, what is the percent of people with your level of knowledge who ever lived are alive now?

REFERENCES

Bauer, F.L., Bolliet, L., and Helms, H.J. (1968), Nato software engineering conference 1968, http://homepages.cs.ncl.ac.uk/brian.randell/NATO/nato1968.PDF, last accessed on May 19, 2010.

Booch, G. (2009), *Handbook of Software Architecture,* http://www.booch.com/architecture/index.jsp, last accessed on May 19, 2010.

Brooks, F.P. (1987), 'No silver bullet: Essence and accidents of software engineering', *Computer*, 20(4):10–19.

CSC-NCSU (2009), 'Bibliography on mining software engineering data', http://ase.csc.ncsu.edu/dmse/, last accessed on May 19, 2010.

Fowler, M. (2003), *Patterns of Enterprise Application Architecture*, Addison-Wesley.

Gabriel, R.P. and Goldman, R. (2006), 'Conscientious software', In *OOPSLA* *'06: Proceedings of the 21st annual ACM SIGPLAN conference on Object-oriented programming systems, languages, and applications*, pp. 433–450, ACM, New York, USA.

Hamming, R.R. (1997), *Art of Doing Science and Engineering: Learning to Learn*, CRC.

Kruchten, P. (2008), 'Licensing software engineers?', *Software, IEEE*, 25(6):35–37.

Weinberg, G.M. (1975), *An Introduction to General Systems Thinking*, John Wiley and Sons Inc.

Weinberg, G.M. (1988), *Rethinking Systems Analysis and Design*, Dorset House Publishing Company, Inc.

Weinberg, G.M. and Weinberg, D. (1988), *General Principles of Systems Design*, Dorset House Publishing Company, Inc.

Index